AF389440

New Trends on Vanadium Chemistry, Biochemistry, and Medicinal Chemistry

New Trends on Vanadium Chemistry, Biochemistry, and Medicinal Chemistry

Editor

Dinorah Gambino

MDPI • Basel • Beijing • Wuhan • Barcelona • Belgrade • Manchester • Tokyo • Cluj • Tianjin

Editor
Dinorah Gambino
Área Química Inorgánica,
Facultad de Química
Universidad de la República
Montevideo
Uruguay

Editorial Office
MDPI
St. Alban-Anlage 66
4052 Basel, Switzerland

This is a reprint of articles from the Special Issue published online in the open access journal *Inorganics* (ISSN 2304-6740) (available at: www.mdpi.com/journal/inorganics/special_issues/vanadium_chem_bio_med).

For citation purposes, cite each article independently as indicated on the article page online and as indicated below:

LastName, A.A.; LastName, B.B.; LastName, C.C. Article Title. *Journal Name* **Year**, *Volume Number*, Page Range.

ISBN 978-3-0365-5766-3 (Hbk)
ISBN 978-3-0365-5765-6 (PDF)

Contents

About the Editor

Dinorah Gambino

Dinorah Gambino is a Full Professor of Inorganic Chemistry at the Faculty of Chemistry of Universidad de la República, Montevideo, Uruguay. She got both Doctor in Pharmaceutical Technology and Doctor in Chemistry degrees from this Faculty. She leads a group working on Medicinal Inorganic Chemistry whose research is focused on the development of multiple-target organometallic and classical coordination compounds as prospective drugs for the treatment of diseases that seriously affect Latin American countries: Chagas disease or American Trypanosomiasis, Leishmaniasis, Tuberculosis and Cancer. Currently, the research of this group focuses on the development of broad-spectrum multi-functional metal compounds that could affect more than one trypanosomatid parasite: *Trypanosoma cruzi* (Chagas disease), *Leishmania infantum* (visceral Leishmaniasis) and *Trypanosoma brucei* (African Trypanosomiasis).

In this context, her group works intensively on the rational design and study of new vanadium compounds as prospective agents against *Trypanosoma cruzi*.

inorganics

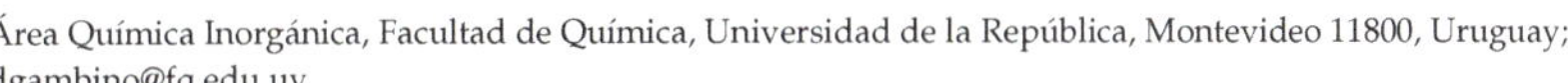

Editorial

New Trends on Vanadium Chemistry, Biochemistry, and Medicinal Chemistry

Dinorah Gambino

Área Química Inorgánica, Facultad de Química, Universidad de la República, Montevideo 11800, Uruguay; dgambino@fq.edu.uy

Citation: Gambino, D. New Trends on Vanadium Chemistry, Biochemistry, and Medicinal Chemistry. *Inorganics* **2022**, *10*, 68. https://doi.org/10.3390/inorganics10050068

Received: 5 May 2022
Accepted: 18 May 2022
Published: 22 May 2022

Publisher's Note: MDPI stays neutral with regard to jurisdictional claims in published maps and institutional affiliations.

Vanadium was discovered twice. Andrés Manuel del Rio, a Professor of Mineralogy in Mexico, discovered it for the first time in a vanadinite ore. Nevertheless, at that time, French chemists dismissed the discovery, concluding that this ore was a chromium mineral. Vanadium was rediscovered in 1830 by Nils Gabriel Sefström, a Swedish chemist, while analyzing samples of iron from a mine in Sweden. He named the new element after Vanadis, the Swedish goddess of beauty and fertility, because of the brilliant and attractive colors of the chemical compounds in which it was first found. Currently, the originally suggested name of the element could equally be associated with the fertile work of scientists over the last 190 years developing new compounds and materials based on vanadium.

In particular, the recognition of the exceptional chemical and biological properties of vanadium compounds has led, in recent decades, to vast research in order to explore their chemistry, biochemistry, and medicinal chemistry. Although the benefits of using vanadium compounds in medicine are still inconclusive, their prospective application as therapeutic agents against diseases, such as diabetes, cancer and those provoked by parasites and bacteria, has led to extensive research. Researchers from all over the world are dedicating their efforts to vanadium research that is related to the potential therapeutic applications of vanadium compounds and to obtain an insight into their modes of action and beneficial effects on health. This Special Issue collects research contributions focused on recent advances in vanadium chemistry, biochemistry, medicinal chemistry, and toxicology. In the following introduction, the contributions covered in this themed issue are summarized in alphabetical order of the family name of corresponding author.

Debbie Crans and colleagues studied the acute toxicity in mice of a previously developed heteroleptic oxidovanadium (V) schiff base complex that had demonstrated anticancer properties against human ovarian, prostate and brain cells as well as enhancing effects of oncolytic viruses. The compound showed low oral toxicity which encourages the design of oxidovanadium (V) complexes with low toxicity for potential applications in cancer therapy [1].

Joao Costa Pessoa and Isabel Correia presented a review work on how the interactions of vanadium complexes with proteins and other biological targets can be misinterpreted. The review emphasizes the fact that in the case of studying biochemical interactions or effects in order to determine binding constants or active species, or propose mechanisms of action, it is essential to evaluate the speciation of the vanadium compound in the media where it is acting. When vanadium (IV) and vanadium (V) compounds are dissolved in biological media, they undergo chemical transformations, particularly at the low concentrations used in biological experiments [2].

Rupam Dinda and colleagues synthesized and characterized two new oxidovanadium (V) complexes that manifested in vitro cytotoxic potential comparable with that of clinically used drugs, which caused cell death by apoptosis [3].

Teresa Fortoult and colleagues explored the use of inhaled vanadium (V) as an option for lung cancer treatment. Aerosol delivery increased apoptosis and growth arrest of the tumors with no respiratory clinical changes in mice [4].

Enrique Gónzalez-Vergara, Antonio Rodríguez-Diéguez and colleagues described the synthesis and characterization of a new member of a family of compounds based on decavanadate and a 2-aminopyrimidine ligand, together with its interaction with RNA as potential target for cancer treatment [5].

Anastasios Keramidas, Yannis Simos and colleagues reported novel vanadium (V) complexes with the siderophore vitamin E-hydroxylamino-triazine ligands and their chemical and biological properties. Instead of exhibiting antioxidant activity, the compounds were radical initiators and did not exert significant cytotoxic activity against tumor cell lines [6].

Ignacio León and colleagues performed a review of the activity of vanadium compounds over cell signaling pathways on cancer cells and of the underlying mechanisms, thereby providing insight into the role of these proteins as potential new molecular targets of vanadium complexes [7].

Irma Sánchez-Lombardo and colleagues performed kinetic studies of sodium and metforminium decavanadates and unraveled the nature of their decomposition products. Cytotoxic activity studies using non-tumorigenic HEK293 cell line and human liver cancer HEPG2 cells showed that decavanadate compounds exhibited selective action toward HEPG2 cells after 24 h. In addition, an insulin release assay in βTC-6 cells showed that metforminium decavanadate enhanced insulin release [8].

Patricia Williams and colleagues reported the antioxidant and anticancer activities and the bovine serum albumin interaction of the oxidovanadium (IV) complex with the flavonoid naringin. The complex generated typical effects shown by apoptotic pathways, such as the generation of intracellular reactive oxygen species (ROS), depletion of reduced glutathione and depolarization of mitochondrial membrane potential, producing cell death by an oxidative stress mechanism. Although the oxidovanadium (IV) naringin complex showed a greater affinity to serum albumin than free naringin, it could still be transported and delivered by it [9].

In a second contribution to the issue, Williams and colleagues reported the synthesis and characterization of an oxidovanadium (IV) heteroleptic complex with the polyphenol chrysin and 1,10-phenanthroline as ligands. The cytotoxic effect of this complex proved to be higher in the human lung cancer A549 cell line than that of the oxidovanadium (IV) chrysin homoleptic complex. The probable mechanism of action proved to involve the production of ROS, the decrease in the natural antioxidant compound glutathione (GSH) and the ratio GSH/GSSG (GSSG, oxidized GSH), and mitochondrial membrane damage. Cytotoxicity studies using the non-tumorigenic HEK293 cell line showed that the new heteroleptic compound exhibited selectivity towards tumor cells [10].

As a concluding remark, I thank all the contributors to this Special Issue which covers recent progress in vanadium science across a range of subject areas.

Funding: This research received no external funding.

Conflicts of Interest: The author declares no conflict of interest.

References

1. Lima, L.; Murakami, H.; Gaebler, D.; Silva, W.; Belian, M.; Lira, E.; Crans, D. Acute Toxicity Evaluation of Non-Innocent Oxidovanadium(V) Schiff Base Complex. *Inorganics* **2021**, *9*, 42. [CrossRef]
2. Pessoa, J.; Correia, I. Misinterpretations in Evaluating Interactions of Vanadium Complexes with Proteins and Other Biological Targets. *Inorganics* **2021**, *9*, 17. [CrossRef]
3. Sahu, G.; Tiekink, E.; Dinda, R. Study of DNA Interaction and Cytotoxicity Activity of Oxidovanadium(V) Complexes with ONO Donor Schiff Base Ligands. *Inorganics* **2021**, *9*, 66. [CrossRef]
4. López-Valdez, N.; Rojas-Lemus, M.; Fortoul, T. The Effect of Vanadium Inhalation on the Tumor Progression of Urethane-Induced Lung Adenomas in a Mice Model. *Inorganics* **2021**, *9*, 78. [CrossRef]
5. García-García, A.; Noriega, L.; Meléndez-Bustamante, F.; Castro, M.; Sánchez-Gaytán, B.; Choquesillo-Lazarte, D.; González-Vergara, E.; Rodríguez-Diéguez, A. 2-Aminopyrimidinium Decavanadate: Experimental and Theoretical Characterization, Molecular Docking, and Potential Antineoplastic Activity. *Inorganics* **2021**, *9*, 67. [CrossRef]
6. Loizou, M.; Hadjiadamou, I.; Drouza, C.; Keramidas, A.; Simos, Y.; Peschos, D. Vanadium(V) Complexes with Siderophore Vitamin E-Hydroxylamino-Triazine Ligands. *Inorganics* **2021**, *9*, 73. [CrossRef]

7. Ferretti, V.; León, I. An Overview of Vanadium and Cell Signaling in Potential Cancer Treatments. *Inorganics* **2022**, *10*, 47. [CrossRef]
8. Silva-Nolasco, A.; Camacho, L.; Saavedra-Díaz, R.; Hernández-Abreu, O.; León, I.; Sánchez-Lombardo, I. Kinetic Studies of Sodium and Metforminium Decavanadates Decomposition and In Vitro Cytotoxicity and Insulin- Like Activity. *Inorganics* **2020**, *8*, 67. [CrossRef]
9. Restrepo-Guerrero, A.; Goitia-Semenco, H.; Naso, L.; Rey, M.; Gonzalez, P.; Ferrer, E.; Williams, P. Antioxidant and Anticancer Activities and Protein Interaction of the Oxidovanadium(IV) Naringin Complex. *Inorganics* **2022**, *10*, 13. [CrossRef]
10. Actis Dato, A.; Naso, L.; Rey, M.; Gonzalez, P.; Ferrer, E.; Williams, P. Phenanthroline Complexation Enhances the Cytotoxic Activity of the VO-Chrysin System. *Inorganics* **2022**, *10*, 4. [CrossRef]

 *inorganics*

Review

An Overview of Vanadium and Cell Signaling in Potential Cancer Treatments

Valeria Alejandra Ferretti [1] and Ignacio Esteban León [1,2,*]

1   Cequinor (UNLP, CCT-CONICET La Plata, Asociado a CIC), Departamento de Química, Facultad de Ciencias Exactas, Universidad Nacional de La Plata, Blvd. 120 N° 1465, La Plata 1900, Argentina; vferretti@med.unlp.edu.ar
2   Cátedra de Fisiopatología, Departamento de Ciencias Biológicas, Facultad de Ciencias Exactas, Universidad Nacional de La Plata, 47 y 115, La Plata 1900, Argentina
*   Correspondence: ileon@biol.unlp.edu.ar

Abstract: Vanadium is an ultratrace element present in higher plants, animals, algae, and bacteria. In recent years, vanadium complexes have been studied to be considered as a representative of a new class of nonplatinum metal anticancer drugs. Nevertheless, the study of cell signaling pathways related to vanadium compounds has scarcely been reported on and reviewed thus far; this information is highly critical for identifying novel targets that play a key role in the anticancer activity of these compounds. Here, we perform a review of the activity of vanadium compounds over cell signaling pathways on cancer cells and of the underlying mechanisms, thereby providing insight into the role of these proteins as potential new molecular targets of vanadium complexes.

Keywords: vanadium biochemistry; cell signaling; cancer; anticancer agents

Citation: Ferretti, V.A.; León, I.E. An Overview of Vanadium and Cell Signaling in Potential Cancer Treatments. *Inorganics* 2022, 10, 47. https://doi.org/10.3390/inorganics10040047

Academic Editor: Isabel Correia

Received: 26 February 2022
Accepted: 29 March 2022
Published: 1 April 2022

Publisher's Note: MDPI stays neutral with regard to jurisdictional claims in published maps and institutional affiliations.

1. Introduction

Metallodrugs have a wide field of therapeutical activities towards several pathologies, including infections, neurodegenerative diseases, diabetes, and cancer [1–4]. Platinum compounds, in special cisplatin (CDDP), carboplatin, and oxaliplatin, are the most relevant and effective metallodrugs [5]. Nevertheless, the lack of specificity, poor absorption, and chemoresistance limit the current use in the clinic. Therefore, medicinal inorganic chemistry focuses on the design and synthesis of novel metal-based drugs aiming to overcome these complications [6,7]. In this sense, the anticancer activity of vanadium complexes has been extensively in vitro and in vivo tested on several types of cancer cell lines. VO (oxidovanadium) flavonoids are an important group of compounds with selective antitumor effects on bone cancer cells [8–10]. Metvan ($V^{IV}O(SO_4)(4,7\text{-Mephen})_2$) is a vanadium complex with anticancer activity on different human tumor cell lines, such as leukemia cells, multiple myeloma cells, and solid tumor cells (brain, prostate, breast, ovarian, etc.) [11–14]. Moreover, vanadocene derivatives have shown important anticancer effects on human cancer cell lines, mainly involving liver and testicular tumors. Another interesting group with antitumor properties is vanadium compounds with heterocycles and Schiff bases. Their complexes have shown antitumor actions on bone, breast, and colon cancer cells [15–17].

The role of vanadium in the regulation of cell signaling pathways converts it into a prospective therapeutic agent to be employed in various pathologies. However, the known activation pathways targeted by vanadium compounds are narrowly reported in the literature and, thus far, these data for the discovery of novel intracellular targets in cancer have not been widely analyzed.

In this review, we present an outline of the cell pathways activated or inactivated by vanadium complexes on cancer cells and the relationship with the anticancer activity of these compounds. This overview is expected to achieve a better understanding of these intracellular signaling pathways and, thereby, may facilitate the design of vanadium complexes with promising therapeutic applications as well as the comprehension of side effects derived from the use of the vanadium compounds as therapeutic agents.

2. Vanadium in Cancer Therapeutics

Vanadium is a transition metal that exists in different oxidation states ranging from -1 to $+5$. At pharmacological doses, compounds with vanadium III, IV, and V show biologically significant effects such as insulin imitators [18,19], growth factor-like activity [20], and antitumoral properties [9,21–24].

Various effects of vanadium derivatives have been observed on the activity of several enzymes, especially those related to phosphate reactions. Vanadium inhibits various ATPases with different effectiveness [25]. Furthermore, vanadium compounds also inhibit several phosphatases, such as alkaline phosphatase, acid phosphatase, and tyrosine–protein phosphatases (PTPases) [26]. The PTPases activate or inhibit intracellular signaling pathways, triggering different biological events in a cascade manner, among these being cancer signaling pathways.

3. Cancer-Related Signaling Pathways Activated by Vanadium Complexes

Diverse assays performed in vitro or in vivo demonstrated that vanadium compounds can activate different cancer signaling pathways, and so exert their antitumoral action.

3.1. MAPK (Mitogen-Activated Protein Kinases)/ERK (Extracellular Signal-Regulated Kinase) Signaling Pathway

The MAPK/ERK pathway is one of the early signaling pathways for cell cycle progression [27,28]. An essential role in cancer development is attributable to alterations regarding different molecular pathways such as the MAPK involved in regulating cell growth. The uncontrolled activation of MAPKs is due to diverse gene mutations, some of which regulate the constitutive activation of the B-Raf protein kinase (cytoplasmic protein) that induces the activation of the mitogen-activated protein kinase (MEK), which in turn activates the extracellular signal-regulated kinase (ERK), the final effector of the pathway, inducing the transcription of target genes that generate the cell entering the cell cycle.

Bis(acetylacetonate)-oxidovanadium(IV) (VO(acac)$_2$) and sodium metavanadate (NaVO$_3$), two well-known antidiabetic compounds, have shown an antiproliferative effect through inducing a G_2/M cell cycle arrest and an elevation in reactive oxygen species (ROS) levels in human pancreatic cancer cell line AsPC-1. It is important to highlight that NaVO$_3$ converts to $H_2VO_4^-$ at physiological conditions, in which the cellular assays were carried out.

ROS are fundamental agents in cell fate. Their intracellular accumulation in normal cells means the oxidation of cellular components, such as nucleic acids, proteins, and lipids. These oxidative reactions cause extensive damage and in cases of irreparable damages, they stimulate apoptosis [29]. In this sense, it has been found that these two vanadium compounds prompt the activation of the MAPK/ERK signaling pathway in a dose- and time-dependent manner. Both compounds generate an increase in the phosphorylated ERK levels; therefore, these vanadium compounds could cause a cell cycle arrest and a high elevation in the ROS levels by positively modulating the MAPK/ERK signaling pathway and, thus, causing tumor suppression [30].

Another study revealed an antiproliferative activity of the inorganic anion vanadate(V) (VN) and the oxidovanadium (IV) complex (VO(1,2-dimethyl-3-hydroxy-4(1H)-pyridinonate)$_2$) (VS2) on the melanoma A375 cell line. The authors demonstrated that both vanadium (IV, V) species displayed an antitumoral activity by arresting the cell cycle and causing apoptosis across intracellular ROS production, ERK, and retinoblastoma protein (Rb) dephosphorylation and p21Cip1 overexpression [31]. The retinoblastoma protein (Rb) constitutes an essential control point for the switch from the G_1 phase to the S phase, whereas the cyclin/CDK complex inhibitor p21Cip1 is involved in the cell cycle blockade.

Vanadium compounds are characterized by their ability to regulate stem cell differentiation [24]. Recently, N,N-bis(salicylidene)-o-phenylenediamine vanadium(IV) oxide was reported to upmodulate osteoblast differentiation. Thus, the V(IV) species seem to stimulate the differentiation and mineralization of the mesenchymal stem cells via the activation of the ERK signaling pathway and the subsequent improvement of the NF-κB (nuclear factor kappa-light-chain-enhancer of activated B cells) mediated action. Furthermore, it has been established that ERK is involved in the increase in the transcriptional activity of NF-κB. Thus, V(IV) may modulate both ERK and NF-κB pathways, and both pathways would act jointly to encourage osteoblasts [32].

In summary, different research suggests that some vanadium compounds can affect the MAPK/ERK signaling pathway, prompting a cell cycle arrest, an increase in the apoptosis, and, thus, causing tumor reduction.

3.2. PI3K (Phosphatidylinositol 3-Kinase)/AKT (Protein Kinase B) Signaling Pathway

Oncogenic RET/PTC1 (receptor tyrosine kinase/type two C phosphatase 1) chromosomal rearrangements are hallmarks of thyroid papillary carcinoma. The resulting protein, mainly, through tyrosine 451, is responsible for the activation of pathways controlling cell survival, including the PI3K/Akt cascade. The PI3K/Akt signaling cascade has an important role in the control of cell survival, metabolism, and motility, with unsuitable signals through this pathway occurring habitually in cancer [33]. Vanadium compounds were revealed to have antitumoral potential in thyroid papillary carcinoma, among others [34]. In this sense, a study realized utilizing papillary thyroid carcinoma-derived TPC-1 cells revealed that a low dose of orthovanadate (OV) (100 nM) induces a pro-proliferative response. In contrast, treatment with inhibitory amounts of the compound (10 μM) resulted in a greater phosphorylation of tyrosine 451 of RET/PTC1, triggering the mTOR/S6R branch of the PI3K/Akt signaling pathway. These concentrations of the drug also generate typical features of apoptosis, including DNA fragmentation, the loss of mitochondrial membrane potential, production of ROS, and activation of caspase-3 [35].

Another study realized utilizing MCF7 human breast cancer epithelial and A549 lung adenocarcinoma cells revealed that vanadium produces a significant decline in cancer cell viability, decreasing H-ras signaling and metalloproteinase-2 (MMP-2) expression by raising ROS-mediated apoptosis [36]. On the other hand, it is well known that vanadium compounds are effective in diabetes treatment due to their insulin-mimetic behavior and the stimulation of glucose catchment [21,37]. Pandey et al. [38] demonstrated that vanadyl sulfate stimulates the ras-ERK pathway through the activation of PI3K, and they presumed that the stimulation of the PI3-K/ras/ERK cascade plays an essential role in mediating the insulin-mimetic effects of the vanadium compounds (Figure 1).

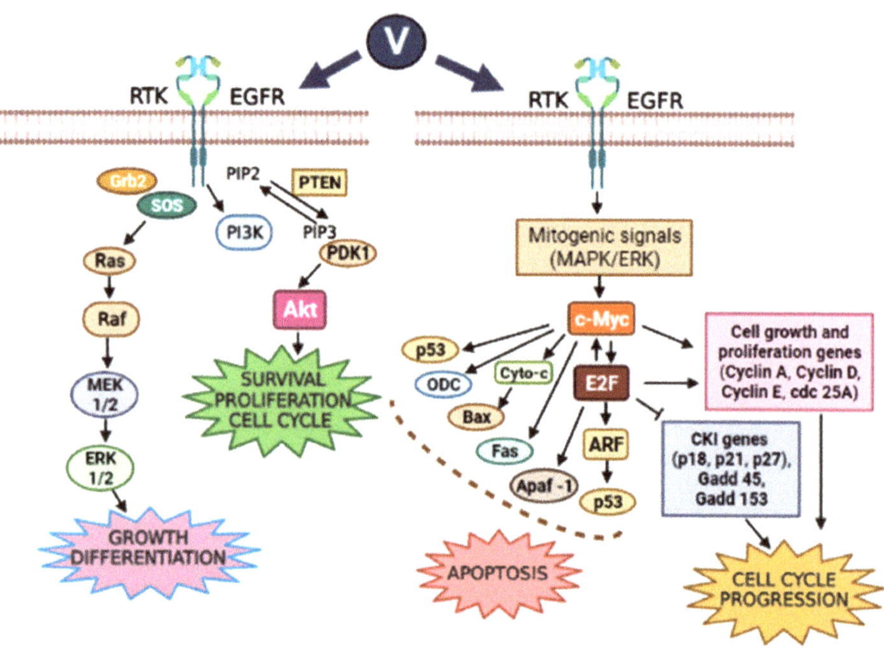

Figure 1. MAPK/ERK cell signaling pathways affected by vanadium compounds. Adapted from [39].

In consequence, it was observed that the vanadium compounds can lead to an increase in the ROS-mediated apoptosis and a decline in cancer cell viability by stimulating the PI3K/Akt signaling pathway.

3.3. Caspase Signaling Pathway

Vanadium is characterized by its capacity to stimulate the apoptotic machinery in cancer cells through the upregulation of the primary apoptosis proteins. In relation to this, we found that the oxidovanadium(IV) flavonoids caused a cell cycle arrest and activated caspase-3, triggering apoptosis in a human osteosarcoma cell line MG-63 [8,15]. In another report, we studied the mechanism of action of the oxidovanadium(IV) complexes with the flavonoids silibinin $Na_2[VO(silibinin)_2]\cdot6H_2O$ (VOsil) and chrysin $[VO(chrysin)_2EtOH]_2$ (VOchrys), utilizing human colon adenocarcinoma-derived cell line HT-29. In this work, we found that the VOchrys caused a cell cycle arrest in the G_2/M phase, while VOsil activated caspase-3, triggering the cells directly into apoptosis [40] (Figure 2). Moreover, VOsil diminished the NF-κB activation via increasing the sensitivity of cells to apoptosis [37]. As mentioned before, orthovanadate also generates an activation of caspase-3 in papillary thyroid carcinoma cells [35]. Moreover, it has been demonstrated that vanadium compounds induce apoptosis and the expression of caspase-3, Bcl-2, and Bax, which regulate cell apoptosis in neuronal cells [41,42].

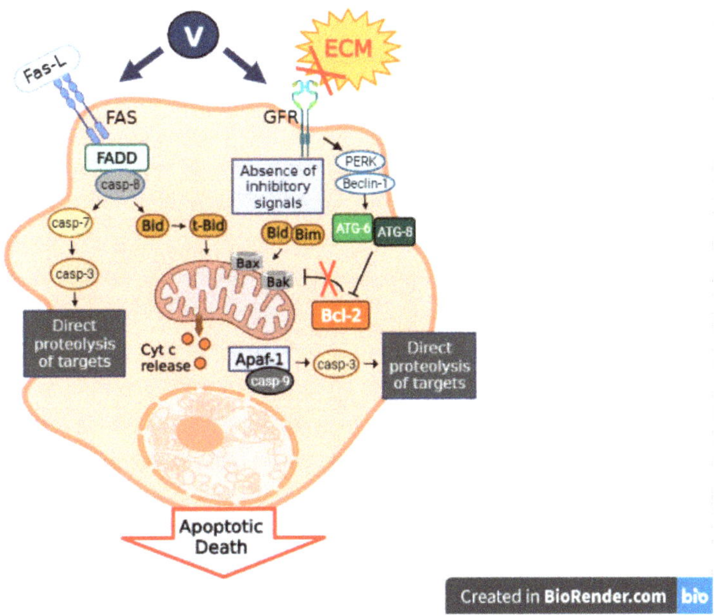

Figure 2. Apoptosis cell pathways activated by vanadium compounds. Adapted from [43].

Briefly, several vanadium compounds can exert their effects on the caspase signaling pathway, generating a rise in cancer cell apoptosis and, thus, tumor suppression.

3.4. JAK (Janus Kinase Protein)/STAT (Signal Transducer and Activator of Transcription Proteins) Signaling Pathway

Vanadium is also considered an air contaminant released into the atmosphere by burning fossil fuels. Moreover, its carcinogenic potential has been assessed to establish permissible limits of exposure at workplaces. Gonzalez—Villalva et al. [44,45] detected a growth in the number and size of platelets and their precursor cells and megakaryocytes in bone marrow and spleen as a consequence of the vanadium exposure. In another study, performed on mice exposed to vanadium pentoxide, they found an increase in JAK2 ph (phosphorylated Janus kinase 2 protein) and STAT3 ph (phosphorylated STAT3), but a decline in the Mpl (myeloproliferative leukemia protein) receptor [45]. In consequence, they concluded that the vanadium pentoxide could activate the JAK/STAT pathway and decrease the Mpl receptor; thus, leading to a condition analogous to essential thrombocythemia. They also proposed that the reduction in Mpl was a negative feedback mechanism after the JAK/STAT activation.

The Mpl receptor does not have tyrosine kinase activity, but is constitutively linked to JAK2, which has tyrosine kinase activity that phosphorylates the signal transducer and activates the transcription of STAT3. STAT3 ph translocates to the nucleus to function as a transcription factor, which activates genes that stimulate survival and apoptosis inhibition, such as Bcl-xl, p27, p21, and cyclin D1. Since megakaryocytes are platelet precursors, their modification affects platelet morphology and function, which might have consequences in hemostasis; therefore, it is imperative to continue assessing the effects of chemicals and contaminants on megakaryocytes and platelets.

Apparently, some vanadium compounds in certain doses can activate the JAK/STAT signaling pathway and, hence, prompt an increase in the number and size of the platelets, a condition analogous to essential thrombocythemia.

3.5. Nrf2 (Nuclear Factor Erythroid 2-Related Factor 2)/HO-1 (Heme Oxygenase-1) Signaling Pathway

CDDP is one of the first-line anticancer treatments; however, the main limitation of CDDP therapy is the development of nephrotoxicity (25–35% cases), whose specific mechanism primarily includes oxidative stress, inflammation, and cell death. Thus, looking for a potential chemo-protectant, Basu et al. [46] assessed an organo–vanadium complex, vanadium(III)-L-cysteine (VC-III), against CDDP-induced nephropathy in mice.

The VC-III treatment significantly avoided the CDDP-induced generation of ROS, reactive nitrogen species, and the beginning of lipid peroxidation in kidney tissues of the experimental mice. Furthermore, VC-III also extensively returned CDDP-induced depleted activities of the renal antioxidant enzymes, such as superoxide dismutase, catalase, glutathione peroxidase, glutathione- S-transferase, and glutathione (reduced) levels. In addition, the VC-III treatment also quite successfully reduced the expression of proinflammatory mediators, such as NFκβ, COX-2, and IL-6, and activated the Nrf2-mediated antioxidant defense system through the promotion of downstream antioxidant enzymes (HO-1). The treatment with VC-III considerably improved CDDP-mediated cytotoxicity in MCF-7 and NCI-H520 human cancer cell lines. Therefore, VC-III could function as a proper chemo-protectant via stimulating the Nrf2/HO-1 signaling pathway and enhancing the therapeutic window of CDDP in cancer patients (Figure 3).

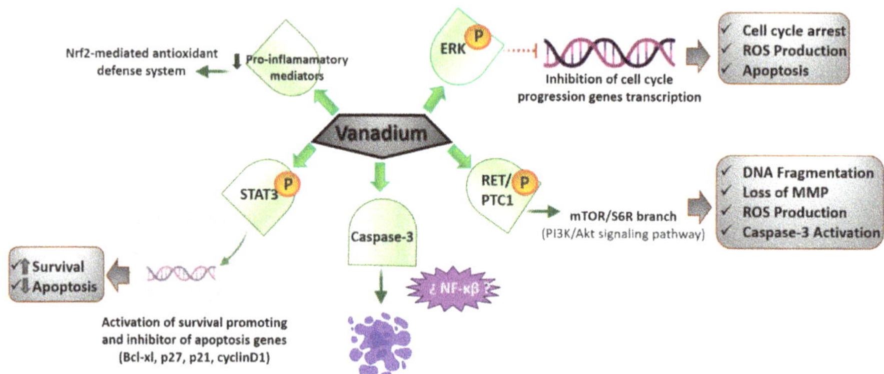

Figure 3. Upregulated proteins involved in survival and cell death induced by vanadium complexes.

4. Cancer-Related Signaling Pathways Inactivated by Vanadium Complexes

Interestingly, many works have revealed that vanadium compounds are also capable of inhibiting several cancer signaling pathways, and so work as antitumor and antimetastatic agents.

4.1. FAK Signaling Pathway

Recently, we demonstrated that oxidovanadium (IV)–chrysin and oxidovanadium (IV)–clioquinol (VO(CQ)$_2$) complexes prevent the activation of focal adhesion kinase (FAK), reducing the cell proliferation in human osteosarcoma cells [10,47]. Our results showed that VO(CQ)$_2$ is located near the activation loop of the kinase domain and establishes interactions with residues in the ATP binding site. Particularly, vanadium–clioquinol exhibited a dual behavior at 2.5 μM, since the Tyr576 and Tyr577 sites were upmodulated; however, at 10 μM, the phosphorylation of Tyr576 and Tyr577 declined 14-fold [48]. Likewise, in another work, we reported that the oxidovanadium (IV)–chrysin complex ([VO(chrysin)$_2$EtOH]$_2$) upmodulated the Tyr577 site of phosphorylation, but downmodulated Tyr397 [10]. The Tyr397 site of tyrosine phosphorylation is the most active and common site for the autocatalytic function of FAK [49]. Thus, these results would indicate that [VO(chrysin)$_2$EtOH]$_2$

selectively repressed the autophosphorylation activity of FAK kinase, directly affecting the Tyr397 site.

On the other hand, FAK is a tyrosine kinase that carries out a crucial role in the adhesion, survival, motility, angiogenesis, and metastasis of cancer cells [50]. Moreover, FAK is overexpressed in numerous kinds of solid and nonsolid tumors [51], so FAK has been suggested as a therapeutic target [52].

Furthermore, vanadium compounds diminished the cell migration in 2D and 3D human bone cancer cell models. Additionally, $VO(CQ)_2$ considerably reduced the activity of MMP-2 and MMP-9 in a dose-dependent way, advising the direct relationship between FAK inhibition and the inactivation of MMPs (matrix metalloproteinases).

4.2. Autophagy Signaling Pathway

Recent studies have demonstrated the great ability of vanadium to regulate the process of autophagy. Inorganic sodium orthovanadate (SOV) converts to an active agent at physiological conditions in $H_2VO_4^-$, potentially prompting cell apoptosis and the prevention of autophagy in human hepatocellular carcinoma (HCC) cells, in vitro and in vivo, concurrently. Additionally, a further decrease in autophagy by 3-methyladenine (3MA) considerably improves SOV-induced apoptosis in HCC cells, while rapamycin could reverse such autophagy inhibition and decrease the apoptosis-stimulating effect of SOV in HCC cells, both in vitro and in vivo. The results showed that such an autophagy inhibition effect plays a pro-death role [53]. Likewise, nano-sized paramontroseite VO2 nanocrystals (P-VO2) prompted cyto-protective, rather than death-inducing, autophagy in cultured HeLa cells. P-VO2 also prompted the upmodulation of hemeoxygenase-1 (HO-1), a cellular protein with a proved role in protecting cells against death under stress conditions. The autophagy inhibitor 3-methyladenine considerably repressed HO-1 upregulation and augmented the rate of cell death in cells treated with P-VO2, while the HO-1 inhibitor protoporphyrin IX Zn(II) (ZnPP) improved the existence of cell death in the P-VO2-treated cells, while displaying no effect on the autophagic response induced by P-VO2. Likewise, Y_2O_3 nanocrystals, a control nanomaterial, prompted death-inducing autophagy without affecting the level of expression of HO-1 and the pro-death effect of the autophagy prompted by Y_2O_3. These data represent the first report on a novel nanomaterial-induced cyto-protective autophagy, possibly through the upregulation of HO-1, and potentially leading to new opportunities for taking advantage of nanomaterial-induced autophagy for cancer therapeutic applications [54].

4.3. TGFβ (Transforming Growth Factor-β)- EMT (Epithelial to Mesenchymal Transition) Signaling Pathway

The EMT plays a crucial role in tumor advancement and metastasis as an essential event for cancer cells to generate the metastatic niche. TGF-β has been revealed to play a significant role as an EMT inducer in several stages of carcinogenesis. Some studies have revealed that vanadium inhibits the metastatic potential of tumor cells by decreasing MMP-2 expression and prompting ROS-dependent apoptosis. Petanidis et al. [55] described, for the first time, the inhibitory effects of vanadium on (TGF-β)-mediated EMT, followed by the downmodulation of cancer stem cell markers in human lung cancer adenocarcinoma A549 and breast cancer MDA-MB-231 epithelial cell models. The results showed a blockage of TGF-β-mediated EMT by vanadium and a decay in the mitochondrial potential of cancer cells related to EMT and cancer metabolism. Moreover, they reported that the combination of vanadium and carboplatin results in a G_0/G_1 cell cycle arrest and the sensitization of cancer cells to carboplatin-induced apoptosis. This knowledge could be valuable for targeting the cancer cell metabolism and cancer stem cell-mediated metastasis in aggressive chemoresistant tumors.

4.4. Notch-1 Signaling Pathway

In recent work, it was found that vanadium compounds can suppress the growth of the MDA-MB-231 cell line, a model of the most aggressive and therapy-resistant triple-negative breast cancer. The [VO(bpy)$_2$Cl]Cl complex (bpy = bipyridyl) generated a rise in caspase-3 levels and, thus, an induction of the apoptotic cell death [56]. Moreover, the authors found a decrease in the Notch1 gene expression, thereby inhibiting the Notch-1 pathway. The Notch signaling pathway is a greatly conserved cell signaling system, which plays a major role in the regulation of embryonic development and is dysregulated in numerous types of cancers, such as lung and breast cancers [57–59]. Additionally, the inactivation of Notch signaling has been demonstrated to have antiproliferative effects on T-cell acute lymphoblastic leukemia in cultured cells and a mouse model [59] (Figure 4).

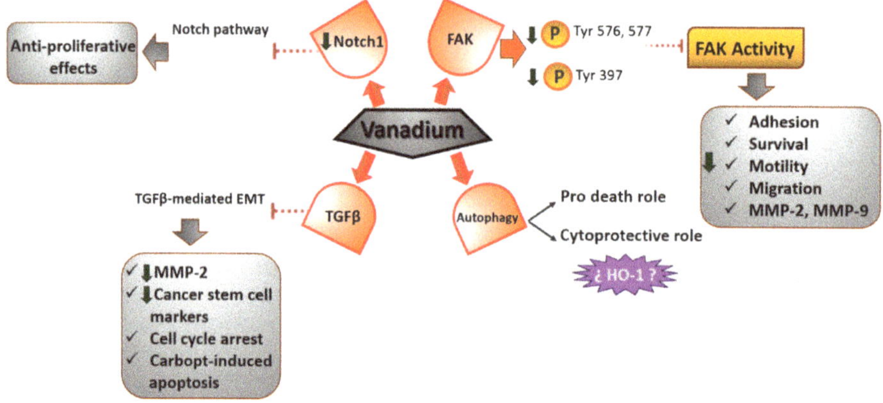

Figure 4. Downregulated proteins involved in survival and cell death induced by vanadium complexes.

5. Conclusions

Over recent years, vanadium compounds have been studied and considered as representatives of a new class of nonplatinum metal antitumor agents. Nevertheless, knowledge surrounding cell signaling pathways related to vanadium drugs is scarce. In this review, we presented a brief overview of the cell pathways activated or inactivated by vanadium complexes on cancer cells and the relationship with the anticancer activity of these compounds. The MAPK/Erk, PI3K/Akt, and caspase family and JAK/Stat signaling pathways were stimulated by the vanadium compounds, prompting a cell cycle arrest, ROS production, and apoptosis towards different types of cancer cells. The Nrf-2 was also activated by the vanadium; however, in this case, it seemed to enhance the defense system and functioned as a chemoprotective. On the other hand, the vanadium complexes would also stimulate the JAK/Stat signaling pathway, generating a growth effect in platelets and their precursors.

Likewise, the FAK, TGF-B/EMT, Notch-1, and autophagy signaling pathways were inactivated by vanadium compounds, potentially leading to an increase in cell cycle arrest and apoptosis, and a decrease in cellular migration and adhesion, generating tumor suppressor effects.

Taken together, hopefully this review will generate further understanding of the activity of vanadium compounds over cell signaling pathways on cancer cells and of the underlying mechanisms, and may thereby facilitate the design of vanadium complexes with promising therapeutic applications to improved cancer treatments.

Author Contributions: Conceptualization, V.A.F. and I.E.L. investigation, V.A.F. and I.E.L.; writing—original draft preparation, V.A.F.; writing—review and editing, I.E.L.; supervision, I.E.L.; funding acquisition, I.E.L. All authors have read and agreed to the published version of the manuscript.

Funding: This work was partly supported by UNLP (PPID X053) CONICET (PIP 2051), and ANPCyT (PICT 2019-2322) from Argentina.

Acknowledgments: The authors thank Consejo Nacional de Investigaciones Científicas y Técnicas (CONICET) since I.E.L. is a member of the Researcher Career.

Conflicts of Interest: The authors declare no conflict of interest.

References

1. Zhang, C.X.; Lippard, S.J. New metal complexes as potential therapeutics. *Curr. Opin. Chem. Biol.* **2003**, *7*, 481–489. [CrossRef]
2. McQuitty, R.J. Metal-based drugs. *Sci. Prog.* **2014**, *97*, 1–19. Available online: https://pubmed.ncbi.nlm.nih.gov/24800466/ (accessed on 25 February 2022). [CrossRef] [PubMed]
3. Leon, I.; Cadavid-Vargas, J.; Di Virgilio, A.; Etcheverry, S. Vanadium, Ruthenium and Copper Compounds: A New Class of Nonplatinum Metallodrugs with Anticancer Activity. *Curr. Med. Chem.* **2017**, *24*, 112–148. Available online: https://pubmed.ncbi.nlm.nih.gov/27554807/ (accessed on 25 February 2022). [CrossRef] [PubMed]
4. Tisato, F.; Marzano, C.; Porchia, M.; Pellei, M.; Santini, C. Copper in diseases and treatments, and copper-based anticancer strategies. *Med. Res. Rev.* **2010**, *30*, 708–749. Available online: https://pubmed.ncbi.nlm.nih.gov/19626597/ (accessed on 25 February 2022). [CrossRef]
5. Apps, M.; Choi, E.; Wheate, N. The state-of-play and future of platin drugs. *Endocr. Relat. Cancer* **2015**, *22*, R219–R233. [CrossRef]
6. Butler, J.S.; Sadler, P.J. Targeted delivery of platinum-based anticancer complexes. *Curr. Opin. Chem. Biol.* **2013**, *17*, 175–188. [CrossRef]
7. Barry, N.P.E.; Sadler, P.J. Exploration of the medical periodic table: Towards new targets. *Chem. Commun.* **2013**, *49*, 5106–5131. [CrossRef]
8. León, I.E.; di Virgilio, A.L.; Porro, V.; Muglia, C.I.; Naso, L.G.; Williams, P.A.; Bollati-Fogolin, M.; Etcheverry, S.B. Antitumor properties of a vanadyl(IV) complex with the flavonoid chrysin [VO(chrysin)2EtOH]2 in a human osteosarcoma model: The role of oxidative stress and apoptosis. *Dalton Trans.* **2013**, *42*, 11868–11880. Available online: https://pubmed.ncbi.nlm.nih.gov/2376 0674/ (accessed on 25 February 2022). [CrossRef]
9. León, I.E.; Porro, V.; di Virgilio, A.L.; Naso, L.G.; Williams, P.A.; Bollati-Fogolin, M.; Etcheverry, S.B. Antiproliferative and apoptosis-inducing activity of an oxidovanadium(IV) complex with the flavonoid silibinin against osteosarcoma cells. *J. Biol. Inorg. Chem.* **2014**, *19*, 59–74. Available online: https://pubmed.ncbi.nlm.nih.gov/24233155/ (accessed on 25 February 2022). [CrossRef]
10. Lón, I.E.; Díez, P.; Etcheverry, S.B.; Fuentes, M. Deciphering the effect of an oxovanadium(iv) complex with the flavonoid chrysin (VOChrys) on intracellular cell signalling pathways in an osteosarcoma cell line. *Metallomics* **2016**, *8*, 739–749. Available online: https://pubmed.ncbi.nlm.nih.gov/27175625/ (accessed on 25 February 2022). [CrossRef]
11. Dong, Y.; Narla, R.K.; Sudbeck, E.; Uckun, F.M. Synthesis, X-ray structure, and anti-leukemic activity of oxovanadium(IV) complexes. *J. Inorg. Biochem.* **2000**, *78*, 321–330. [CrossRef]
12. Narla, R.K.; Dong, Y.; Klis, D.; Uckun, F.M. Bis(4,7-dimethyl-1,10-phenanthroline) sulfatooxovanadium(I.V.) as a novel antileukemic agent with matrix metalloproteinase inhibitory activity. *Clin. Cancer Res.* **2001**, *7*, 1094–1101. [PubMed]
13. Evangelou, A.M. Vanadium in cancer treatment. *Crit. Rev. Oncol. Hematol.* **2002**, *42*, 249–265. [CrossRef]
14. León, I.E.; Ruiz, M.C.; Franca, C.A.; Parajón-Costa, B.S.; Baran, E.J. Metvan, bis(4,7-Dimethyl-1,10-phenanthroline)sulfatooxidova nadium(IV): DFT and Spectroscopic Study-Antitumor Action on Human Bone and Colorectal Cancer Cell Lines. *Biol. Trace Elem. Res.* **2019**, *191*, 81–87. Available online: https://pubmed.ncbi.nlm.nih.gov/30519799/ (accessed on 25 February 2022). [CrossRef]
15. León, I.E.; Butenko, N.; Di Virgilio, A.L.; Muglia, C.I.; Baran, E.J.; Cavaco, I.; Etcheverry, S.B. Vanadium and cancer treatment: Antitumoral mechanisms of three oxidovanadium(IV) complexes on a human osteosarcoma cell line. *J. Inorg. Biochem.* **2014**, *134*, 106–117. Available online: https://pubmed.ncbi.nlm.nih.gov/24199985/ (accessed on 25 February 2022). [CrossRef]
16. Rodríguez, M.R.; Balsa, L.M.; Del Pla, J.; García-Tojal, J.; Pis-Diez, R.; Parajón-Costa, B.S.; León, I.E.; González-Baró, A.C. Synthesis, characterization, DFT calculations and anticancer activity of an new Oxidovanadium(IV) complex with a ligand derived from o-vanillin and thiophene. *New J. Chem.* **2019**, *43*, 11784–11794. [CrossRef]
17. Lewis, N.A.; Liu, F.; Seymour, L.; Magnusen, A.; Erves, T.R.; Arca, J.F.; Beckford, F.A.; Venkatraman, R.; González-Sarrías, A.; Fronczek, F.R.; et al. Synthesis, characterization, and preliminary in vitro studies of vanadium(IV) complexes with a Schiff base and thiosemicarbazones as mixed-ligands. *Eur. J. Inorg. Chem.* **2012**, *2012*, 664–677. Available online: https://pubmed.ncbi.nlm.nih.gov/23904789/ (accessed on 25 February 2022). [CrossRef]
18. Bruck, R.; Halpern, Z.; Aeed, H.; Shechter, Y.; Karlish, S.J.D. Vanadyl ions stimulate K+ uptake into isolated perfused rat liver via the Na+/K+-pump by a tyrosine kinase-dependent mechanism. *Pflug. Arch.* **1998**, *435*, 610–616. Available online: https://pubmed.ncbi.nlm.nih.gov/9479013/ (accessed on 25 February 2022). [CrossRef]
19. Thompson, K.H.; Liboiron, B.D.; Sun, Y.; Bellman, K.D.D.; Setyawati, I.A.; Patrick, B.O.; Karunaratne, V.; Rawji, G.; Wheeler, J.; Sutton, K.; et al. Preparation and characterization of vanadyl complexes with bidentate maltol-type ligands; in vivo comparisons of anti-diabetic therapeutic potential. *J. Biol. Inorg. Chem.* **2003**, *8*, 66–74. Available online: https://pubmed.ncbi.nlm.nih.gov/12 459900/ (accessed on 25 February 2022). [CrossRef]

20. Cortizo, A.M.; Etcheverry, S.B. Vanadium derivatives act as growth factor–mimetic compounds upon differentiation and proliferation of osteoblast-like UMR106 cells. *Mol. Cell Biochem.* **1995**, *145*, 97–102. Available online: https://pubmed.ncbi.nlm.nih.gov/7675039/ (accessed on 25 February 2022). [CrossRef]
21. Barrio, D.A.; Etcheverry, S.B. Vanadium and bone development: Putative signaling pathways. *Can. J. Physiol. Pharmacol.* **2006**, *84*, 677–686. [CrossRef] [PubMed]
22. León, I.E.; Cadavid-Vargas, J.F.; Resasco, A.; Maschi, F.; Ayala, M.A.; Carbone, C.; Etcheverry, S.B. In vitro and in vivo antitumor effects of the VO-chrysin complex on a new three-dimensional osteosarcoma spheroids model and a xenograft tumor in mice. *J. Biol. Inorg. Chem.* **2016**, *21*, 1009–1020. [CrossRef] [PubMed]
23. Galluzzi, L.; Baehrecke, E.H.; Ballabio, A.; Boya, P.; Bravo-San Pedro, J.M.; Cecconi, F.; Choi, A.M.; Chu, C.T.; Codogno, P.; Colombo, M.I.; et al. Molecular definitions of autophagy and related processes. *EMBO J.* **2017**, *36*, 1811–1836. [CrossRef] [PubMed]
24. Kioseoglou, E.; Petanidis, S.; Gabriel, C.; Salifoglou, A. The chemistry and biology of vanadium compounds in cancer therapeutics. *Coord Chem. Rev.* **2015**, *301–302*, 87–105. [CrossRef]
25. Nechay, B.R.; Saunders, J.P. Inhibition by vanadium of sodium and potassium dependent adenosinetriphosphatase derived from animal and human tissues. *J. Environ. Pathol. Toxicol.* **1978**, *2*, 247–262. Available online: https://pubmed.ncbi.nlm.nih.gov/216760/ (accessed on 25 February 2022).
26. McLauchlan, C.C.; Hooker, J.D.; Jones, M.A.; Dymon, Z.; Backhus, E.A.; Greiner, B.A.; Dorner, N.A.; Youkhana, M.A.; Manus, L.M. Inhibition of acid, alkaline, and tyrosine (PTP1B) phosphatases by novel vanadium complexes. *J. Inorg. Biochem.* **2010**, *104*, 274–281. Available online: https://pubmed.ncbi.nlm.nih.gov/20071031/ (accessed on 25 February 2022). [CrossRef]
27. García, Z.; Kumar, A.; Marqués, M.; Cortés, I.; Carrera, A.C. Phosphoinositide 3-kinase controls early and late events in mammalian cell division. *EMBO J.* **2006**, *25*, 655–661. Available online: https://pubmed.ncbi.nlm.nih.gov/16437156/ (accessed on 25 February 2022). [CrossRef]
28. Meloche, S.; Pouysségur, J. The ERK1/2 mitogen-activated protein kinase pathway as a master regulator of the G1- to S-phase transition. *Oncogene* **2007**, *26*, 3227–3239. Available online: https://pubmed.ncbi.nlm.nih.gov/17496918/ (accessed on 25 February 2022). [CrossRef]
29. Wu, J.X.; Hong, Y.H.; Yang, X.G. Bis(acetylacetonato)-oxidovanadium(IV) and sodium metavanadate inhibit cell proliferation via ROS-induced sustained MAPK/ERK activation but with elevated AKT activity in human pancreatic cancer AsPC-1 cells. *J. Biol. Inorg. Chem.* **2016**, *21*, 919–929. [CrossRef]
30. Pisano, M.; Arru, C.; Serra, M.; Galleri, G.; Sanna, D.; Garribba, E.; Palmieri, G.; Rozzo, C. Antiproliferative activity of vanadium compounds: Effects on the major malignant melanoma molecular pathways. *Metallomics* **2019**, *11*, 1687–1699. [CrossRef]
31. Srivastava, S.; Kumar, N.; Roy, P. Role of ERK/NFκB in vanadium (IV) oxide mediated osteoblast differentiation in C3H10t1/2 cells. *Biochimie* **2014**, *101*, 132–144. Available online: https://pubmed.ncbi.nlm.nih.gov/24440756/ (accessed on 25 February 2022). [CrossRef] [PubMed]
32. Courtney, K.D.; Corcoran, R.B.; Engelman, J.A. The PI3K pathway as drug target in human cancer. *J. Clin. Oncol.* **2010**, *28*, 1075–1083. Available online: https://pubmed.ncbi.nlm.nih.gov/20085938/ (accessed on 25 February 2022). [CrossRef] [PubMed]
33. Yu, Q.; Jiang, W.; Li, D.; Gu, M.; Liu, K.; Dong, L.; Wang, C.; Jiang, H.; Dai, W. Sodium orthovanadate inhibits growth and triggers apoptosis of human anaplastic thyroid carcinoma cells in vitro and in vivo. *Oncol. Lett.* **2019**, *17*, 4255–4262. Available online: https://pubmed.ncbi.nlm.nih.gov/30944619/ (accessed on 25 February 2022). [CrossRef] [PubMed]
34. Gonçalves, A.P.; Videira, A.; Soares, P.; Máximo, V. Orthovanadate-induced cell death in RET/PTC1-harboring cancer cells involves the activation of caspases and altered signaling through PI3K/Akt/mTOR. *Life Sci.* **2011**, *89*, 371–377. [CrossRef]
35. Petanidis, S.; Kioseoglou, E.; Hadzopoulou-Cladaras, M.; Salifoglou, A. Novel ternary vanadium-betaine-peroxido species suppresses H-ras and matrix metalloproteinase-2 expression by increasing reactive oxygen species-mediated apoptosis in cancer cells. *Cancer Lett.* **2013**, *335*, 387–396. [CrossRef]
36. Etcheverry, S.B.; Crans, D.C.; Keramidas, A.D.; Cortizo, A.M. Insulin-mimetic action of vanadium compounds on osteoblast-like cells in culture. *Arch. Biochem. Biophys.* **1997**, *338*, 7–14. Available online: https://pubmed.ncbi.nlm.nih.gov/9015381/ (accessed on 25 February 2022). [CrossRef]
37. Pandey, S.K.; Anand-Srivastava, M.B.; Srivastava, A.K. Vanadyl sulfate-stimulated glycogen synthesis is associated with activation of phosphatidylinositol 3-kinase and is independent of insulin receptor tyrosine phosphorylation. *Biochemistry* **1998**, *37*, 7006–7014. Available online: https://pubmed.ncbi.nlm.nih.gov/9578588/ (accessed on 25 February 2022). [CrossRef]
38. Khan, Z.; Bisen, P.S. Oncoapoptotic signaling and deregulated target genes in cancers: Special reference to oral cancer. *Biochim. Biophys. Acta* **2013**, *1836*, 123–145. Available online: https://pubmed.ncbi.nlm.nih.gov/23602834/ (accessed on 25 February 2022). [CrossRef]
39. León, I.E.; Cadavid-Vargas, J.F.; Tiscornia, I.; Porro, V.; Castelli, S.; Katkar, P.; Desideri, A.; Bollati-Fogolin, M.; Etcheverry, S.B. Oxidovanadium(IV) complexes with chrysin and silibinin: Anticancer activity and mechanisms of action in a human colon adenocarcinoma model. *J. Biol. Inorg. Chem.* **2015**, *20*, 1175–1191. Available online: https://pubmed.ncbi.nlm.nih.gov/26404080/ (accessed on 25 February 2022). [CrossRef]
40. Zhao, J.; Wu, J.X.; Yang, W. Expression of caspase-3, Bcl-2, and Bax in pentavalent vanadium-induced neuronal apoptosis. *Zhonghua Lao Dong Wei Sheng Zhi Ye Bing Za Zhi* **2013**, *31*, 589–592. Available online: https://pubmed.ncbi.nlm.nih.gov/24053958/ (accessed on 25 February 2022).

41. Zhao, J.; Wang, J.; Wu, J. Roles of cytochrome c, caspase-9, and caspase-3 in pentavalent vanadium-induced neuronal apoptosis. *Zhonghua Lao Dong Wei Sheng Zhi Ye Bing Za Zhi* **2014**, *32*, 664–667. Available online: https://pubmed.ncbi.nlm.nih.gov/25511266/ (accessed on 25 February 2022). [PubMed]
42. Taddei, M.L.; Giannoni, E.; Fiaschi, T.; Chiarugi, P. Anoikis: An emerging hallmark in health and diseases. *J. Pathol.* **2012**, *226*, 380–393. Available online: https://pubmed.ncbi.nlm.nih.gov/21953325/ (accessed on 25 February 2022). [CrossRef] [PubMed]
43. González-Villalva, A.; Fortoul, T.I.; Avila-Costa, M.R.; Piñón-Zarate, G.; Rodriguez-Lara, V.; Martínez-Levy, G.; Rojas-Lemus, M.; Bizarro-Nevarez, P.; Díaz-Bech, P.; Mussali-Galante, P.; et al. Thrombocytosis induced in mice after subacute and subchronic V2O5 inhalation. *Toxicol. Ind. Health* **2006**, *22*, 113–116. Available online: https://pubmed.ncbi.nlm.nih.gov/16716040/ (accessed on 25 February 2022). [CrossRef] [PubMed]
44. Gonzalez-Villalva, A.; Piñon-Zarate, G.; Falcon-Rodriguez, C.; Lopez-Valdez, N.; Bizarro-Nevares, P.; Rojas-Lemus, M.; Rendon-Huerta, E.; Colin-Barenque, L.; Fortoul, T.I. Activation of Janus kinase/signal transducers and activators of transcription pathway involved in megakaryocyte proliferation induced by vanadium resembles some aspects of essential thrombocythemia. *Toxicol. Ind. Health* **2016**, *32*, 908–918. [CrossRef] [PubMed]
45. Basu, A.; Singha Roy, S.; Bhattacharjee, A.; Bhuniya, A.; Baral, R.; Biswas, J.; Bhattacharya, S. Vanadium(III)- L -cysteine protects cisplatin-induced nephropathy through activation of Nrf2/HO-1 pathway. *Free Radic. Res.* **2016**, *50*, 39–55. [CrossRef] [PubMed]
46. León, I.E.; Díez, P.; Baran, E.J.; Etcheverry, S.B.; Fuentes, M. Decoding the anticancer activity of VO-clioquinol compound: The mechanism of action and cell death pathways in human osteosarcoma cells. *Metallomics* **2017**, *9*, 891–901. Available online: https://pubmed.ncbi.nlm.nih.gov/28581009/ (accessed on 25 February 2022). [CrossRef]
47. Balsa, L.M.; Quispe, P.; Baran, E.J.; Lavecchia, M.J.; León, I.E. In silico and in vitro analysis of FAK/MMP signaling axis inhibition by VO-clioquinol in 2D and 3D human osteosarcoma cancer cells. *Metallomics* **2020**, *12*, 1931–1940. Available online: https://pubmed.ncbi.nlm.nih.gov/33107537/ (accessed on 25 February 2022). [CrossRef]
48. Ciccimaro, E.; Hevko, J.; Blair, I.A. Analysis of phosphorylation sites on focal adhesion kinase using nanospray liquid chromatography/multiple reaction monitoring mass spectrometry. *Rapid Commun. Mass Spectrom.* **2006**, *20*, 3681–3692. Available online: https://pubmed.ncbi.nlm.nih.gov/17117420/ (accessed on 25 February 2022). [CrossRef]
49. Luo, M.; Fan, H.; Nagy, T.; Wei, H.; Wang, C.; Liu, S.; Wicha, M.S.; Guan, J.-L. Mammary epithelial-specific ablation of the focal adhesion kinase suppresses mammary tumorigenesis by affecting mammary cancer stem/progenitor cells. *Cancer Res.* **2009**, *69*, 466–474. Available online: https://pubmed.ncbi.nlm.nih.gov/19147559/ (accessed on 25 February 2022). [CrossRef]
50. Golubovskaya, V.M. Targeting FAK in human cancer: From finding to first clinical trials. *Front. Biosci.* **2014**, *19*, 687–706. Available online: https://pubmed.ncbi.nlm.nih.gov/24389213/ (accessed on 25 February 2022). [CrossRef]
51. McLean, G.W.; Avizienyte, E.; Frame, M.C. Focal adhesion kinase as a potential target in oncology. *Expert Opin. Pharm.* **2003**, *4*, 227–234. Available online: https://pubmed.ncbi.nlm.nih.gov/12562313/ (accessed on 25 February 2022). [CrossRef] [PubMed]
52. Wu, Y.; Ma, Y.; Xu, Z.; Wang, D.; Zhao, B.; Pan, H.; Wang, J.; Xu, D.; Zhao, X.; Pan, S.; et al. Sodium orthovanadate inhibits growth of human hepatocellular carcinoma cells in vitro and in an orthotopic model in vivo. *Cancer Lett.* **2014**, *351*, 108–116. Available online: https://pubmed.ncbi.nlm.nih.gov/24858025/ (accessed on 25 February 2022). [CrossRef] [PubMed]
53. Zhou, W.; Miao, Y.; Zhang, Y.; Liu, L.; Lin, J.; Yang, J.Y.; Xie, Y.; Wen, L. Induction of cyto-protective autophagy by paramontroseite VO2 nanocrystals. *Nanotechnology* **2013**, *24*, 165102. Available online: https://pubmed.ncbi.nlm.nih.gov/23535229/ (accessed on 25 February 2022). [CrossRef]
54. Petanidis, S.; Kioseoglou, E.; Domvri, K.; Zarogoulidis, P.; Carthy, J.M.; Anestakis, D.; Moustakas, A.; Salifoglou, A. In vitro and ex vivo vanadium antitumor activity in (TGF-β)-induced EMT. Synergistic activity with carboplatin and correlation with tumor metastasis in cancer patients. *Int. J. Biochem. Cell Biol.* **2016**, *74*, 121–134. [CrossRef] [PubMed]
55. El-Shafey, E.S.; Elsherbiny, E.S. Possible Selective Cytotoxicity of Vanadium Complex on Breast Cancer Cells Involving Pathophysiological Pathways. *Anticancer Agents Med. Chem.* **2019**, *19*, 2130–2139. Available online: https://pubmed.ncbi.nlm.nih.gov/31696812/ (accessed on 25 February 2022). [CrossRef] [PubMed]
56. Kumar, R.; Juillerat-Jeanneret, L.; Golshayan, D. Notch Antagonists: Potential Modulators of Cancer and Inflammatory Diseases. *J. Med. Chem.* **2016**, *59*, 7719–7737. Available online: https://pubmed.ncbi.nlm.nih.gov/27045975/ (accessed on 25 February 2022). [CrossRef]
57. Lim, J.S.; Ibaseta, A.; Fischer, M.M.; Cancilla, B.; O'Young, G.; Cristea, S.; Luca, V.C.; Yang, D.; Jahchan, N.S.; Hamard, C.; et al. Intratumoural heterogeneity generated by Notch signalling promotes small-cell lung cancer. *Nature* **2017**, *545*, 360–364. Available online: https://pubmed.ncbi.nlm.nih.gov/28489825/ (accessed on 25 February 2022). [CrossRef]
58. Shen, S.; Chen, X.; Hu, X.; Huo, J.; Luo, L.; Zhou, X. Predicting the immune landscape of invasive breast carcinoma based on the novel signature of immune-related lncRNA. *Cancer Med.* **2021**, *10*, 6561–6575. [CrossRef]
59. Moellering, R.E.; Cornejo, M.; Davis, T.N.; Bianco, C.D.; Aster, J.C.; Blacklow, S.C.; Kung, A.L.; Gilliland, D.G.; Verdine, G.L.; Bradner, J.E. Direct inhibition of the NOTCH transcription factor complex. *Nature* **2009**, *462*, 182–188. Available online: https://pubmed.ncbi.nlm.nih.gov/19907488/ (accessed on 25 February 2022). [CrossRef]

Article

Antioxidant and Anticancer Activities and Protein Interaction of the Oxidovanadium(IV) Naringin Complex

Andrés Gonzalo Restrepo-Guerrero [1], Helen Goitia-Semenco [1], Luciana G. Naso [1], Marilin Rey [2], Pablo J. Gonzalez [2], Evelina G. Ferrer [1] and Patricia A. M. Williams [1,*]

[1] Centro de Química Inorgánica (CEQUINOR, UNLP, CONICET, Asociado a CICPBA), Departamento de Química, Facultad de Ciencias Exactas, Universidad Nacional de La Plata, Bv. 120 N° 1465, La Plata B1906, Argentina; gonzalo.restrepo@quimica.unlp.edu.ar (A.G.R.-G.); helengoitia@quimica.unlp.edu.ar (H.G.-S.); luciananaso504@hotmail.com (L.G.N.); evelina@quimica.unlp.edu.ar (E.G.F.)

[2] Departamento de Física, Facultad de Bioquímica y Ciencias Biológicas, Universidad Nacional del Litoral and CONICET, Santa Fe S3000, Argentina; mrey@fbcb.unl.edu.ar (M.R.); pablogonzalez1979@gmail.com (P.J.G.)

* Correspondence: williams@quimica.unlp.edu.ar

Abstract: The complex of oxidovanadium(IV) with naringin (Narg) [VO(Narg)$_2$] 8H$_2$O (VONarg) was prepared according to the literature improving the synthetic procedure and physicochemical characterization. In addition, biological activities (cytotoxic, antioxidant, and BSA interaction) were determined. The metal coordinated through the 5-hydroxy and 4-carbonyl groups of rings A and C of naringin, respectively. The antioxidant activity of VONarg, determined in vitro, was higher than those of the flavonoid against superoxide and peroxyl reactive oxygen species (ROS) and DPPH radical. The cytotoxic properties were determined by a MTT assay on adenocarcinoma human alveolar basal epithelial cells (A549). VONarg exerted a 20% decrease in cancer cells viability at 24 h incubation, while naringin and oxidovanadium(IV) cation did not show cytotoxicity. Measurements with the normal HEK293 cell line showed that the inhibitory action of the complex is selective. VONarg generated intracellular reactive oxygen species (ROS), depletion of reduced glutathione and depolarization of mitochondrial membrane potential, typical for apoptotic pathway, producing cell death by oxidative stress mechanism. Moreover, naringin interacted with bovine serum albumin (BSA) through hydrophobic interactions in a spontaneous process, and VONarg showed greater affinity for the protein but can still be transported and delivered by it (K$_a$ 10^4 L·mol^{-1} order).

Keywords: glycosylated flavonoid; oxidovanadium(IV) complex; antitumoral; antioxidant

Citation: Restrepo-Guerrero, A.G.; Goitia-Semenco, H.; Naso, L.G.; Rey, M.; Gonzalez, P.J.; Ferrer, E.G.; Williams, P.A.M. Antioxidant and Anticancer Activities and Protein Interaction of the Oxidovanadium(IV) Naringin Complex. *Inorganics* **2022**, *10*, 13. https://doi.org/10.3390/inorganics10010013

Academic Editor: Dinorah Gambino

Received: 18 November 2021
Accepted: 13 January 2022
Published: 15 January 2022

Publisher's Note: MDPI stays neutral with regard to jurisdictional claims in published maps and institutional affiliations.

1. Introduction

Flavonoids are an important group of natural substances, containing phenolic structures, which can be found in fruits, vegetables, and some beverages [1]. They are benzo-γ-pyrone derivatives consisting of aromatic (A and B) and pyrene (C) rings (Figure 1A) and are classified according to structural changes in hydroxylation pattern, conjugation between the aromatic rings, glycosidic moieties, and methoxy group [2,3]. These secondary metabolites have shown many and different applications, such as antiviral activity against zika virus infection [4], anti-hyperglycemic, liver protective [5], and anti-rheumatoid arthritis effects [6], antigenotoxic activity [7], and modulation of cardiovascular K$^+$ channels [8], just to mention some cutting-edge studies from the last few years. It is also well known these compounds have anti-oxidative, anti-inflammatory, anti-mutagenic, and anti-carcinogenic properties coupled with their capacity to modulate the cellular key and enzyme function [1].

Another important feature of flavonoids is their capacity to chelate metal ions and form coordination complexes through hydroxyl and keto groups [9]. There are three main chelating sites for metal ions in flavonoids, e.g., the 3′,4′-dihydroxy group located on the B ring, the 3-hydroxy or 5-hydroxy, and the 4-carbonyl group in the C ring [3]. In glycosylated flavonoids, metal coordination can also take place through hydroxy groups of the saccharide moiety [10].

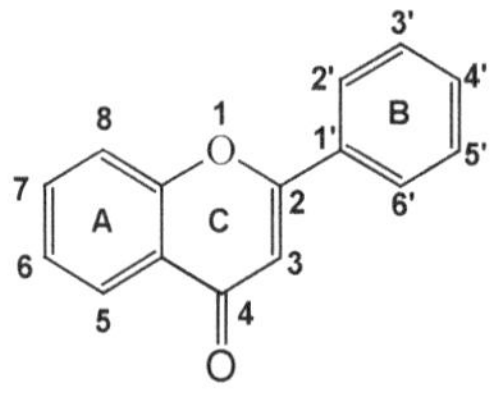

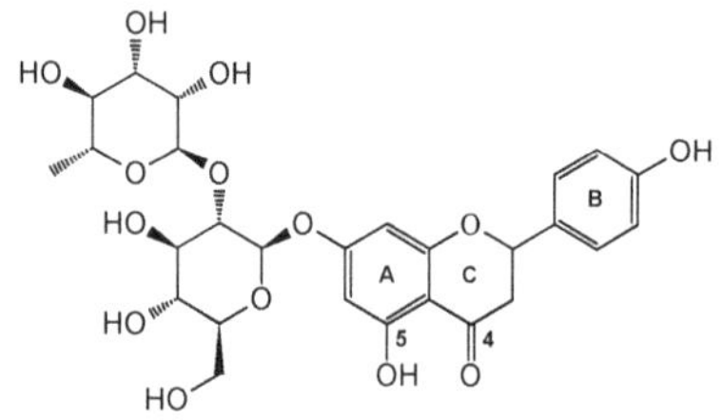

(A)   (B)

Figure 1. (**A**) Flavone structure. (**B**) Naringin structure.

Naringin (Figure 1B) is a natural flavanone glycoside biosynthesized via the phenylpropanoid pathway from shikimic acid [11]. It is mainly found in citrus fruits, such as grapefruits, lemons and oranges [12], also apples, onions, and tea [13]. Recent studies have demonstrated its numerous in vivo and in vitro properties exerting potential therapeutic effects by modulating various protein and enzyme expressions [14]. It possesses properties such as anti-inflammatory effects [15,16], oxidative stress reduction [17,18], and it has proven to be one potential anticancer agent against different types of cancer and also acted synergistically enhancing the anticancer activities of antitumor drugs in combination therapies [19]. Molecular docking studies showed that naringin interacted with COVID-19 main protease and displayed higher binding affinity than other flavonoids [20].

Metal-flavonoid chelates have a considerably better biological activity than their flavonoid and metal precursors on their own and we have reported this behavior in previous works with other flavonoids, such as hesperidin, diosmin, and rutin and their oxidovanadium(IV) complexes [21–23].

Vanadium derivatives have shown insulin-mimetic, antidiabetic properties, and antitumor properties; vanadium compounds are promising non-platinum anticancer agents due to their low IC_{50}, low toxicity, along with antiproliferative, genotoxic, and proapoptotic effects by modulation of important protein function and reactive oxygen species (ROS) action [24].

Naringin is capable of binding the oxidovanadium(IV) cation through two different sites (the flavonoid or the saccharide moiety). At the experimental conditions selected herein, we have modified the structure of the ligand at the flavonoid moiety (pH 9) and compared the physicochemical, antioxidant, cytotoxic, and protein interaction properties of both the complex and the ligand.

2. Results

2.1. Characterization of V(IV)O–Naringin Complex

2.1.1. Fourier Transformed Infrared Spectroscopy

The FTIR main vibrational bands of naringin and the oxidovanadium(IV) complex are presented in Table 1. The assignments were performed according to general references of VOflavonoid systems [21]. Similar broad bands for the ligand and the complex and assigned to the OH stretching modes (including OH groups of crystallization H_2O for the complex) could be seen at the 3400–3200 cm^{-1} region. The band due to C=O stretching at 1645 cm^{-1} shifted in the complex to 1637 cm^{-1} because of the coordination of this group to the metal ion that produced a decrease in bond order and an increase in bond length. The bands related to the C=C stretching modes of the aromatic group (1630–1500 cm^{-1}) showed the expected changes due to the resonance established by coordination with the oxidovanadium(IV) cation. Modes involving OH stretching and COH bending (1360–1000 cm^{-1})

were modified and/or reduced the intensities due to deprotonation or coordination of the C_5-OH group, indicating the interaction of the metal center through this group: the bending COH modes at 1355 cm^{-1} and 1341 cm^{-1} appeared at 1357 cm^{-1}, the band at 1295 cm^{-1} decreased its intensity and the bands at 1281 cm^{-1} and 1265 cm^{-1} shifted to 1256 cm^{-1}. Bands related to C-O stretching also decreased their intensities. However, the band related to C-O stretching of the sugar moiety at 1074 cm^{-1} did not change upon metal coordination to naringin, indicating that the glycoside group did not participate in the coordination. Moreover, the V=O stretching appeared at 980 cm^{-1} as with naringenin (VONar), so the typical coordination through 5-O$^-$ and 4-C=O groups of A and C rings of naringin, respectively, can be assumed and the coordination of the metal center through the sugar moiety can be discarded. In this latter case, the V=O stretching appeared at low energies (ca. 920 cm^{-1}) [25]. This band appeared overlapped with the band of the ligand at 987 cm^{-1}, previously assigned to V=O stretching.

Table 1. Assignment of the FTIR spectra of naringin and the oxidovanadium(IV) complex (band positions in cm^{-1}) [a].

Naringin	VONarg	Vibrational Modes-Functional Groups
3422 br 3231 sh	3395 br 3205 sh	ν O-H
1645 vs	1637 sh	ν C=O ring C
1629 sh 1615 sh	1614 vs	ν C=C
1582 m	1574 vs 1537 m	ν C=C
1520 m 1504 m	1520 m	ν C=C
1355 sh 1341 m	1357 sh	δ COH
1295 m	1292 w	δ HOC
1281 w 1265 w	1256 w	δ HOC, ν (C-O-C)
1134 s	1134 m	ν C-O secondary alcohol
1074 vs	1076 vs	ν O-C sugar
1062 vs	1060 sh	ν C-O primary alcohol
1041 vs	1040 sh	ν O-C
987 m	987 m	ν O-C
	980 m	ν V=O
822 m	814 w	ν C-C, ν O-C

[a] vs, very strong; s, strong; m, medium; w, weak br, broad; sh, shoulder.

2.1.2. Powder and Frozen Solution EPR Spectra

The powder EPR spectrum of the VONarg complex obtained at 120 K is shown in Figure 2A. The spectrum consisted of a weak—although detectable—signal with a hyperfine structure produced by the I = 7/2 nuclear spin from magnetically isolated V^{4+} ions (inset of Figure 2A), superimposed with a broad resonance line attributed to metal sites with spin–spin exchange interactions between neighboring V^{4+} sites.

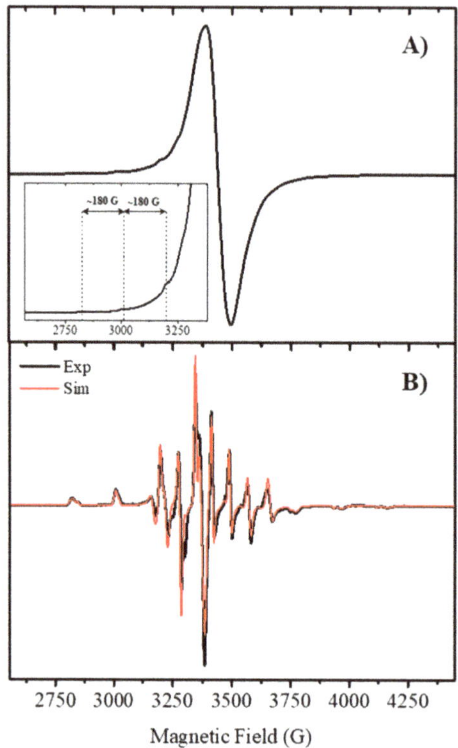

Figure 2. EPR spectrum of VONarg: (**A**) Powdered sample at 120 K, (**B**) Experimental spectrum (black) and simulation (red) of frozen pure DMSO solution at 120 K.

The EPR spectra of the frozen solutions in both pure DMSO and in H_2O/DMSO mixtures (ratios 50/50, 90/10, and 99/1, pH ca. 7) measured at 120 K showed minor differences than that obtained in pure DMSO (see Figures 2B and S1). They showed the typical axial spectrum with a hyperfine structure produced by the [51] V nucleus. The spectrum seemed to be composed of two species (labeled as S1 and S2) with different contributions (1:0.2). Simulations yielded the spin Hamiltonian parameters shown in Table 2.

Table 2. Spin Hamiltonian parameters obtained from simulations.

	$g_\parallel$	$g_\perp$	$^{(1)}A_\parallel$	$^{(1)}A_\perp$	$\Delta g_\parallel/\Delta g_\perp$	$^{(1)}P$	k	$^{(1)}P \times k$
S1	1.9332	1.9717	167.0	62.8	2.26	121.8	0.72	87.3
S2	1.9398	1.9751	159.4	51.6	2.30	125.8	0.62	78.2

$^{(1)}$ A, P and $P \times k$ are given in $\times 10^{-4}$ cm^{-1}.

For both S1 and S2 the $g_\parallel < g_\perp < g_e = 2.0023$ and $|A_\parallel| > |A_\perp|$, indicating an octahedral geometry with tetragonal compression with a d_{xy} ground state, and the ratio $\Delta g_\parallel/\Delta g_\perp$ indicated considerable tetragonal distortion. According to the empirical relationship $A_z = \sum(n_i \cdot A_{z,i})$ used to determine the identity of the equatorial ligands in V(IV) complexes (accuracy of $\pm 3 \times 10^{-4}$ cm^{-1}) (n_i, number of equatorial ligands of type i and $A_{z,i}$, the contribution to the parallel hyperfine coupling from each of them) [26], and considering the contributions to the parallel hyperfine coupling constant of the different coordination modes (CO = 44.7×10^{-4} cm^{-1}, ArO$^-$ = 38.6×10^{-4} cm^{-1}), S1 was consistent with the proposed structure, with 2 C=O and 2 ArO$^-$ moieties in the equatorial plane (A$_\parallel$ = 166.6 cm^{-1}) that should correspond to a square pyramidal geometry. However, a decrease in A$_{II}$ (as in Table 2) was found by Gorelsky et al. [27] when the solvent coordinated to V=O in the axial position, so it could be proposed that DMSO (or the solvent H$_2$O) may coordinate to the sixth position, *trans-* to the metal center, in S2. However, vanadium coordination to the sugar moieties with (O$^-$O$^-$) donor set (A$_{II}$ ca. 155–157 × 10^{-4} cm^{-1}), could not be ruled out [22,28].

Using the relations developed by Kivelson and Lee [29],

$$A_{\parallel} = -P\left[k + \frac{4}{7} - \Delta g_{\parallel} - \frac{3}{7}\Delta g_{\perp}\right] \tag{1}$$

$$A_{\perp} = -P\left[k - \frac{2}{7} - \frac{11}{4}\Delta g_{\perp}\right] \tag{2}$$

the *p*-value (dipolar hyperfine coupling parameter) that represents the dipole–dipole interaction of the electronic and nuclear moments can be obtained. *p*-values range from 100 to 160 × 10^{-4} cm^{-1} in oxidovanadium(IV) compounds [30] and are calculated as P = g$_e$g$_N$μ$_B$μ$_N$‹r^{-3}›, where g$_N$ is the nuclear g-factor, g$_e$ is the g-factor of the free electron, μ$_N$ the nuclear magneton, and ‹r^{-3}› can be calculated for the vanadium 3d orbitals. The parameter k is the dimensionless Fermi contact interaction constant, ranging from 0.6 to 0.9 [31], is very sensitive to deformations of the metal orbitals, and indicates the isotropic Fermi contact contribution to the hyperfine coupling. The calculated values of *p* = 120–125 × 10^{-4} cm^{-1} (Table 2) were considerably reduced when compared to the value of the free ion (160 × 10^{-4} cm^{-1}) and indicated a considerable amount of covalent bonding in the complex. The values of k = 0.62–0.72 indicated a moderate contribution to the hyperfine constant by the unpaired s-electron.

2.1.3. Spectrophotometric Titrations

The spectral bands for the visible part of the spectrum of the complex in DMSO, located at 598 nm and 810 nm are typical for VOflavonoid complexes interacting by the 4-carbonyl and 5-hydroxy groups of rings A and C (Figure 1) [32]. Spectrophotometric titration in DMSO was performed at different ligand-to-metal ratios from 0.5 to 10.0 and pH 9. The absorbance of the band at 810 nm in the electronic spectra was monitored at each molar ratio. From Figure 3, a L:M 2:1 stoichiometry was determined.

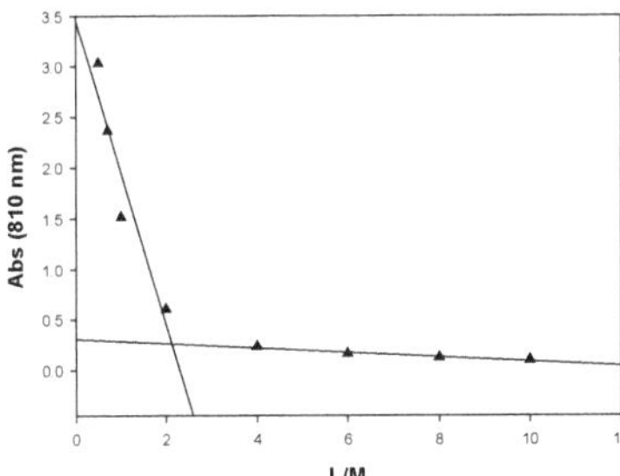

Figure 3. Spectrophotometric titration of V^{IV}O^{2+} with naringin at pH = 9, under nitrogen atmosphere. L/M, naringin/ V^{IV}O^{2+} molar ratio. Absorbance at 810 nm.

2.2. Antiradical Behavior

For biological studies, it is important to measure, in the first instance, the stability of the complex. The stability of VONarg in DMSO solution was followed by UV-vis spectroscopy and conductivity measurements over time. Both the spectral pattern (Figure S2A) and the molar conductivities (data not shown) did not change at least during the first 15 min of manipulation (for biological studies), demonstrating that the compounds were stable in DMSO. Although a spectral shift due to solvatochromic effects was observed between DMSO and DMSO/water (1/99) solutions, the complex behaved in the same way as DMSO during the first 15 min (Figure S2B).

From Table 3 an enhancement on the superoxide dismutase (SOD) activity of the complex concerning naringin can be observed, but the effect was still low ($IC_{50} = 870$ μM), hence, the complex could not be considered as a good SOD simil agent. The scavenging activity for peroxyl radical (ROO$^\bullet$) was enhanced by complexation. The complex caused a delay in pyranine consumption (lag phase) by the ROO$^\bullet$ radical of 3.8 min at 100 μM. However, this delay was not so effective as that produced by the reference compound Trolox. No antiradical effect against hydroxyl (OH$^\bullet$) was measured for the complex, though naringin produced a 30% scavenge of hydroxyl radical at 100 μM concentration. Oxidovanadium(IV) complexation showed much stronger 2,2′-diphenyl-1-picrylhydrazyl radical (DPPH$^\bullet$) scavenging than the ligand, and the same effect has previously been reported for the copper/naringin metal complex [33]. In summary, metal complexation improved the antioxidant behavior of the flavonoid against the ROS superoxide and peroxyl and also against the DPPH$^\bullet$ radical. These results were obtained when the coordination, as in VONarg, occurred through C=O and deprotonated O moieties (from 4-carbonyl and 5-hydroxyl, respectively). This acetylacetonate-type coordination enhanced π delocalization between V=O and the ligand, which conferred a high capacity to stabilize unpaired electrons and then scavenge free radicals. However, the antioxidant behavior of naringin was moderate because of the lack of a double bond between C2 and C3 and the absence of this double bond also forbids the electronic resonance of the generated radicals with ring B and only ring A was involved in π delocalization with the metal center.

Table 3. Percentage of free radical scavenging of naringin, VONarg, and oxidovanadium(IV) cation. Values are expressed as the mean ± standard error of at least three independent experiments.

Radical	% Scavenging		
	Naringin	VONarg	V(IV)O^{2+}
SOD (IC_{50}, μM)	>1000	870 ± 5.2	15 ± 0.2
ROO$^\bullet$, lag (min), 100 μM	0	3.8 ± 0.8	6.4 ± 1.1
OH$^\bullet$, 100 μM	28 ± 0.4	0	38 ± 2
DPPH$^\bullet$, 100 μM	2.0 ± 0.7	45.0 ± 5.0	37.0 ± 2

IC_{50} SOD$_{native}$ = 0.21 μM; lag phase Trolox, 37.28 min.

2.3. Anticancer Effects

Numerous studies have shown that naringin inhibits cell proliferation and induces apoptosis in a majority of tumor cells, including breast cancer (TNBC), human cervical cancer (SiHa), and 5637 bladder cancer cells [19]. The cytotoxicity of the compounds was investigated by the MTT assay. Results are expressed as a percent with respect to control values. In particular, in the human lung cancer A549 cells naringin induced no cytotoxicity at the tested concentrations (up to 500 μM at 24 h incubation). On the contrary, VONarg reduced cell viability in a dose response manner with a 20% reduction at 100 μM (Figure 4) and 40% at 500 μM. In addition, the oxidovanadium(IV) cation did not show significant cytotoxic effects up to 100 μM, 24 h incubation [34]. Moreover, the complex did not reduce the cell viability of the non-tumorigenic HEK293 cell line (0–100 μM) (Figure S3).

In addition, alterations in cell morphology as a consequence of the exposition to VONarg were observed using Giemsa staining (Figure 5). Control A549 cells displayed a typical epithelial-like morphology. Upon treatment with VONarg, 100 μM concentration for 24 and 48 h, different degrees of morphological changes were observed. The complex caused cytoplasmic shrinkage, elongated lamellipodia, and moderate cell population decrease at 24 h incubation. The formation of apoptotic bodies was observed at 48 h incubation. It can be seen that the cytotoxic effect of the complex was greatly increased with the incubation time.

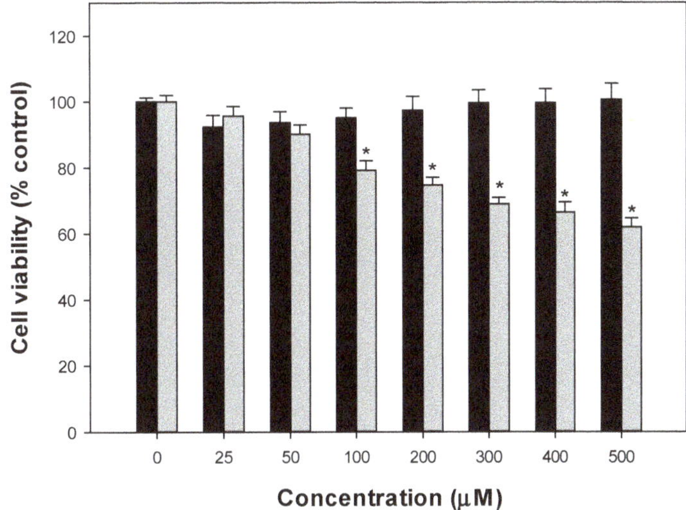

Concentration (µM)

Figure 4. Effects of naringin (black) and VONarg (grey) on A549 cell line viability. Cells were incubated in serum-free Dulbecco's modified Eagle's medium (DMEM) alone (control) or with different concentrations of the compounds at 37 °C for 24 h. The results are expressed as the percentage of the control level and represent the mean ± SEM. * Significant difference with control at the 0.05 level.

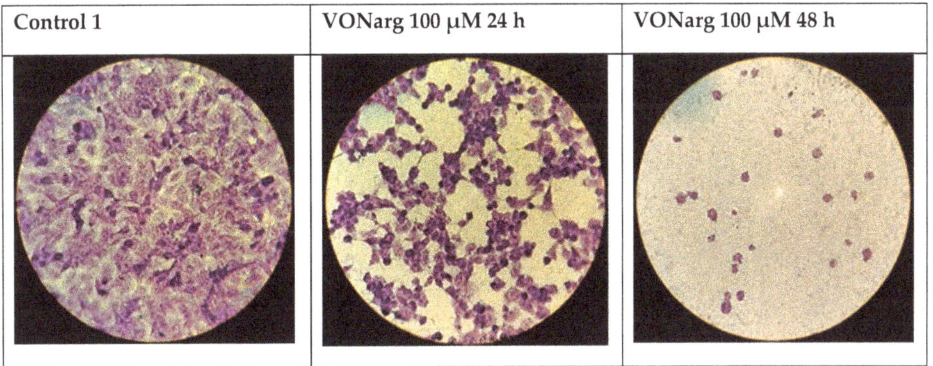

Figure 5. Effect on cell morphology for the treatment of A549 cell line with VONarg. Cells were incubated for 24 and 48 h without drug additions (control) and with the complex (100 µM, 400× *g* magnification).

It is known that cell growth and death are regulated by the redox status of the environment and excessive production of ROS and low GSH levels induce apoptosis or necrosis. To evaluate if the toxic mechanism of action occurs through the induction of intracellular ROS production, A549 cells incubated with different doses of the compounds were assessed using the oxidant-sensitive dye H_2DCFDA. From Figure 6, it can be seen that while naringin did not generate excessive cellular ROS levels, VONarg produced an increase in ROS (250% of the control level at 100 µM and 24 h incubation). Furthermore, the oxidovanadium(IV) cation did not produce ROS in the A549 cell line, 24 h incubation [34].

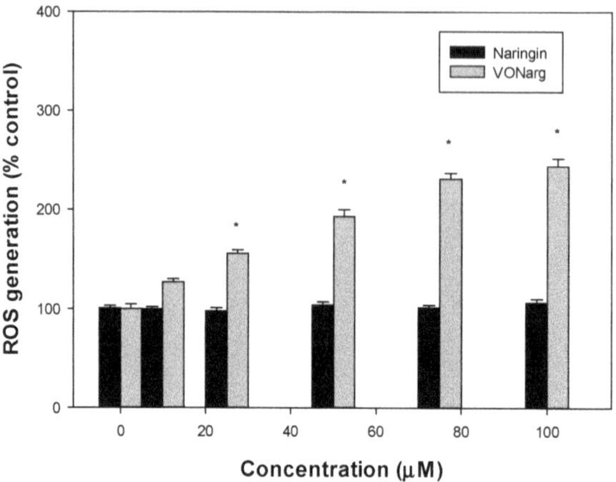

Figure 6. Effects of naringin (black) and VONarg (grey) on H_2DCFDA oxidation to DCF. A549 cells were incubated at 37 °C in the presence of 10 µM H_2DCFDA. The values are expressed as the percentage of the control level and represent the mean ± SEM. * significant difference with control at the 0.05 level.

GSH is one of the natural antioxidants that prevent the oxidation of cellular components. Cellular redox imbalance is also demonstrated by the reduction in endogenous GSH production at the same time that cells are producing increased levels of ROS. When the tripeptide non-protein thiol compound, GSH, acted as a defense system against oxidative stress it oxidized to GSSG. Therefore, GSH/GSSG ratio is a better signal of oxidative stress generation. Treatment with VONarg caused a decrease in GSH and the GSH/GSSG ratio levels in a dose–response manner (Figure 7).

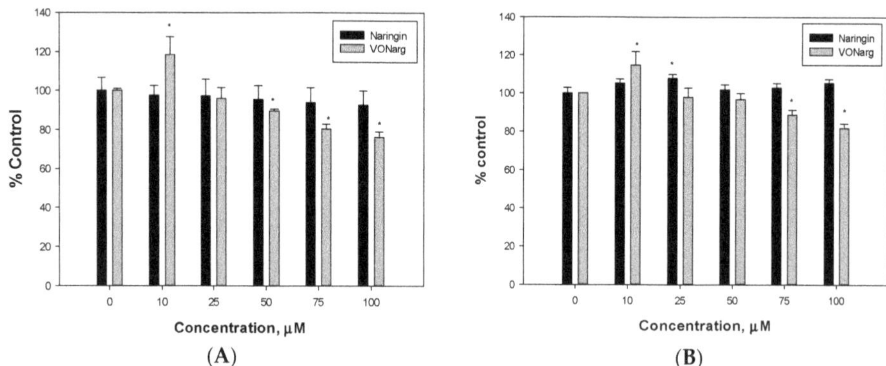

Figure 7. Effect of naringin and VONarg on GSH cellular levels (**A**) and GSH/GSSG ratio (**B**) in A549 cells. Results are expressed as mean ± SEM of three independent experiments, * significant differences in comparison with the control level ($p < 0.05$).

To elucidate the mechanism of action of the complex under oxidative stress conditions, the mitochondrial membrane potential at different concentrations of VONarg was measured (Figure 8). $DioC_6$ is a lipophilic cationic probe that is able to enter into the mitochondria and when depolarization occurs, it will accumulate less dye. VONarg caused mitochondrial depolarization as evidenced by the decrease in membrane potential in response to the damage produced by ROS increment. Then, VONarg could induce apoptosis of A549 cells by activating mitochondrial death pathways related to ROS and GSSG generation.

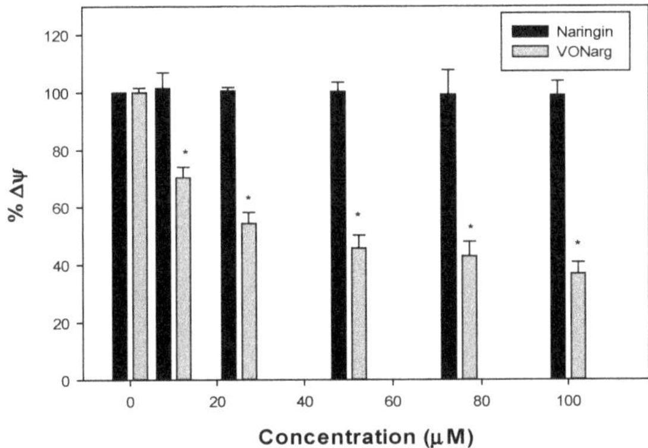

Figure 8. Changes of the mitochondrial membrane potential (% $\Delta\psi$) in A549 cells treated with increasing concentrations naringin (black) and VONarg (grey) for 24 h. Each point represents the mean $\pm$ S.E.M of three measurements in three independent experiments. Significant differences: * $p < 0.05$ compared to the control.

2.4. Bovine Serum Albumin (BSA) Interactions

Many drugs circulate reversibly in the bloodstream bound to plasma protein. Human serum albumin is the best carrier for acidic and neutral drugs and influences their pharmacokinetics and pharmacodynamics. It can be used in cosmetics, food technology, crop protection, industrial chemical processes, having great potential in pharmacology because of its ability to increase solubility for lipophilic drugs, allowing their interaction with the site of action. Thus, the interaction of a drug with plasma proteins is a relevant factor influencing the distribution, elimination, duration, and intensity of their pharmacological activities [35]. Consequently, the albumin-binding assay may provide some information for the pathway for bioactive compound transportation. In this sense, bovine serum albumin (BSA) is often used as a model to study in vitro interactions because of its similarity to HSA, stability, low cost, and intrinsic fluorescence properties originated by two tryptophan residues (Trp-134 and Trp-212). BSA is very sensitive to the alteration of its tertiary structure, and this effect can be followed by fluorescence spectroscopy.

To analyze the binding affinity of naringin and VONarg the fluorescence spectra of BSA were recorded in the presence of increasing quantities of the flavonoid and the complex in the 5–100 μM range. As shown in Figure 9, the increment of the concentration of naringin and VONarg at 298 K produced a quenching process on the intrinsic albumin fluorescence. The reduction in fluorescence intensity suggests BSA-compound interactions and that, at the same concentration value, the effect of the quenching process is higher for the VONarg complex than the flavonoid. The observed intensity of fluorescence (F_{obs}) was corrected because the UV absorbance of the samples can decrease the observed values (inner-filter effect) [36,37]. Considering the absorbances at the exciting and emitted radiation A_{ex} and A_{em}, respectively, the corrected fluorescence intensity (F_{corr}) was determined using the following equation: $F_{corr} = F_{obs} \times e^{(A_{ex} + A_{em})/2}$.

Fluorescence quenching can be static or dynamic. In the static one, there exists a complex formation between the quencher and the fluorophore, while a collision process is described for the dynamic one. The Stern–Volmer equation was applied to determine the type of fluorescence quenching:

$$F_0/F = 1 + K_q \tau_0 [Q] = 1 + K_{sv}[Q] \tag{3}$$

where F_0 and F are the fluorescence intensities in the absence and presence of the quencher, respectively; K_q is the bimolecular quenching constant; τ_0 is the lifetime of the fluorophore in the absence of the quencher (usually taken as 1×10^{-8} s); [Q] is the concentration of the quencher and K_{sv} is the Stern–Volmer quenching constant. The calculated curves of F_0/F vs. [Q], depicted in Figure S4, showed a linear dependence that suggests a single quenching process, either static or dynamic. The K_{sv} values presented in Table 4 showed a slight increase correlated with temperatures and the calculated quenching rate constants (K_q) were higher than the limiting diffusion rate constant (2.0×10^{10} L·mol^{-1}·s^{-1}). Thus, the most probable quenching mechanism was initiated by the ground-state complex formation (static quenching), but the possible existence of combined quenching cannot be ruled out [38].

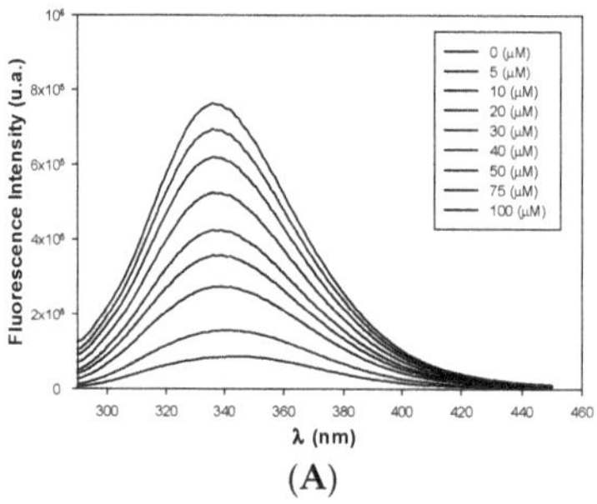
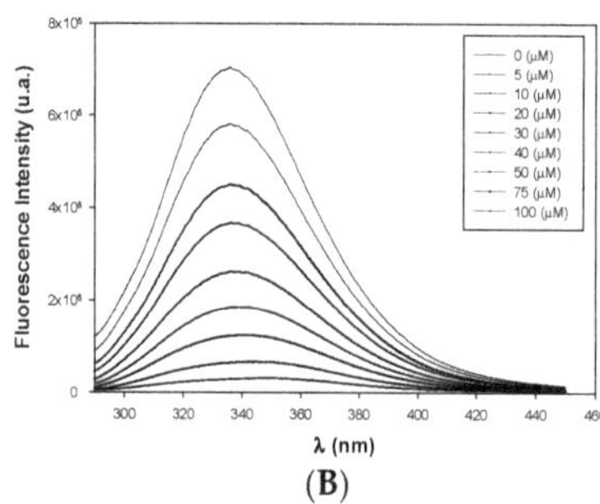

Figure 9. Fluorescence spectra of BSA in the presence of (**A**) naringin and (**B**) VONarg in Tris–HCl buffer of pH 7.4 at 298 K. BSA (6 µM), naringin and VONarg (0–1.00 µM).

For a fluorescence quenching of protein with a static quenching process, the free and bound molecule equilibrium can be analyzed by the following equation:

$$\log[(F_0 - F)/F] = \log K_b + n \log [Q] \tag{4}$$

where K_b is the binding constant (displaying the interaction degree of BSA and the quencher) and n is the number of binding sites (Table 4).

Figure 10 shows the resulting $\log [(F_o - F)/F]$ vs. $\log [Q]$ plots. The same tendency of the K_b constant with temperature increments can be seen and higher K_b values of VONarg in comparison with naringin support the strong interaction with the albumin for the complex at the same temperature. In addition, reversible binding to BSA is inferred based on the obtained values lying in the 10^4 to 10^6 M^{-1} range and it is possible that VONarg remained stable in the biological systems and could be carried by albumin [39]. The n value (approximately equal to 1) suggests the existence of one independent class of binding sites on BSA for both compounds. Moreover, it has been determined that naringin interacted with BSA at the named Sudlow's site I within the hydrophobic pocket of subdomain IIA, mainly through hydrophobic interactions [40].

Table 5 shows the K_b binding values for naringin, its aglycon naringenin and their oxidovanadium(IV) and palladium(II) complexes. It is known that glycosylation of flavonoids lowered the binding affinity for BSA. The change in the planar structure due to the replacement of the hydroxyl group by the disaccharide neohesperidose may cause steric hindrance diminishing affinity for BSA [42]. Indeed, according to the obtained K_b values, the affinity of naringenin for BSA was about 7-fold higher than naringin. The complexes showed a different behavior. The reported Pd complex and VONarg, both containing naringin displayed greater K_b values than the oxidovanadium(IV) complex of naringenin. Therefore, the presence of naringin favors the interaction with albumin in these complexes.

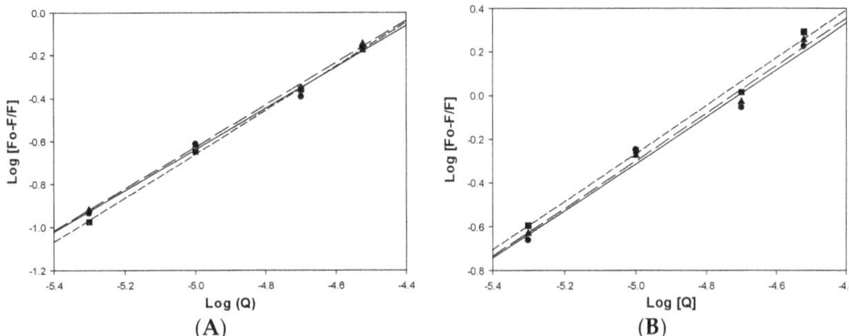

Figure 10. Plots of log $(F_o - F)/F$ vs. log [Q] for the naringin-BSA system (**A**) and VONarg-BSA system (**B**): (●) 298 K; (▲) 303 K; (■) 310 K, [BSA] = 6 µM, λ_{ex} = 280 nm.

Table 4. Stern–Volmer constant (K_{sv}), quenching rate constant (K_q), binding constant (K_b), and n binding sites for the interaction of naringin and VONarg with BSA (6 µM) in Tris–HCl buffer (pH 7.4).

	T (K)	K_{sv} (10^4) $(L \cdot mol^{-1})$	K_q (10^{12}) $(L \cdot mol^{-1}s^{-1})$	K_b (10^4) $(L \cdot mol^{-1})$	n
Naringin *	298	2.18 ± 0.15	2.18 ± 0.15	1.45 ± 0.23	0.96 ± 0.03
	303	2.28 ± 0.09	2.28 ± 0.09	1.81 ± 0.44	0.98 ± 0.10
	310	2.32 ± 0.13	2.32 ± 0.13	2.81 ± 0.38	1.02 ± 0.09
VONarg	298	5.67 ± 0.33	5.67 ± 0.33	11.6 ± 0.19	1.08 ± 0.04
	303	5.88 ± 0.31	5.88 ± 0.31	14.1 ± 0.13	1.09 ± 0.03
	310	5.99 ± 0.36	5.99 ± 0.36	16.9 ± 0.19	1.10 ± 0.04

r^2 (correlation coefficient) values are all ranged from 0.977 to 0.999 for the K_{sv}, K_q, and K_b constants. * K_{sv} and K_b values are in the same order of magnitude as those from literature [40,41].

Table 5. K_b binding values for naringenin, naringin, VO(naringenin)$_2$·H$_2$O, VONarg and [Pd{(C,N)-C$_6$H$_4$CH$_2$NH(Et)(Narg)]] complexes at 298 K.

Compounds	$K_b \times 10^4$ M^{-1}
Naringenin *	10.20
naringin	1.45
VO(naringenin)$_2$·H$_2$O *	0.31
VONarg	11.6
[Pd{(C,N)-C$_6$H$_4$CH$_2$NH(Et)(Narg)]] *	18

* Data taken from references [25,43].

The thermodynamics of the process for the formation of a complex between BSA and flavonoid and/or its metal complex can be determined from the dependence of the binding constant with temperature. Some non-covalent interactions may contribute to these interactions: hydrogen bonds, Van der Waals force, electrostatic and hydrophobic interactions. The changes in the thermodynamic parameters for the systems, free energy ΔG, enthalpy ΔH, and entropy ΔS, were analyzed to estimate the types of complex–protein binding at different temperatures. The enthalpy and entropy values were resolved from the relationship between ln K_b and 1/T. The $\Delta H/R$ values were obtained from the slope, and from the intersection with the *y* axis, the $\Delta S/R$ values. Values of ΔG were then found as $\Delta G = -RT \ln K_b$.

The spontaneity of the processes was determined by the value of ΔG, while the combinations of the signs associated with ΔH and ΔS suggest the bonding strength. If

ΔH and ΔS values are negative they give evidence to the existence of hydrogen bonds and/or van der Waals forces. Electrostatic interactions are suggested for negative ΔH and positive ΔS values and hydrophobic interactions occur when both parameters display a positive sign.

The average values of the thermodynamic parameters are depicted in Table 6 where positive ΔH and ΔS values were found. Thus, the sign of ΔG implies spontaneity, and the ΔS value is the major contribution (entropically driven) in an endothermic process. However, the comparison of the values for both compounds suggests that the complex shifted to an environment of reduced hydrophobicity (-56% in ΔH value and -79% in a ΔS value).

Table 6. Thermodynamic parameters for the systems naringin-BSA and VONarg-BSA.

Compounds	ΔH (KJ/mol)	ΔS (J/mol)	ΔG (KJ/mol)
Naringin	42.7	222.7	-23.7 (298 K) -24.8 (303 K) -26.3 (310 K)
VONarg	23.9	177.0	-28.8 (298 K) -29.7 (303 K) -31.0 (310 K)

3. Discussion

We have synthesized the oxidovanadium(IV) complex of naringin using different techniques, as previously reported [44] and extended the physicochemical characterization. The vibrational spectral characterization of the naringin complex with the oxidovanadium(IV) cation showed an acetylacetonate coordination mode (C_4=O, C_5-O$^-$). EPR determinations in solution (DMSO and H_2O/DMSO mixtures) support this coordination mode with square pyramidal geometry (90%), and most likely, a distorted octahedral with DMSO in *trans*-position (10%). It was shown by electronic spectroscopy that the complex was stable in DMSO and DMSO/water (1/99), at least during the manipulation time. The stability of the complex in DMSO was also shown by molar conductivity determinations.

Natural polyphenols showed antioxidant properties that were enhanced by metal complexation (see Section 2.2) because of the increased stability of the free radical generated in the first step of their oxidation by electron delocalization with the V=O orbitals. This behavior was also displayed by VONarg. Previously, we have studied the biological improvement of naringenin (aglycone of naringin) by oxidovanadium(IV) complexation (VONar) [25]. Antioxidant determinations showed that VONar is somehow a better antioxidant compound (only the peroxyl radical scavenging activity is low, lag phase VONar = 2.9 min, a 100 µM). Glycosylation on C7-OH may disrupt the electron delocalization of the free radical generated when that H atom is abstracted in the aglycone form.

Cancer cells, with a pro-oxidant status, seem to be more susceptible to treatment with agents that cause oxidative stress than normal cells. Excessive ROS production, exceeding a threshold level, can lead to cellular damage and induce cell death and it has been shown that many compounds exert anticancer activity by cellular ROS generation [45]. Accordingly, in the present study, we studied cellular ROS production and GSH/GSSG depletion in the human lung A549 cancer cell line, showing that the ROS generation by the complex was accompanied by the oxidation of the natural antioxidant agent GSH to give GSSG, being this behavior different from what occurred in the in vitro determinations (Section 2.2). It has to be noted that naringin at 100 µM did not show cytotoxic activity towards A549 cancer cells, in agreement with reported results [46,47] nor did it produce or deplete cellular ROS, and it did not generate oxidized glutathione. On the other hand, the oxidovanadium(IV) cation did not generate ROS in the A549 cell line (100 µM, 24 h incubation) [35] nor deplete GSH, showing that the anticancer effect is only due to the effect of metal complexation of the flavonoid. A similar enhancement of the decrease in cell viability upon complexation was shown for naringin-Ag(I) and -Y(III) in A549 cell line, 48 h incubation [48].

Because one of the major events during apoptosis is the depolarization of the mitochondrial membrane, which subsequently induces the release of the proapoptotic proteins [49], we measured if the complex could damage the mitochondrial membrane. Treatment of A549 cells with VONarg induced a significant loss of $\Delta\Psi$m and may cause A549 cell apoptosis. Then, it has been demonstrated that the complex can be toxic to cancer cells inducing ROS production and GSH depletion and the loss of $\Delta\Psi$m. On the other hand, the complex of naringenin, VONar, significantly decreased the viability of lung cancer cells in a dose–response manner (35% at 100 μM) but in both cases, the IC_{50} values were higher than 100 μM [25]. Flavonoids with C2-C3 double bond exert better anticancer effects [50]. The low anticancer activities of both compounds could be supported by the structure of the flavonoid: the lack of the C2-C3 double bond forbids the electronic delocalization between rings B and C. However, we found an improvement in the cytotoxic effects of both flavonoids upon metal complexation, due to the planar structure produced by the resonance of ring B with the V=O moiety. The interaction with BSA was spontaneous, entropically driven, and the obtained value of the binding constant suggests that albumin could act as a carrier for the transport in plasma and delivery to target cells of VONarg.

4. Material and Methods

4.1. Materials and Instrumentation

Naringin (Sigma, St. Louis, MO, USA), oxidovanadium(IV) chloride (50% aqueous solution, Carlo Erba, Buenos Aires, Argentina), bis(acetylacetonato)oxidovanadium(IV) (Fluka, Munich, Germany), and oxidovanadium(IV) sulfate pentahydrate (Merck) were used as supplied. Corning or Falcon provided cell culture materials. Dulbecco's modified Eagle's medium (DMEM) was purchased from Gibco (Gaithersburg, MD, USA), Tryple^{TM} from Invitrogen (Buenos Aires, Argentina SRL), and fetal bovine serum (FBS) from Internegocios, Buenos Aires, Argentina. All other chemicals used were of analytical grade. Elemental analysis for carbon and hydrogen was performed using a Carlo Erba EA1108 analyzer. Vanadium content was determined by the tungstophosphovanadic method [51]. Thermogravimetric analysis (TGA) was performed with a Shimadzu thermobalance (model TG-50), working in an oxygen flow of 50 mL/min, and at a heating rate of 10 °C/min. Sample quantities ranged from 10 to 20 mg. UV-vis and diffuse reflectance ($BaSO_4$ as standard) spectra determinations were recorded with a Shimadzu UV-2600/2700 spectrophotometer. Infrared spectra were measured using the KBr pellet technique with a Bruker IFS 66 FTIR spectrophotometer from 4000 to 400 cm^{-1}. Fluorescence spectra were obtained with a Shimadzu RF-6000 spectrophotometer equipped with a pulsed xenon lamp. The molar conductance of the complex was measured using a Conductivity TDS Probe-850084, Sper Scientific Direct, using 10^{-3} M DMSO solutions. X-band CW-EPR spectra of both powdered samples and DMSO or DMSO/water solutions (pH ca. 7) were obtained at 120 K on a Bruker EMX-Plus spectrometer, equipped with a rectangular cavity with 100 kHz field modulation. EPR spectra were baseline corrected (when necessary) using the WinEPR Processing software (Bruker, Inc. Billerica, MA, USA). *g*- and *A*-values were obtained by simulations using the EasySpin 5.2.3 toolbox based on MATLAB [26], assuming an axial spin-Hamiltonian of the form: $H = \mu_B \left[g_\parallel B_z S_z + g_\perp \left(B_x S_x + B_y S_y \right) \right] + \left[A_\parallel S_z I_z + A_\perp \left(S_x I_x + S_y I_y \right) \right]$, where μ_B is the Bohr magneton, and $g_\parallel, g_\perp, A_\parallel, A_\perp$ are the components of the axial **g** and **A** tensors, respectively. $B_{x/y/z}, S_{x/y/z}$, and $I_{x/y/z}$ are the components of the magnetic field, and the spin operators of the electron and V nucleus, respectively.

4.2. Synthesis of [VO(Narg)$_2$]·8H$_2$O (VONarg)

The solid complex was prepared using different preparative methods than the reported one [44]. A solution of oxidovanadium(IV) sulfate pentahydrate in methanol (0.25 mmol, 5 mL) was dropwise added to a naringin methanolic solution (0.5 mmol, 20 mL). pH was then adjusted to 9 by the addition of 0.2 M NaOH drops in methanol. After 1 h of stirring (under nitrogen atmosphere), a yellow-greenish solid was precipitated by the addition of 100 mL of bidistilled water. The solid was filtered off and washed with plenty of bidistilled water and dried at 60 °C. It was possible to obtain the same complex using a different synthetic method by adding 0.25 mmol of bis(acetylacetonato)oxidovanadium(IV) over a hot aqueous solution of naringin (80 °C). After 30 min of stirring (under nitrogen atmosphere), a yellow-greenish precipitate formed, it was filtered off, washed several times with hot bidistilled water, and dried at 60 °C. Due to the reduction in costs and waste production as well as its simplicity, the synthesis using hot water was employed to prepare the solid compound that was used to conduct all the other physiochemical characterizations. Anal calcd. For $C_{54}H_{78}O_{37}V$: C 47.33%; H 5.73%; V 3.71%; Found: C 47.1%; H 5.6%; V 3.9%; Yield: 51.62%. UV-Vis data for 1:2 VO:Naringin, DMSO: 598 nm ($3d_{xy} \rightarrow 3d_{x2-y2}$, ε = 50.4 M^{-1} cm^{-1}); 810 nm ($3dxy \rightarrow 3d_{xz}$, $3d_{yz}$, ε = 60.3 M^{-1} cm^{-1}). Thermal decomposition measurements gave the same results as reported by Badea et al. [44]. A first weight loss at 105 °C of 10.4% corresponded to the loss of 8 molecules of water ($\Delta\omega\%$ calc = 10.5%). The presence of V_2O_5 in the residue was confirmed by FTIR spectroscopy. The molar conductance of the complex measured in DMSO, Λm = 0 ($\Omega^{-1} \cdot cm^2 \cdot mol^{-1}$), suggested that it had a non-electrolytic nature. These results allowed us to postulate a L:M stoichiometry of 2:1 [52], which was also supported by spectral determinations (see below).

4.3. Spectrophotometric Titration

In order to establish VONarg stoichiometry, the molar ratio method was applied. A DMSO solution of naringin (2×10^{-2} M) was prepared and its electronic spectrum was recorded. The absorption spectra of different solutions of naringin (2×10^{-2} M) and oxidovanadium(IV) chloride in ligand-to-metal molar ratios from 10 to 0.5 (pH 9) were measured.

4.4. Antioxidant Properties

The antioxidant activities of the complex and the ligand were determined by the Superoxide Dismutase (SOD) mimetic assay, scavenging power of the hydroxyl radical, inhibitory action on peroxyl radical, and 1,1-Diphenyl-2-picrylhydrazyl radical (DPPH) assay. Reported protocols were used in these determinations [53]. Briefly, the nitrobluetetrazolium (NBT) assay was used to determine indirectly the SOD activity. The activity of naringin and VONarg for their ability to inhibit the reduction in NBT by the superoxide anion generated by the phenazinemethosulfate (PMS) and reduced nicotinamide adenine dinucleotide (NADH) system was measured. The 50% inhibition (IC_{50}) for naringin or VONarg was estimated by plotting the percentage of reduction in NBT versus the negative log of the concentration of the sample solution. The capacity of naringin and VONarg to scavenge hydroxyl radicals (generated by the ascorbate-Fe-H_2O_2 system) was measured by the determination of the extent of deoxyribose degradation by hydroxyl radical at 535 nm. The inhibitory activity against peroxyl radicals was determined by measuring the delay produced by the compounds on pyranine consumption (lag phase) from the peroxyl radicals. The DPPH radical scavenging activity was measured in the presence and absence of the compounds and this later value was assigned to 100%. Each experiment was performed in triplicate and at least three independent experiments were evaluated in each case.

4.5. Biological Assays
4.5.1. Cell Culture

Adenocarcinomic human alveolar basal epithelial cells (A549 cell line) were obtained from ABAC (Argentinean Cell Bank Association INEVH, Pergamino, Buenos Aires, Argentina). DMEM supplemented with 100 $U \cdot mL^{-1}$ penicillin, 100 $\mu g \cdot mL^{-1}$ streptomycin, and 10% (v/v) fetal bovine serum was used as culture medium. When 70–80% confluence was reached, cells were washed with phosphate-buffered saline (PBS) (11 mM KH_2PO_4, 26 mM Na_2HPO_4, 115 mM NaCl, pH 7.4) and were subcultured using TrypLE™. All reagents were from Sigma–Aldrich (St Louis, MO, USA). Cultures were maintained at 37 °C in a humidified atmosphere with 5% CO_2 and passaged, according to the manufacturer's instructions.

4.5.2. MTT Assay

Stock complex solutions of compounds were prepared in DMSO with a manipulation time of 15 min. A549 cells and HEK293 (human embryonic kidney) were seeded at a density of 15,000 cells/well in 96 well plates, grown overnight, and treated with either vehicle (DMSO), naringin, VONarg, or oxidovanadium(IV) in DMSO at different concentrations in DMEM supplemented with 100 U·mL^{-1} penicillin and 100 µg·mL^{-1} streptomycin. The dissolution vehicle yielded a maximum final concentration of 0.5% in the treated well. After 24 h of incubation, 3-(4,5-methyl-thiazol-2-yl)-2,5diphenyl tetrazolium bromide (MTT) (Sigma–Aldrich, St. Louis, MO, USA) was added at 100 µg/well for 2 h. The formazan products generated by cellular reduction of MTT were dissolved in DMSO and the optical density was measured at 450 nm using Sinergy 2 Multi Mode Microplater reader Biotek (Wisnooski, VT, USA). All experiments were done in triplicate. Data were presented as proportional viability (%) comparing the treated group with the untreated cells, for which the viability was assumed to be 100%.

4.5.3. Stress Oxidative Determinations

Generation of intracellular reactive oxygen species (ROS) was measured using a 2′,7′-dichlorodihydrofluorescein diacetate probe (H2DCF-DA, Sigma–Aldrich). The probe hydrolyzed by intracellular esterases to the non-fluorescent 2′,7′-dichlorodihydrofuorescein (H2DCF), and the cytosolic cellular oxidants oxidized H2DCF to the fluorescent product, dichlorofluorescin (DCF). Hence, DCF fluorescence intensity is proportional to the amount of ROS. Here, after 24 h incubation with the complex, the cells were incubated with 10 mM DCFH-DA (dissolved in PBS with 5% DMSO) at 37 °C for 60 min under light protection and then lysed with 1 mL Tryton X-100 0.1%. The intracellular ROS were detected at 485 nm for excitation and 520 nm for emission.

The effect of the complex on the intracellular glutathione (GSH) level was assessed using o-phthaldialdehyde (OPT) [54] that reacted with the amino and sulfhydryl groups of glutathione emitting intense fluorescence at pH 8 and 12. For the GSH assay, 100 mL aliquots of the treated cells were incubated with OPT (0.1% MeOH in cold phosphate buffer (Na$_2$HPO$_4$ 0.1 M, M-EDTA 0.005 M, pH 8)). For the oxidized glutathione (GSSG) determinations, 100 mL aliquots of the cellular lysate were incubated with 0.04 M N-ethylmaleimide (NEM) to block the thiol groups in GSH, followed by the addition of OPT solution and the incubation in NaOH 0.1 M solution. The fluorescence intensity was detected at 350 nm excitation and 420 nm emission. The data were normalized against the total protein content measured by the Bradford method [55]. The GSH and GSSG contents (Sigma–Aldrich) were calculated from a calibration curve from standard solutions.

The mitochondrial membrane potential analysis was assessed using the DiOC6 fluorescent probe (3,3′-dihexyloxacarbocyanine iodide), which was added to the wells (400 nM concentration) and incubated for 30 min at 37 °C. The cells were resuspended in PBS and measured at 485 nm excitation and 535 nm emission [56].

4.6. Data Analysis

All the experiments were performed in triplicate, and each one was repeated at least three times. All the values were expressed as the mean ± standard error. Differences between the mean values were analyzed for significance using the one-way ANOVA test on Origin 9.1 (OriginLab Corporation, Northampton, MA, USA). $p < 0.05$ values were considered to be statistically significant.

4.7. Bovine Serum Albumin Interactions

BSA was dissolved in Tris–HCl (pH 7.4) buffer to attain a final concentration of 6 µM, 10 mL. Naringin and VONarg, dissolved in 0.5% DMSO/buffer solution, were added dropwise to the BSA solution to obtain the desired concentration of 5 to 100 µM. The solutions were left into a thermostatic water bath for 1 h at 298, 303, and 310 K, to get homogeneous solutions and allow the interaction between compound and protein.

The compounds did not show any fluorescence intensity that could interfere with the measurements. BSA 6 μM was titrated by successive additions of naringin and VONarg solutions and the fluorescence intensity was measured on a luminescence spectrometer in the range from 290 to 450 nm. For each sample and concentration, three independent replicates were performed. All the fluorescence quenching data were analyzed according to previous studies performed in our laboratory applying the well-known Stern–Volmer equation and other traditional mathematical procedures were applied to obtain the Stern–Volmer quenching constant and the binding constant (K_b) and the binding site value (n) [25].

5. Conclusions

The V(IV)O-naringin complex was prepared and characterized in the solid state (EPR measurements) and solution (EPR spectrum, conductance, spectrophotometric titration, and stability studies). Metal coordination to positions 4 and 5 of the glycosylated flavonoid enhanced the moderate antioxidant activity of the ligand against superoxide and peroxyl ROS, which may be due to the higher electron delocalization induced by the resonance of the π bond of the oxidovanadium(IV) cation with ring A. In addition, the metal complex showed higher anticancer activity when compared to naringin through ROS elevation and GSH/GSSG decrease, enhancing the oxidative stress that led to mitochondrial damage and cell death. A selective anticancer behavior was found. BSA interaction studies showed a static quenching process and the binding constant values were indicative that the complex could be transported and delivered by the protein. The experimental results showed the potential of flavonoids' complexation with oxidovanadium(IV) cation to improve their biological actions.

Supplementary Materials: The following supporting information can be downloaded at: https://www.mdpi.com/article/10.3390/inorganics10010013/s1, **Figure S1.** EPR spectrum of frozen solutions of VONarg at 120 K with different water:DMSO ratios. Black: 100% DMSO, Red: 50% water pH 7, Blue: 90% water pH 7. **Figure S2.** UV-Vis spectra recorded after 4 h of 0.005 M solutions of [VO(Narg)$_2$]·8H$_2$O in: (A) DMSO; (B) DMSO/H$_2$O 1/99. **Figure S3.** Cell viability assay at different VONarg concentrations after treatment for 24 h on HEK293 cells. The results are expressed as a percentage of the control level and represent the mean $\pm$ the standard error of the mean (SEM) from three separate experiments. **Figure S4.** Stern–Volmer plot of the fluorescence quenching of BSA with different concentrations of naringin (A) and VONarg (B) systems: ($\bullet$) 298 K; ($\blacktriangle$) 303 K; ($\blacksquare$) 310 K, [BSA] = 6 μM, λ_{ex} = 280 nm.

Author Contributions: Conceptualization, P.A.M.W.; validation, E.G.F., L.G.N., and P.A.M.W.; formal analysis, A.G.R.-G. and L.G.N.; investigation, A.G.R.-G. and H.G.-S.; EPR measurements, P.J.G. and M.R.; resources, P.A.M.W.; writing-original draft preparation, A.G.R.-G., H.G.-S., E.G.F. and P.J.G.; writing-review and editing, P.A.M.W.; visualization, P.A.M.W.; supervision, P.A.M.W. and L.G.N.; project administration, P.A.M.W.; funding acquisition, P.A.M.W. All authors have read and agreed to the published version of the manuscript.

Funding: This research was funded by ANPCyT, 2019-0945 and UNLP X871.

Institutional Review Board Statement: Not applicable.

Informed Consent Statement: Not applicable.

Data Availability Statement: The data presented in this study are available in the Supplementary Materials.

Acknowledgments: This work was supported by CICPBA, ANPCyT, CONICET, and UNLP, Argentina. L.G.N., P.J.G. and E.G.F. are members of the Research Career, CONICET, Argentina. P.A.M.W. is a member of the Research Career, CICPBA. G.R. is a fellowship holder from ANPCyT.

Conflicts of Interest: The authors declare no conflict of interest. The funders had no role in the design of the study; in the collection, analyses, or interpretation of data; in the writing of the manuscript, or in the decision to publish the results.

References

1. Panche, A.N.; Diwan, A.D.; Chandra, S.R. Flavonoids: An overview. *J. Nutr. Sci.* **2016**, *5*, E47. [CrossRef] [PubMed]
2. Wang, T.Y.; Li, Q.; Bi, K.S. Bioactive flavonoids in medicinal plants: Structure, activity and biological fate. *Asian J. Pharm. Sci.* **2018**, *13*, 12–23. [CrossRef]
3. Dias, M.C.; Pinto, D.C.G.A.; Silva, A.M.S. Plant Flavonoids: Chemical Characteristics and Biological Activity. *Molecules* **2021**, *26*, 5377. [CrossRef]
4. Zou, M.; Liu, H.; Li, J.; Yao, X.; Chen, Y.; Ke, C.; Liu, S. Structure-activity relationship of flavonoid bifunctional inhibitors against zika virus infection. *Biochem. Pharmacol.* **2020**, *177*, 113962. [CrossRef]
5. Zhu, X.; Ouyang, W.; Lan, X.; Xiao, H.; Tang, L.; Liu, G.; Feng, K.; Zhang, L.; Song, M.; Cao, Y. Anti-hyperglycemic and liver protective effects of flavonoids from *Psidium guajava* L. (guava) leaf in diabetic mice. *Food Biosci.* **2020**, *135*, 100574. [CrossRef]
6. Sun, Y.W.; Bao, Y.; Yu, H.; Chen, Q.J.; Lu, F.; Zhai, S.; Zhang, C.F.; Li, F.; Wang, C.Z.; Yuan, C.S. Anti-rheumatoid arthritis effects of flavonoids from Daphne genkwa. *Int. Immunopharmacol.* **2020**, *83*, 106384. [CrossRef]
7. Tuentera, E.; Creylman, J.; Verheyen, G.; Pieters, L.; Van Miert, S. Development of a classification model for the antigenotoxic activity of flavonoids. *Bioorg. Chem.* **2020**, *98*, 103705. [CrossRef] [PubMed]
8. Fusia, F.; Trezza, A.; Tramaglino, M.; Sgaragli, G.; Saponara, S.; Spiga, O. The beneficial health effects of flavonoids on the cardiovascular system: Focus on K^+ channels. *Pharmacol. Res.* **2020**, *152*, 104625. [CrossRef]
9. Khater, M.; Ravishankar, D.; Greco, F.; Osborn, H.M.I. Metal complexes of flavonoids: Their synthesis, characterization and enhanced antioxidant and anticancer activities. *Future Med. Chem.* **2019**, *11*, 2845–2867. [CrossRef]
10. Allscher, T.; Klüfers, P.; Mayer, P. Carbohydrate-metal complexes: Structural Chemistry of Stable Solution Species. In *Glycoscience*; Springer: Berlin/Heidelberg, Germany, 2008; pp. 1077–1139. [CrossRef]
11. Sharma, P.; Kumar, V.; Guleria, P. Naringin: Biosynthesis and pharmaceutical applications. *Indian J. Pharm. Sci.* **2019**, *89*, 988–999. [CrossRef]
12. Kumar, S.; Pandey, A.K. Chemistry and biological activities of flavonoids: An overview. *Sci. World J.* **2013**, *2013*, 162750. [CrossRef]
13. Sharma, M.; Dwivedi, P.; Singh Rawat, A.K.; Dwivedi, A.K. Nutrition nutraceuticals: A proactive approach for healthcare. In *Nutraceuticals*; Elsevier Inc.: Amsterdam, The Netherlands, 2016; pp. 79–116. [CrossRef]
14. Chen, R.; Qi, Q.L.; Wang, M.T.; Li, Q.Y. Therapeutic potential of naringin: An overview. *Pharm. Biol.* **2016**, *54*, 3203–3210. [CrossRef]
15. Zhao, Y.; Liu, S. Bioactivity of naringin and related mechanisms. *Pharmazie* **2021**, *76*, 359–363. [CrossRef] [PubMed]
16. Zaragoz, C.; Villaescusa, L.; Monserrat, J.; Zaragoz, F.; Melchor, A. Potential therapeutic anti-inflammatory and immunomodulatory effects of dihydroflavones, flavones, and flavonols. *Molecules* **2020**, *25*, 1017. [CrossRef]
17. Qi, Z.; Xu, Y.; Liang, Z.; Li, S.; Wang, J.; Wei, Y.; Dong, B. Naringin ameliorates cognitive deficits via oxidative stress, proinflammatory factors and the PPARγ signaling pathway in a type 2 diabetic rat model. *Mol. Med. Rep.* **2015**, *12*, 7093–7101. [CrossRef]
18. Viswanatha, G.L.; Shylaja, H.; Moolemath, Y. The beneficial role of Naringin- a citrus bioflavonoid, against oxidative stress-induced neurobehavioral disorders and cognitive dysfunction in rodents: A systematic review and meta-analysis. *Biomed. Pharmacother.* **2017**, *94*, 909–929. [CrossRef] [PubMed]
19. Memariani, Z.; Qamar Abbas, S.; Shams ul Hassan, S.; Ahmadi, A.; Chabra, A. Naringin and naringenin as anticancer agents and adjuvants in cancer combination therapy: Efficacy and molecular mechanisms of action, a comprehensive narrative review. *Pharmacol. Res.* **2021**, *171*, 105264. [CrossRef]
20. Amin Hussen, N.H. Docking Study of Naringin Binding with COVID-19 Main Protease Enzyme. *Iraqi J. Pharm. Sci.* **2020**, *29*, 231–238. [CrossRef]
21. Etcheverry, S.B.; Ferrer, E.G.; Naso, L.; Rivadeneira, J.; Salinas, V.; Williams, P.A.M. Antioxidant effects of the VO(IV) hesperidin complex and its role in cancer chemoprevention. *J. Biol. Inorg. Chem.* **2008**, *13*, 435–447. [CrossRef]
22. Naso, L.; Martínez, V.R.; Lezama, L.; Salado, C.; Valcarcel, M.; Ferrer, E.G.; Williams, P.A.M. Antioxidant, anticancer activities and mechanistic studies of the flavone glycoside diosmin and its oxidovanadium(IV) complex. Interactions with bovine serum albumin. *Bioorg. Med. Chem.* **2016**, *24*, 4108–4119. [CrossRef] [PubMed]
23. Goitia, H.; Quispe, P.; Naso, L.G.; Martínez, V.R.; Rey, M.; Rizzi, A.C.; Ferrer, E.G.; Williams, P.A.M. Interactions of rutin with the oxidovanadium(IV) cation. Anticancer improvement effects of glycosylated flavonoids. *New J. Chem.* **2019**, *43*, 17636–17646. [CrossRef]
24. Pessoa, J.C.; Etcheverry, S.; Gambino, D. Vanadium compounds in medicine. *Chem. Rev.* **2015**, *301–302*, 24–48. [CrossRef]
25. Islas, M.S.; Naso, L.G.; Lezama, L.; Valcarcel, M.; Salado, C.; Roura-Ferrer, M.; Ferrer, E.G.; Williams, P.A.M. Insights into the mechanisms underlying the antitumor activity of the new coordination complex with oxidovanadium(IV) and naringenin. Albumin binding studies. *J. Inorg. Biochem.* **2015**, *149*, 12–24. [CrossRef]
26. Chasteen, N.D. *Biological Magnetic Resonance*; Berliner, L.J., Reuben, J., Eds.; Plenum: New York, NY, USA, 1981; p. 3.
27. Gorelsky, S.; Micera, G.; Garribba, E. The Equilibrium Between the Octahedral and Square Pyramidal Form and the Influence of an Axial Ligand on the Molecular Properties of $V^{IV}O$ Complexes: A Spectroscopic and DFT Study. *Chem. Eur. J.* **2010**, *16*, 8167–8180. [CrossRef] [PubMed]

28. Sanna, D.; Ugone, V.; Lubinu, G.; Micera, G.; Garribba, E. Behavior of the potential antitumor VIVO complexes formed by flavonoid ligands. 1. Coordination modes and geometry in solution and at the physiological pH. *J. Inorg. Biochem.* **2014**, *140*, 173–184. [CrossRef] [PubMed]
29. Kivelson, D.; Lee, S. ESR Studies and the Electronic Structure of Vanadyl Ion Complexes. *J. Chem. Phys.* **1964**, *41*, 1896–1903. [CrossRef]
30. Chand, P.; Murali Krishna, R.; Lakshamana Rao, J.; Lakshaman, S.V.J. EPR and optical studies of vanadyl complexes in two host-crystals of tutton salts of thallium. *Rad. Eff. Def. Solids* **1993**, *127*, 245–254. [CrossRef]
31. Liu, K.T.; Yu, J.T.; Lou, S.H.; Lee, C.H.; Huang, Y.; Lii, K.H.J. Electron paramagnetic resonance study of V^{4+}-doped $KTiOPO_4$ single crystals. *Phys. Chem. Solids* **1994**, *55*, 1221–1226.
32. Ferrer, E.G.; Salinas, M.V.; Correa, M.J.; Naso, L.; Barrio, D.A.; Etcheverry, S.B.; Lezama, L.; Rojo, T.; Williams, P.A.M. Synthesis, characterization, antitumoral and osteogenic activities of quercetin vanadyl(IV) complexes. *J. Biol. Inorg. Chem.* **2006**, *11*, 791–801. [CrossRef]
33. Pereira, R.M.S.; Andrades, N.E.D.; Paulino, N.; Sawaya, A.C.H.F.; Eberlin, M.N.; Marcucci, M.C.; Marino Favero, G.; Novak, E.M.; Bydlowski, S.P. Synthesis and Characterization of a Metal Complex Containing Naringin and Cu, and its Antioxidant, Antimicrobial, Antiinflammatory and Tumor Cell Cytotoxicity. *Molecules* **2007**, *12*, 1352–1366. [CrossRef]
34. Naso, L.G.; Lezama, L.; Valcarcel, M.; Salado, C.; Villacé, P.; Kortazar, D.; Ferrer, E.G.; Williams, P.A.M. Bovine serum albumin binding, antioxidant and anticancer properties of an oxidovanadium(IV) complex with luteolin. *J. Inorg. Biochem.* **2016**, *157*, 80–93. [CrossRef] [PubMed]
35. Fanali, G.; di Masi, A.; Trezza, V.; Marino, M.; Fasano, M.; Ascenzi, P. Human serum albumin: From bench to bedside. *Mol. Asp. Med.* **2012**, *33*, 209–290. [CrossRef]
36. Weinryb, I.; Steiner, R.F. The Luminescence of the Aromatic Amino Acids. In *Excited States of Proteins and Nucleic Acids*; Steiner, R.F., Weinryb, I., Eds.; Springer: Boston, MA, USA, 1971. [CrossRef]
37. Lakowicz, J.R. *Principles of Fluorescence Spectroscopy*; Springer Science & Business Media: New York, NY, USA, 2013.
38. Sun, Y.; Zhang, H.; Sun, Y.; Zhang, Y.; Liu, H.; Cheng, J.; Bi, S.; Zhang, H. Study of interaction between protein and main active components in *Citrus aurantium* L. by optical spectroscopy. *J. Luminescence* **2010**, *130*, 270–279. [CrossRef]
39. Kragh-Hansen, U.; Chuang, V.T.G.; Otagiri, M. Practical aspects of the ligand-binding and enzymatic properties of human serum albumin. *Biol. Pharm. Bull.* **2002**, *25*, 695–704. [CrossRef]
40. Zhang, X.; Li, L.; Xu, Z.; Liang, Z.; Su, J. Investigation of the Interaction of Naringin Palmitate with Bovine Serum Albumin: Spectroscopic Analysis and Molecular Docking. *PLoS ONE.* **2013**, *8*, e59106. [CrossRef]
41. Roy, A.S.; Tripathy, D.R.; Chatterjee, A.; Dasgupta, S. A spectroscopic study of the interaction of the antioxidant naringin with bovine serum albumin. *J. Biophys. Chem.* **2010**, *1*, 141–152. [CrossRef]
42. Shi, J.; Cao, H. Molecular structure-affinity relationship of dietary flavonoids for bovine serum albumin. *Braz. J. Pharmacogn.* **2011**, *21*, 594–600. [CrossRef]
43. Karami, K.; Mehri Lighvan, Z.; Farrokhpour, H.; Dehdashti Jahromi, M.; Momtazi-borojeni, A.A. Synthesis and spectroscopic characterization study of new palladium complexes containing bioactive O,O-chelated ligands: Evaluation of the DNA/protein BSA interaction, in vitro antitumoural activity and molecular docking. *J. Biomol. Struct. Dyn.* **2018**, *36*, 3324–3340. [CrossRef] [PubMed]
44. Badea, M.; Olar, R.; Uivarosi, V.; Marinescu, V.; Aldea, V. Synthesis and characterization of some vanadyl complexes with flavonoid derivatives as potential insulin-mimetic agents. *J. Therm. Anal. Calorim.* **2012**, *107*, 279–285. [CrossRef]
45. Perillo, B.; Di Donato, M.; Pezone, A.; Di Zazzo, E.; Giovannelli, P.; Galasso, G.; Castoria, G.; Migliaccio, A. ROS in cancer therapy: The bright side of the moon. *Exp. Mol. Med.* **2020**, *52*, 192–203. [CrossRef]
46. Yoshinaga, A.; Kajiya, N.; Oishi, K.; Kamada, Y.; Ikeda, A.; Chigwechokha, P.K.; Kibe, T.; Kishida, M.; Kishida, S.; Komatsu, M.; et al. NEU3 inhibitory effect of naringin suppresses cancer cell growth by attenuation of EGFR signaling through GM3 ganglioside accumulation. *Eur. J. Pharmacol.* **2016**, *782*, 21–29. [CrossRef] [PubMed]
47. Nie, Y.; Wu, H.; Li, P.; Xie, L.; Luo, Y.; Shen, J.; Su, W. Naringin attenuates EGF-induced MUC5AC secretion in A549 cells by suppressing the cooperative activities of MAPKs-AP-1 and IKKs-IkB-NF-kB signaling pathways. *Eur. J. Pharmacol.* **2012**, *690*, 207–213. [CrossRef]
48. Atta, E.M.; Hegab, K.H.; Abdelgawad, A.A.M.; Youssef, A.A. Synthesis, characterization and cytotoxic activity of naturally isolated naringin-metal complexes. *Saudi Pharm. J.* **2019**, *27*, 584–592. [CrossRef] [PubMed]
49. Qian, J.; Li, J.; Ding, J.; Wang, Z.; Zhang, W.; Hu, G. Erlotinib activates mitochondrial death pathways related to the production of reactive oxygen species in the human non-small cell lung cancer cell line A549. *Clin. Exp. Pharm. Physiol.* **2009**, *36*, 487–494. [CrossRef]
50. Chidambara Murthy, K.; Kim, J.; Vikram, A.; Patil, B.S. Differential inhibition of human colon cancer cells by structurally similar flavonoids of citrus. *Food Chem.* **2012**, *132*, 27–34. [CrossRef] [PubMed]
51. Onishi, H. *Photometric Determination of Traces of Metals*, 4th ed.; John Wiley and Sons, Inc.: New York, NY, USA, 1986.
52. Ali, I.; Wani, W.A.; Saleem, K. Empirical formulae to molecular structures of metal complexes by molar conductance. *Synth. React. Inorg. Met. Nano-Metal Chem.* **2013**, *43*, 1162–1170. [CrossRef]

53. Martínez Medina, J.J.; Naso, L.G.; Pérez, A.L.; Rizzi, A.; Okulik, N.B.; Ferrer, E.G.; Williams, P.A.M. Apigenin oxidovanadium(IV) cation interactions. Synthesis, spectral, bovine serum albumin binding, antioxidant and anticancer studies. *J. Photochem. Photobiol. A Chem.* **2017**, *344*, 84–100. [CrossRef]
54. Hissin, P.J.; Hilf, R. A fluorometric method for determination of oxidized and reduced glutathione in tissues. *Anal. Biochem.* **1976**, *74*, 214–226. [CrossRef]
55. Bradford, M.M. A rapid and sensitive method for the quantitation of microgram quantities of protein utilizing the principle of protein-dye binding. *Anal. Biochem.* **1976**, *72*, 248–254. [CrossRef]
56. Zamzami, N.; Métivier, D.; Kroemer, G. Quantitation of mitochondrial transmembrane potential in cells and in isolated mitochondria. *Methods Enzymol.* **2000**, *322*, 208–213. [CrossRef]

 *inorganics*

MDPI

Article

Phenanthroline Complexation Enhances the Cytotoxic Activity of the VO-Chrysin System

Agustin Actis Dato [1], Luciana G. Naso [1], Marilin Rey [2], Pablo J. Gonzalez [2], Evelina G. Ferrer [1] and Patricia A. M. Williams [1,*]

[1] Centro de Química Inorgánica (CEQUINOR, UNLP, CONICET), Departamento de Química, Facultad de Ciencias Exactas, Universidad Nacional de La Plata, Bv. 120 No 1465, La Plata CP B1906CXS, Argentina; a.actisdato@hotmail.com (A.A.D.); lucianananaso504@hotmail.com (L.G.N.); evelina@quimica.unlp.edu.ar (E.G.F.)

[2] Departamento de Física, Facultad de Bioquímica y Ciencias Biológicas, Universidad Nacional del Litoral and CONICET, Santa Fe S3000ZAA, Argentina; mrey@fbcb.unl.edu.ar (M.R.); pablogonzalez1979@gmail.com (P.J.G.)

* Correspondence: williams@quimica.unlp.edu.ar

Abstract: Metal complexation in general improves the biological properties of ligands. We have previously measured the anticancer effects of the oxidovanadium(IV) cation with chrysin complex, $VO(chrys)_2$. In the present study, we synthesized and characterized a new complex generated by the replacement of one chrysin ligand by phenanthroline (phen), VO(chrys)phenCl, to confer high planarity for DNA chain intercalation and more lipophilicity, giving rise to a better cellular uptake. In effect, the uptake of vanadium has been increased in the complex with phen and the cytotoxic effect of this complex proved higher in the human lung cancer A549 cell line, being involved in its mechanisms of action, the production of cellular reactive oxygen species (ROS), the decrease of the natural antioxidant compound glutathione (GSH) and the ratio GSH/GSSG (GSSG, oxidized GSH), and mitochondrial membrane damage. Cytotoxic activity studies using the non-tumorigenic HEK293 cell line showed that [VO(chrys)phenCl] exhibits selectivity action towards A549 cells after 24 h incubation. The interaction with bovine serum albumin (BSA) by fluorometric determinations showed that the complex could be carried by the protein and that the binding of the complex to BSA occurs through H-bond and van der Waals interactions.

Keywords: oxidovanadium(IV) phenantrholine chrysin; vanadium cellular uptake; anticancer; albumin interaction

Citation: Actis Dato, A.; Naso, L.G.; Rey, M.; Gonzalez, P.J.; Ferrer, E.G.; Williams, P.A.M. Phenanthroline Complexation Enhances the Cytotoxic Activity of the VO-Chrysin System. *Inorganics* **2022**, *10*, 4. https://doi.org/10.3390/inorganics10010004

Academic Editor: Dinorah Gambino

Received: 9 December 2021
Accepted: 23 December 2021
Published: 28 December 2021

Publisher's Note: MDPI stays neutral with regard to jurisdictional claims in published maps and institutional affiliations.

1. Introduction

Chrysin (Scheme 1) is a natural polyphenol with several biological activities, such as antioxidant, anticancer, antiviral, and neuroprotective. It has low solubility at physiological conditions, low bioavailability, quick metabolism, and rapid excretion, limiting its utilization as a chemotherapeutic agent [1]. Hence, its structure has been modified (by functionalization or metal complexation) in order to improve its bioactivities. Metal-based drug development is a promising strategy for the enhancement of the pharmacological action of drugs. In particular, vanadium complexes have been recognized to display biological activities for the treatment of various diseases, such as diabetes, cancer, tuberculosis, and leishmaniasis [2]. Cancer is one of the primary causes of mortality. Moreover, lung cancer is the main cause of cancer deaths. Recently, with the knowledge of the molecular mechanisms related to this disease, it has been found that angiogenesis is one of the causes of its bad prognosis [3,4].

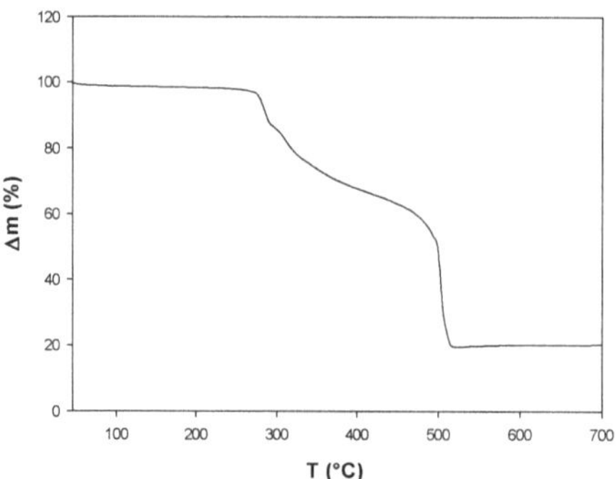

Scheme 1. Draw of the structure of chrysin.

We have previously studied the anticancer behavior of chrysin and chrysin oxidovanadium(IV) metal complex on osteoblast like cells [5], breast cancer cells [6], and human lung A549 cancer cells [7] and showed that complexation improved the biological effects of the polyphenol. The selection of the heterocyclic base phenanthroline (phen) included as a second ligand is related to the fact that planar ligands coordinated to metals could bind DNA through intercalation to base pairs, improving the anticancer action of binary complexes, as well as confer lipophilicity to the compounds [8]. Only a few ternary chrysin metal complexes were reported and the Ga(III), Cu(II), and Ru(II)-chrysin-ancillary aromatic chelator systems proved more cytotoxic than free chrysin in different cancer cell lines [9–11].

In the current work, we design the heteroleptic [VO(chrys)phenCl] complex aiming to enhance the biological behavior of the VO(chrys)$_2$ complex. It was characterized in the solid state and in solution. The biological activity of the complex as an anticancer drug was studied in the human lung cancer cell line A549 and toxicity was evaluated in the cell line derived from human embryonic kidney HEK293. The mechanism of action was studied by means of cellular reactive oxygen species (ROS) generation, natural antioxidant level depletion (glutathione, GSH), mitochondrial membrane damage, and vanadium cellular uptake. The interaction of the complex with BSA was also determined.

2. Results

2.1. Synthesis of [VO(chrys)phenCl]

The complex was prepared by the replacement of one ligand chrysin in the binary [VO(chrys)$_2$EtOH]$_2$ complex by phen. The molar conductance of the complex measured in DMSO, Λ_m = 11 (Ω^{-1} cm^2 mol^{-1}), suggested a non-electrolyte compound. The thermogravimetric analysis (oxygen atmosphere, 50 mL/min) showed that the complex is stable up to 260 °C, indicating that no solvation or coordination solvent molecules are present in the complex (Figure 1). The compound degraded in a series of two consecutive TG steps observed at 295 and 400 °C. Weight constancy is attained at 525 °C and the weight of the remaining solid residue, collected at 700 °C, was 17.2%, in good agreement with the expected value of 17.0%. The presence of V$_2$O$_5$ in the residue was confirmed by FTIR spectroscopy. More details can be found in Section 4.

Figure 1. Thermogravimetric analysis (TGA) curve for the decomposition of [VO(chrys)phenCl].

2.2. FTIR Spectrum

The assignment of the main absorption bands of the FTIR spectrum of the complex was performed in comparison with the binary VO(chrys)$_2$ system [5]. The spectral pattern remained similar to the binary compound regarding the modifications of the vibrational bands of C=O stretching (1635 cm^{-1}) and O-H bendings (1596, 1351, and 1247 cm^{-1}), indicating the coordination of chyrsin to the metal center (Figure 2). The vibrational modes of phen in the 1600–1400 cm^{-1} range (medium intensities) are associated with C=C and C=N stretching modes. These bands appeared to be overlapped with those of chrysin. However, it can be seen that the C=N stretching band of phen at ca. 1646 cm^{-1} shifted to 1635 cm^{-1} upon coordination, but it is masked by the C=O stretching mode (strong intensity) of the ligand chrysin. Main bands of motion of ring hydrogen atoms in phase and out of phase for phen at 852 and 738 cm^{-1}, respectively, shifted to 847 and 725 cm^{-1}, showing that phen is also interacting to the oxidovanadium(IV) cation [12].

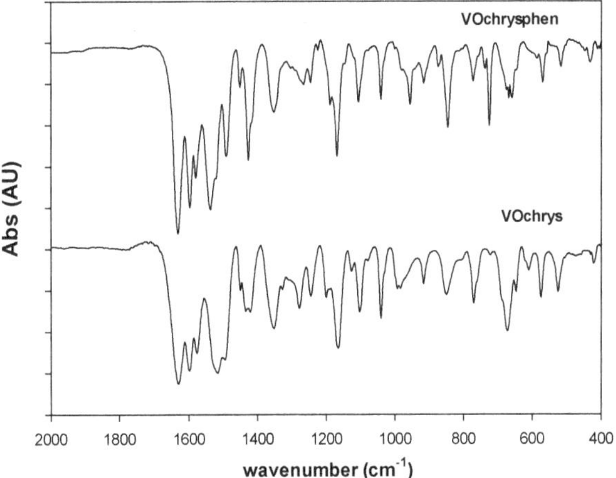

Figure 2. FTIR spectra of [VO(chrys)phenCl] and [VO(chrys)$_2$EtOH]$_2$.

Moreover, the replacement of one chrysin ligand by one phen may produce a decrease of π electron donation to the V=O moiety. The shift of the V=O stretching band from 968 to 957 cm^{-1} is indicative of a decrease of the bond order and an increase of the bond length in agreement with the ligand replacement. These results suggest that the metal ion is interacting with chrysin through C=O and the deprotonated C(5)–O group and with the N atoms of phen.

2.3. EPR Measurements

The powder EPR spectrum of the complex obtained at 120K is shown in Figure 3. A very similar spectrum was obtained at room temperature (Figure S1A). The EPR spectrum of a polycrystalline powder of [VO(chrys)phenCl] gave a unique EPR line, which does not show the typical eight line hyperfine splitting pattern of ^{51}V nucleus ($I = 7/2$), suggesting the presence of extended spin–spin interactions between neighboring oxidovanadium(IV) ions in the solid complex, which collapse the hyperfine interaction into a single line [13]. Similar behavior was observed for oxidovanadium(IV) complexes of apigenin, naringenin, and quercetin [14–16] and it is characteristic of magnetically extended systems of the cation.

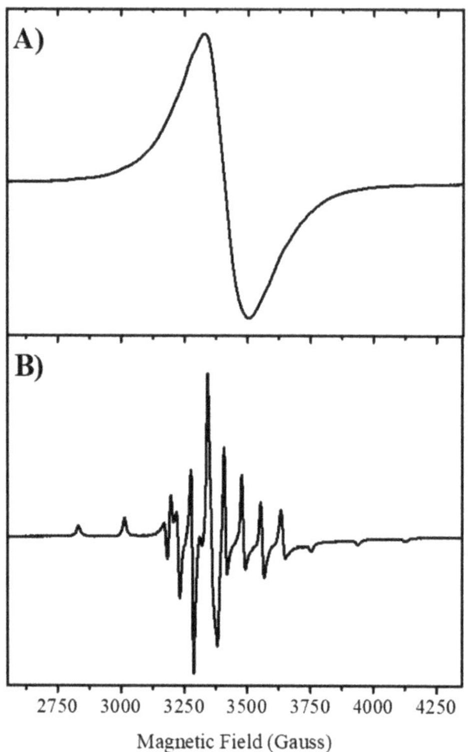

Figure 3. EPR spectrum of [VO(chrys)phenCl] recorded at 120 K. (**A**) Powder sample and (**B**) Frozen DMSO solution. EPR spectra were recorded in a Bruker EMX-Plus spectrometer equipped with a rectangular cavity. Experimental conditions: 100 kHz modulation, 4 Gpp modulation amplitude, 2 mW microwave power.

The EPR spectrum of the DMSO frozen solution measured at 120 K displays the typical eight-line pattern spectrum for axial-V(IV) systems, as shown in Figure 3. The simulation (see Figure S1B) predicted that the observed signal was consistent with the oxidovanadium(IV) ion in a nearly axial or pseudoaxial ligand field. The spin Hamiltonian parameters and the hyperfine coupling constants were $g_{II} = 1.941$; $A_{II} = 162.2 \times 10^{-4}$ cm^{-1}; $g_\perp = 1.977$; $A_\perp = 59.5 \times 10^{-4}$ cm^{-1}. These parameters fit well in the corresponding g_{II} vs. A_{II} diagram for a 2N,2O coordination sphere [17]. Because the parallel component of the hyperfine coupling constant is sensitive to the type of donor atoms on the equatorial positions of the coordination sphere, the empirical relationship $A_z = \sum n_i A_{z,i}$ is frequently used to determine the identity of the equatorial ligands in V(IV) complexes (n_i, number of equatorial ligands of type i and $A_{z,i}$, the contribution to the parallel hyperfine coupling from each of them) [17]. Considering the contributions to the parallel hyperfine coupling constant of the different coordination modes (CO = 44.7×10^{-4}, ArO$^-$ = 38.6×10^{-4}) [14], and N(phen) $\times 2 = 40.4 \times 2 \times 10^{-4}$ [18], the calculated value for A_{II} of 164.1×10^{-4} cm^{-1} agrees with the experimental value. From the EPR parameters, we conclude that the coordination sphere would correspond to a binding mode of (CO, O$^-$, 2N) in the equatorial plane. The chloride ion that also linked to the metal center (see below) may be located in trans-position. Besides, it is observed that $g_{||} < g_\perp < g_e = 2.0023$ and $|A_{||}| > |A_\perp|$, in line with an octahedral site with tetragonal compression and the d_{xy} orbital being the ground state of the V^{4+} ($3d^1$) ion. Moreover, the $\Delta g_{||}/\Delta g_\perp$ ratio ($[g_{||}-g_e]/[g_\perp-g_e]$) proved to be 2.37, showing an octahedral tetragonal distortion.

The hyperfine coupling constants were related to the dipolar hyperfine coupling parameter P, that represents the dipole–dipole interaction of the electronic and nuclear moments, through the relations of Kivelson and Lee [19]: $A_{\parallel} = -P\,[k + 4/7 - \Delta g_{\parallel} - 3/7\,\Delta g_{\perp}]$ and $A_{\perp} = -P\,[k - 2/7 - 11/4\,\Delta g_{\perp}]$. p-value ranges from 100 to 160×10^{-4} cm^{-1} in oxidovanadium(IV) compounds [20] and is calculated as $P = g_e g_N \mu_B \mu_N \langle r^{-3} \rangle$, where g_N is the nuclear g-factor, g_e is the g-factor of the free electron, μ_N the nuclear magneton, and $\langle r^{-3} \rangle$ can be calculated for the vanadium 3d orbitals. The parameter k (between 0.6 to 0.9) is the dimensionless Fermi contact interaction constant [21], is very sensitive to deformations of the metal orbitals, and indicates the isotropic Fermi contact contribution to the hyperfine coupling. The calculated value of $P = 119.6 \times 10^{-4}$ cm^{-1} is considerably reduced when compared to the value of the free ion (160×10^{-4} cm^{-1}) and indicates a considerable amount of covalent bonding in the [VO(chrys)phenCl] complex. The value of $k = 0.71$ indicates a moderate contribution to the hyperfine constant by the unpaired s-electron. Moreover, the product $P \times k = 85.1 \times 10^{-4}$ cm^{-1} represents the anomalous contribution of s-electrons to the $A_{\parallel}$ and $A_{\perp}$ components (the rest being the contribution of $3d_{xy}$ electrons) [22].

2.4. Stability Measurements

The electronic spectral band for the d-d transition of [VO(chrys)phenCl] dissolved in DMSO is shifted to blue regarding the precursor complex (VO(chrys)$_2$), in agreement with the changes in the coordination sphere (2O and 2N atoms) vs. (4O atoms) (766 nm for [VO(chrys)phenCl] vs. 796 nm for the binary compound) [5]. The position of the electronic absorption band proved similar to that of [VO(SO$_4$)(phen)$_2$] in DMSO (ca. 765 nm) [23]). Electronic spectra of a DMSO solution of [VO(chrys)phenCl] (t) and conductivity measurements did not show any significant change during 4 h. Figure S2B shows the spectral changes for a solution of the complex in DMSO/H$_2$O 1/99 (1×10^{-2} M), during 4 h. The complex proved less stable in water solutions, but at least during 15 min (manipulation time for the cellular studies) the complex remained stable in both solutions. It is known that once the complex is added to living cells, it could undergo several chemical interactions with the oxidant and antioxidant cellular systems, including ligand release, but the differences in the anticancer effect between the free ligands and the vanadium complex could demonstrate the efficacy of the vanadium compound.

2.5. Cytotoxic Assays

The cytotoxic effect [VO(chrys)phenCl] on the human lung cancer cell line A549 was determined by the MTT assay, at 24, 48, and 72 h incubation (Table 1). The effects were compared with those of the oxidovanadium(IV) cation [16–24], chrysin [25], VO(chyrs)$_2$ [7], and phen [8], as previously reported.

Table 1. Half maximal inhibitory concentration, IC$_{50}$, values of VO(chrys)phenCl and its components (oxidovanadium(IV) cation (VO), chrysin, phen) on A549 cell line at 24, 48 and 72 h incubation and on HEK cell line for VO(chrys)phenCl (24 h). The IC$_{50}$ values for the binary complex of VO and chrysin was added for comparisons. The results represent the mean $\pm$ the standard error of the mean (SEM) from three separate experiments.

	IC$_{50}$ (μM) 24 h	IC$_{50}$ (μM) 48 h	IC$_{50}$ (μM) 72 h
VO	>100 [a]	>100	15 $\pm$ 1.2 [b]
chrysin	>100	66.4 $\pm$ 4.9 [c]	37.3 $\pm$ 3.5
Phen [d]	66.1 $\pm$ 3.4	23.9 $\pm$ 2.5	1.9 $\pm$ 0.5
VO(chyrs)$_2$ [e]	>100	41.2 $\pm$ 3.9	6.1 $\pm$ 1.2
VO(chyrs)phenCl	28.9 $\pm$ 4.0 >100 (HEK293)	8.3 $\pm$ 1.0	1.7 $\pm$ 0.1

[a] [16], [b] [24], [c] [25], [d] [8], [e] [7].

The replacement of one chyrsin molecule by phen in the coordination sphere of the binary complex greatly enhanced the anticancer effects at 24 and 48 h incubation. As can be seen, cell incubation time had an impact on cell viability in a dose–response manner. Higher incubation times (72 h) result in low IC_{50} values, and most of the cells that are responding at early times after treatment were lost at the time the assay was performed, owing to the disintegration of cells into particulate debris altering the ratio between the live and dead cell populations present at the time of analysis. Therefore, it is difficult to make cell viability comparisons after 72 h incubation and this is the reason why the mechanistic studies have been carried out in the present study at 24 h incubation [26].

It is important to emphasize that the oxidovanadium(IV) complexes may hydrolyze and/or oxidize in aqueous solution. At 24 h incubation, the $V^{IV}O$ cation showed a high IC_{50} value without affecting cell viability up to 100 µM (Table 1), and V(V) showed a decrease of cell viability of 10–20 % at 50 and 100 µM [27]. The anticancer effect of VO(chyrs)phenCl at 24 h incubation proved higher than that exerted by the oxidovanadium(IV) cation, its oxidized species, the binary complex, and the ligands. Hence, we are able to discard that the deleterious effect the complex was due to the metal ion, phen, and/or chrysin ligands generated after decomposition processes (or at least we can assume that these processes were slow at 24 h incubation).

The percentage of non-tumorigenic HEK293 cell viability vs. [VO(chrys)phenCl] concentrations is shown in Figure S3. The complex does not reduce the cellular viability in the range of tested concentrations (0–100 µM), indicating that the toxicity of the compound shows a good correlation in terms of selectivity toward A549 cancer cells within 24 h of tincubation.

2.6. ROS and GSH/GSSG Cellular Levels

Cellular oxidative stress is considered to be inducer of carcinogenesis. Cancer cells show high ROS levels and are more vulnerable to ROS. Hence, ROS generating compounds kill cancer cells selectively. Non-transformed cells have a low basal intracellular ROS level and have a full antioxidant capacity, being less vulnerable to the oxidative stress induced by different compounds in cancer cells. To study the role of the oxidative stress in the cytotoxicity induced by the complex, we measured ROS production in A549 cells at 24 h incubation. From Figure 4, it can be seen that the oxidovanadium(IV) cation did not increase cellular ROS levels, chrysin generated a very low increase of ROS, while phen and the complex elevated the amount of intracellular ROS so that they selectively damage cancer cells. The pro-oxidant nature of the complex and phen improved the anticancer effects of the flavonoid.

Glutathione is an antioxidant, capable to prevent damage produced by ROS in cells, ubiquitously present in all cell types at mM concentration. In mammalian cells under physiological conditions, the GSH redox couple is known to be present with steady-state concentrations of 1–10 mM. The overall ratio of GSH to its oxidized state GSSG in a cell is usually greater than 100:1, and the redox couple GSH/GSSG is used as an indicator of changes in the redox environment and of oxidative stress in the cell. In various models of oxidative stress, this ratio has been demonstrated to decrease to values of 10:1 and even 1:1 [28].

The GSH contents in A549 cell line after incubation with the complex were measured (Figure 5A). As GSH depletion is not a major cause of cytotoxicity, the GSH/GSSG was also calculated (Figure 5B) to demonstrate that the decrease of the GSH/GSSG ratio was due not only to a decrease in the level of GSH, but also to an accumulation of GSSG. Therefore, it can be seen that the cellular damage on the A549 cell line is manifested by the increased levels of ROS that exceed the defense mechanisms inducing GSH oxidation with the resultant reduction of cellular GSH. Therefore, a stress oxidative mechanism could be assumed for the cell-killing action of [VO(chrys)phenCl].

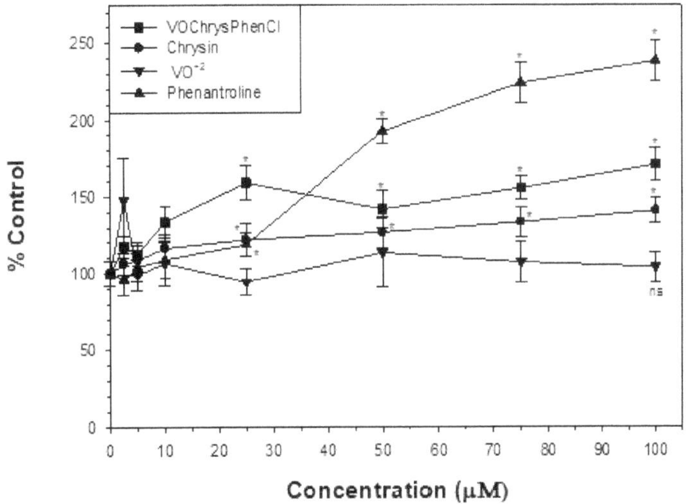

Figure 4. Effect of chrysin, VO(chrys)phenCl, phen and oxidovanadium(IV) cation on H_2DCFDA oxidation to DCF. A549 cells were incubated at 37 °C in the presence of 10 µM H_2DCFDA. The values are expressed as the percentage of the control level and represent the mean $\pm$ SEM. * $p < 0.05$, ns: not significant.

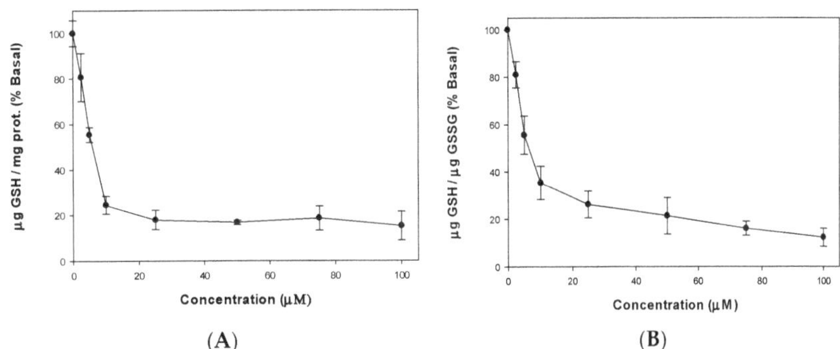

Figure 5. Effect of [VO(chrys)phenCl] on GSH cellular levels (**A**) and GSH/GSSG ratio (**B**) in A549 cells, 24 h incubation. Results are expressed as mean $\pm$ SEM of three independent experiments. All values are statistically significant in comparison with the control.

2.7. Mitochondrial Membrane Potential

The loss of mitochondrial function and the subsequent release of cytochrome C into the intracellular space are some of the mechanisms associated with the apoptosis process [29]. The mitochondrial membrane potential ($\Delta\psi$) was measured to explain the increase of cellular ROS and decrease of the GSH/GSSG ratio on the A549 cell line by incubation of the metal complex. The lipophilic cationic probe DioC6 enters the mitochondria and upon depolarization it will accumulate less dye [30]. Figure 6 shows the effect of mitochondrial dysfunction with a membrane potential loss when A549 cells were treated with increasing concentrations of the metal complex in concordance with a stress oxidative mechanism.

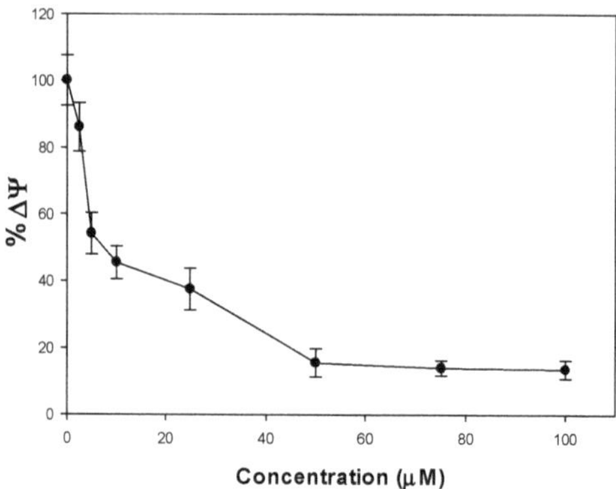

Figure 6. Changes of the mitochondrial membrane potential (% Δψ) in A549 cells treated with increasing concentrations of VO(chrys)phenCl for 24 h. Each point represents the mean ± S.E.M of three measurements in three independent experiments. All values are statistically significant in comparison with the control.

2.8. Cellular Vanadium Uptake Experiments

The vanadium content after 24 h treatment of the compounds at a concentration equivalent to [VO(chrys)phenCl] IC_{50} (28.9 μM) was determined by inductively coupled plasma-mass-spectrometry, ICP-MS (see Table 2). $VO(acac)_2$ exhibited almost the same capacity for cellular uptake as the control. Our results are comparable with data reported for the human A2780 ovarian cancer cells [31]. For [VO(chrys)phenCl], the total amount of V up-taken is ca. five-fold higher than for the binary complex. Our data suggest that its greatest cytotoxicity might be directly correlated with the incorporation of phen into the complex structure, increasing its lipophilicity, which could improve the cellular uptake.

Table 2. Cellular V content (determined by ICP-MS) following cell treatments with 28.9 μM of compounds for 24 h. Results are expressed as mean ± the standard error of the mean (SEM) of two independent experiments.

	Nmol V/Mg Protein
Control	4.6 ± 0.5
$VO(acac)_2$	4.9 ± 0.2
$VO(chrys)_2$	4.7 ± 0.1
[VO(chrys)phenCl]	23.5 ± 1.3

2.9. BSA (Bovine Serum Albumin) Interactions

The binding behavior of drugs with albumin affected their distribution and metabolism in the body. Both human serum albumin (HSA) and BSA are commonly used to determine the binding interactions because they possess near 76% sequence homology and tertiary structure similarity [32]. To study the protein binding ability of the complex, we have selected BSA herein because of its low cost and wide availability. The intrinsic fluorescence of BSA is due to tryptophan residue when excited at 295 nm. Upon titration of BSA with increasing concentrations of the complex, the fluorescence spectra showed a decrease in the intensity or quenching (Figure S4). The Stern–Volmer quenching constant was obtained from the slope of the graph of the fluorescence intensities F_0/F (in the absence and presence of the quencher, respectively, and corrected by the inner-filter effect) vs. the quencher concentration, Q, according to the equation: $F_0/F = 1 + K_q\tau_0[Q] = 1 + K_{SV}[Q]$, where K_q is the bimolecular quenching constant and τ_0 is the lifetime of the fluorophore in the absence of the quencher (considered 1×10^{-8} s for a biopolymer) [33].

An upward curvature of the plots at different temperatures can be seen, showing a combined quenching (static and dynamic). However, at low concentrations, a linear correlation was obtained (Figure 7). The linearity at lower concentrations suggested a single quenching type. Static and dynamic quenching can be differentiated by the analysis of the temperature dependence. Using the data of the linear region, K_{sv} and K_q ($K_q = K_{sv}/\tau_0$, $\tau_0 = 10^{-8}$ s) were calculated (Table 3). The Ksv values showed a decrease with increasing temperature, suggesting a static quenching mechanism. Another criterion for the type of static cooling is that the result of the bimolecular cooling constant should be greater than the maximum dynamic cooling constant for a biopolymer, assumed to be 2.0×10^{10} M^{-1}s^{-1}. As can be seen in Table 3, all the values of K_q obtained were greater than that value, reinforcing the assumption that a static quenching type of interaction between the complex and albumin takes place.

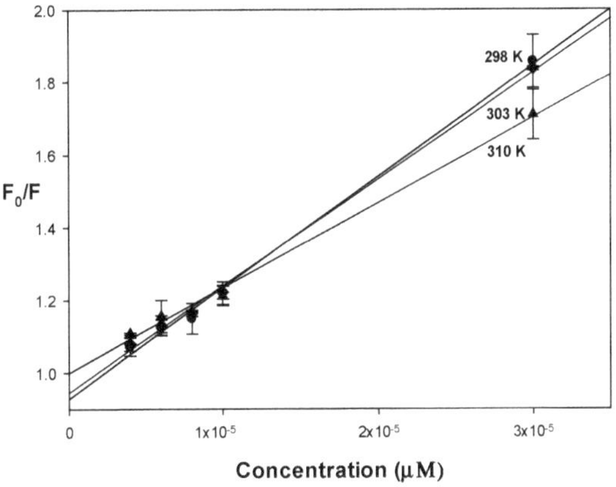

Figure 7. Stern Volmer graph for VO(chrys)phenCl at different temperatures, [BSA] = 6 μM, λ_{ex} = 280 nm.

Table 3. Stern-Volmer constant (K_{sv}), bimolecular quenching constant (K_q), binding constant (K_b) and number of binding sites (n) for the interaction of VO(chrys)phenCl with BSA (6 μM) in Tris-HCl buffer (0.1 M, pH 7.4).

T (K)	K_{sv} ($\times 10^4$) (M^{-1})	r^2	K_q ($\times 10^{12}$) (M^{-1}s^{-1})	K_b ($\times 10^5$) (M^{-1})	n
298	3.06 ± 0.11	0.98	3.06 ± 0.11	3.16 ± 0.24	1.24 ± 0.07
303	2.94 ± 0.07	0.99	2.94 ± 0.07	1.58 ± 0.34	1.17 ± 0.06
310	2.34 ± 0.10	0.97	2.34 ± 0.10	0.16 ± 0.11	0.96 ± 0.06

To determine the binding constant (K_b) and the number of binding sites n (Table 3), the Scatchard equation ($\log[(F_0 - F)/F] = \log K_b + n \log [Q]$) was used (Figure 8). Binding constants prove to be in the order of 10^5–10^6 M^{-1} and decrease at higher temperature, indicating a reduction in stability of the BSA-complex compound, and confirming the involvement of a static quenching. However, the obtained values suggest that the interactions are adequate for the complex to be transported and delivered by BSA. The number of binding sites (*ca.* 1) indicated that the complex may occupy one binding site of the protein.

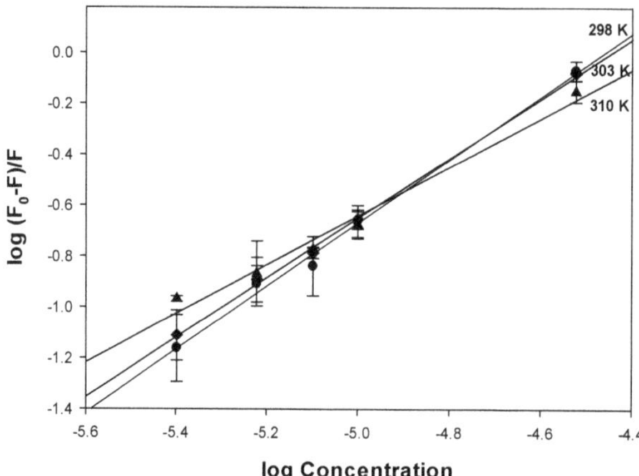

Figure 8. Plots of log [(F$_0$ − F)/F] vs. log [Q] for the VO(chrys)phenCl-BSA system at 298 K, 303 K and 310 K, [BSA] = 6 μM, λ_{ex} = 280 nm.

To determine the main forces operating during the protein binding, thermodynamic parameters were determined, using the van't Hoff equation ln K$_a$ = −ΔH/RT + ΔS/R and ΔG = ΔH − TΔS (Table 4). From the negative values of ΔG, a spontaneous interaction can be confirmed and the decrease of ΔG at higher temperatures indicated a decrease of the binding strength of the complex–protein bond. The negative values of enthalpy and entropy changes indicate that hydrogen bond and van der Waals forces were the major forces operating during the interaction [34].

Table 4. Thermodynamic parameters for the interactions between [VO(chrys)phenCl] with BSA.

ΔH (KJ/Mol)	ΔS (J/Mol)	ΔG (KJ/Mol)
−207.8	−588.9	−32.2 (298 K) −29.3 (303 K) −25.1 (310 K)

3. Discussion

To enhance the biological action of VO(chrys)$_2$ complex, one chrysin ligand was replaced by the planar ligand phen in the metal coordination sphere and the new complex was characterized by common analytical techniques. FTIR studies indicated that the oxidovanadium(IV) ion interacted with C=O and O$^-$ atoms of chrysin and N atoms of phen. Conductivity studies indicated that the chloride ion, which contributed to the electroneutrality of the complex, is also bonded to the V=O moiety, in trans- position (solution EPR determinations). Conductivity and spectral determinations in DMSO and DMSO/H$_2$O showed that the complex did not produce hydrolytic species during the manipulation time of the complex. It has to be noted that the cellular experiments consist in the addition to the cells of a fresh DMSO stock solution of the complex dissolved in the culture media (with a final DMSO concentration of 0.5%) and, therefore, the studies of the stability of the complex must be performed not only on DMSO and the culture media, but the cell components must also be considered (including natural antioxidant compounds that must also been taken under consideration in those kind of studies), as previously mentioned [35]. However, the determination of the speciation in aqueous solution of the complex inside the cell is outside the scope of this study.

In a previous paper, the cytotoxicity of *cis*-[V^{IV}O(OSO$_3$)(phen)$_2$] at 72 h incubation in the A549 cell line was related to its decomposition in cell culture medium generating the ligands and the oxidation of vanadium [36]. Another study showed that up to 24 h incubation of the same compound on different ovarian cancer cell lines displayed different IC_{50} values than the free ligands and inorganic V(IV) and V(V) ions, determining that the biological effects were due to the complex, which is more active than the free ligands [23]. However, at 72 h incubation, similar IC_{50} values were obtained, probably due to chemical changes of the metal complex.

From the data in Table 1, we are able to discard that the cytotoxic effect was due to phen and/or chrys ligands, generated after decomposition processes at least at 24 h incubation, because there were no deleterious effects of the binary complex and chrysin, while the IC_{50} value measured for phen was 66 μM, corresponding to more than twice the IC_{50} value of the complex. Besides, we measured the effect of the 24 h cell incubation with the mixture of each component of the complex in stoichiometric quantities (sodium metavanadate, chrysin and phen, physiological pH). It produced the same inhibitory effect from 10 to 50 μM (*ca.* 50% viable cells, Figure S5) and, hence, the IC_{50} value could not be determined. However, we can conclude that the effect of the mixture proved different from that of the complex (that inhibited cell viability in a dose dependent manner), discarding that the cytotoxic activity of the complex was due to decomposition followed by oxidovanadium(IV) oxidation in the culture media.

It is well known that ROS overproduction is the cause of the development of a number of diseases. Excessive ROS accumulation and the depletion of natural antioxidant compounds such as GSH can cause irreversible cell damage and even cell death [37]. We determined herein that one mechanism involved in the process of cell death is the induction of oxidative stress by the metal complex accompanied by disruption of the mitochondria membrane potential. Meanwhile, the loss of mitochondrial membrane potential is related to activation of the mitochondrial apoptosis pathways. The presence of the lipophilic ligand, phen, favors the cellular uptake of the metal. Phen coordinates to oxidovanadium(IV) cation, improving its transport inside cells, hence behaving as a better cytotoxic agent via the induction of oxidative stress and mitochondrial membrane damage than the VO(chrys)$_2$ complex.

The replacement of one chrys ligand by phen in VOchrysphenCl produced a tighter interaction to BSA than VO(chrys)$_2$ (K_b 0.76×10^5 M^{-1}) [7]. Both compounds bind to BSA in a spontaneous and enthalpy-driven manner and could be transported by albumin (K_b values in the range of 10^4–10^6 M^{-1}).

4. Materials and Methods

4.1. Materials and Instrumentation

Chrysin (Sigma, Buenos Aires, Argentina), phenanthroline hydrochloride (Merck, Buenos Aires, Argentina), and vanadyl acetylacetonate (Fluka Munich, Germany) were used as supplied. Corning or Falcon provided tissue culture materials. Dulbecco's modified Eagle's medium (DMEM, Gibco, Gaithersburg, MD, USA), Tryple^{TM} (Invitrogen, Buenos Aires, Argentina) and fetal bovine serum (FBS, Internegocios, Buenos Aires, Argentina) were used as provided. All other chemicals used were of analytical grade. Elemental analysis for carbon, nitrogen, and hydrogen was performed using a Carlo Erba EA1108 analyzer. Vanadium content was determined by the tungstophosphovanadic method [38]. Thermogravimetric analysis was performed with Shimadzu systems (model TG-50), working in an oxygen flow of 50 mL·min^{-1} and at a heating rate of 10 °C·min^{-1}. Sample quantities ranged between 10 and 20 mg. UV-vis and diffuse reflectance spectra were recorded on a Hewlett-Packard 8453 diode-array and a Shimadzu 2600/2700 spectrophotometer. Infrared spectra were measured with a Bruker IFS 66 FTIR spectrophotometer from 4000 to 400 cm^{-1} using the KBr pellet technique. Fluorescence spectra were obtained with a Shimadzu RF-6000 spectrophotometer equipped with a pulsed xenon lamp. The molar conductance of the complex was measured on a Conductivity TDS Probe-850084, Sper Scientific Direct, using 10^{-3} M DMSO solutions. Exact mass spectra were obtained using a Bruker micrOTOF-Q II mass spectrometer, equipped with an ESI source operating in positive mode.

A Bruker EMX-Plus spectrometer, equipped with a rectangular cavity with 100 kHz field modulation and with standard Oxford Instruments low-temperature devices (ESR900/ITC4), was used to record the EPR spectra of both powdered and DMSO solution spectra. The spectra, obtained at 100 K to room temperature (*ca.* 298 K) were, when necessary, baseline corrected using WinEPR Processing software (Bruker, Inc., Billerica, MA, USA).

The EasySpin 5.2.3. toolbox based on MATLAB was used to simulate g- and A- values [39], assuming an axial spin-Hamiltonian of the form:

$$H = \mu_B \left[g_\parallel B_z S_z + g_\perp \left(B_x S_x + B_y S_y \right) \right] + \left[A_\parallel S_z I_z + A_\perp \left(S_x I_x + S_y I_y \right) \right]$$

where μ_B is the Bohr magneton, and $g_\parallel, g_\perp, A_\parallel, A_\perp$ are the components of the axial **g** and **A** tensors, respectively. $B_{x/y/z}$, $S_{x/y/z}$, and $I_{x/y/z}$ are the components of the magnetic field, and of the spin operators of the electron and V nucleus, respectively.

4.2. Preparative [VO(chrys)phenCl]

The binary complex [VO(chrys)$_2$EtOH]$_2$ (0.1 mmol) prepared as in [5] and phenanthroline hydrochloride (0.2 mmol) in acetone (25 mL) were poured into a round bottomed flask and refluxed for 3 h. The precipitate was filtered and washed three times with hot acetone. The green-yellow solid was dried in an oven at 60 °C. Anal calc for $C_{27}H_{17}ClN_2O_5V$: 535.9 g/mol; C, 60.5; H, 3.2; N, 5.2; V, 9.5%. Exp, C, 60.4; H, 3.3; N, 5.3; V, 9.6. Diffuse reflectance spectrum: 201 nm, 208 nm, 226 nm, 276 nm, 333 nm (sh), 396 nm, 782 nm. UV-Vis (DMSO): 273 nm (105,300 $M^{-1}cm^{-1}$) 288 nm (91,100 $M^{-1}cm^{-1}$) 326 nm (12,341 $M^{-1}cm^{-1}$) 392 nm (6143 $M^{-1}cm^{-1}$) 769 nm (68 $M^{-1}cm^{-1}$). Electrospray ionization mass spectrometry (ESI-MS) analyses for the complex dissolved in DMSO:acetone (1:2): ESI-MS(+) (m/z) (calc for $C_{27}H_{17}ClN_2NaO_5V$): 558.02 [M − Na]$^+$, (found): 558.01, for M, [VO(chrys)phenCl] and [M − Na]$^+$, [VO(chrys)phenCl − Na]$^+$. The mass-to-charge ratio peak detected at (m/z) 500.05 (100%) was due to the presence of [VOchrysphen]$^+$ species: the ligand chloride dissociates under the ESI-MS conditions and the bidentate ligands remained bonded to the metal ion (chelate effect, higher stability), as shown in Figure S6.

4.3. Cell Viability Assay (MTT Assay)

Cell viability was measured by the 3-[4,5-dimethylthylthiazol-2-yl]-2,5 diphenyltetrazolium bromide (MTT) (Sigma-Aldrich, St. Louis, MO, USA) method. Briefly, A549 cells (human lung cancer cell line) and HEK293 (human embryonic kidney) were maintained at 37 °C in a 5% carbon dioxide atmosphere using DMEM supplemented with 100 U/mL penicillin, 100 µg/mL streptomycin, and 10% (*v/v*) fetal bovine serum as the culture medium. When 70–80% confluence was reached, cells were subcultured using TrypLE ™ and free phosphate buffered saline (PBS) (11 mM KH_2PO_4, 26 mM Na_2HPO_4, 115 mM NaCl, pH 7.4). For the treatments, cells were seeded at a density of 1×10^5 per well in 48 well plates, grown overnight, then incubated with the complex, metal ion, and ligands in FBS free medium. After different incubation times, at 37 °C, 100 µg of MTT per well were added and incubated was performed in a CO_2 incubator for 2 h at 37 °C. DMSO was then added to dissolve formazan crystals and the absorbance of each well was measured by a plate reader at a test wavelength of 560 nm. Data are presented as the percentage of cell viability (%) of the treated group with respect to the untreated cells (control), whose viability is assumed to be 100%.

4.4. Oxidative Stress Determinations

Reactive oxygen species (ROS) generation in A549 cell lines was measured by oxidation of 2′,7′-dichlorodihydrofluorescein diacetate (H$_2$DCFDA) to 2′,7′-dichlorofluorescein (DCF). Briefly, 24-well plates were seeded with 5×10^4 cells per well and allowed to adhere overnight. Then, different concentrations of the compounds were added. After 24 h incubation, media was removed, and cells were loaded with 10 µM H$_2$DCFDA diluted in clear media for 30 min at 37 °C. Media was then separated and the cell monolayers rinsed with PBS and lysated into 1 mL 0.1% Triton-X100. The oxidized product DCF was analyzed in the cell extracts using fluorescence spectroscopy (λ_{ex}, 485 nm; λ_{em}, 535 nm) [40].

Natural antioxidant levels of glutathione (GSH) and its oxidized product (GSSG) were determined in A549 cell lines in culture. Confluent cell monolayers from 24 well dishes were incubated with different concentrations of the compounds at 37 °C for 24 h. Then, the monolayers were washed with PBS and harvested by incubating them with 300 µL Triton 0.1% for 30 min. For GSH determinations, 100 µL aliquots were mixed with 1.8 mL of ice cold phosphate buffer (Na_2HPO_4 0.1 M-EDTA 0.005 M, pH 8) and 100 µL o-phthaldialdehyde (OPT) (0.1% in methanol) [41]. For the determination of GSSG, the cellular extracts were incubated with 0.04 M of N-ethylmaleimide (NEM) to avoid GSH oxidation, 100 µL aliquots were mixed with 1.8 mL NaOH 0.1 M and OPT, and the fluorescence was determined (λ_{ex}, 350 nm; λ_{em}, 420 nm). Standard curves with different concentrations of GSH were processed in parallel. The protein content in each cellular extract was quantified using the Bradford assay [42]. The better marker for the cellular redox status, the GSH/GSSG ratio, was calculated as % control for all the experimental conditions.

The mitochondrial membrane potential was assessed to evaluate the mitochondrial function, using the $DiOC_6$ (3,3′-Dihexyloxacarbocyanine Iodide) fluorescent probe. The probe was added to the wells (400 nM concentration) and incubated for 30 min at 37 °C. The cells were resuspended in PBS and measured using fluorescence spectroscopy (λ_{ex}, 485 nm; λ_{em}, 535 nm) [43].

4.5. Cellular Vanadium Uptake Experiments

For vanadium uptake experiments, cells were grown to 80% confluence in 100 mm petri dishes. Incubations with the treatment compounds (28.9 µM of VO(chrys)phenCl, VO(chrys)$_2$ and VO(acac)$_2$ in 0.5% DMSO) were performed for 24 h. Afterwards, V-containing media were removed, and the cell layers were washed twice with PBS. Cells were detached using TrypLE enzyme solution for ca. 15 min at 37 °C. The cell suspensions were collected into centrifuge tubes and pelleted at $4000\times g$ for 2 min. The cell pellets were washed once with phosphate buffered saline (1.0 mL per tube) and lysed with 100 µL 0.10 M NaOH overnight at 4 °C. Each lysate (2 µL) was mixed with 98 µL of Bradford reagent and the absorbance at 560 nm was measured using a plate reader for the determination of protein content. Freshly prepared solutions (0–2.0 mg mL^{-1} in 0.10 M NaOH) of bovine serum albumin were used for calibration. The rest of the lysate was diluted to 1.0 mL with 20% HNO_3 and vanadium contents in the resultant solutions were determined by ICP-MS. Corresponding amounts of NaOH and HNO_3 solutions were used to prepare the blank samples. The content of vanadium in the cell lysates was calculated in nmol *V* per mg protein [44].

4.6. BSA Interactions

The fluorescent technique was used for the measurement of the interactions with BSA. BSA in Tris-HCl (0.1 M, pH = 7) was kept constant at 6 µM and titrated with different concentrations of the complex ranging from 2 to 30 µM with an incubation time of 1 h. For the experiments, the quenching of the emission intensity of BSA (at 336 nm) by the complex was monitored at λ_{ex} = 280 nm with excitation and emission slits of 10 nm. Three different temperatures were selected for thermodynamic determinations (298, 303, and 310 K). Because of the absorption of the BSA-complex near the excitation wavelength, the fluorescence intensities were corrected by the inner-filter effect, using: $F_{corr} = F_{obs} \times e^{(1/2A_{ex} + 1/2A_{em})}$, where F_{corr} and F_{obs} are the fluorescence intensities corrected by inner-filter effect and recorded, respectively, A_{ex} and A_{em} are the electronic absorbances of the solutions at excitation and emission wavelengths, respectively [33]. Three independent replicates were performed for each sample and concentration.

4.7. Statistical Analysis

Statistical differences were analyzed using the analysis of variance method (ANOVA) followed by the test of least significant difference (Fisher). Statistical significance was defined as $p < 0.05$.

5. Conclusions

A new oxidovanadium(IV) metal complex with the flavonoid chrysin and phen has been synthesized and characterized. In vitro cytotoxicity testing showed that the compound exhibit significant cytotoxicity towards A549 cell line at different incubation times, indicating that the compound has the potential to act as an effective metal-based anticancer drug. The induction of intracellular reactive oxygen species (ROS) production, perturbation of mitochondrial membrane potential, and GSH and GSH/GSSG depletion suggested the ability of the complex to induce cell death by the initiation or progression of oxidative stress. It can be seen that while the complexation of the natural antioxidant chrysin by the oxidovanadium(IV) cation increased the anticancer effect of the flavonoid, the addition of phen in the coordination sphere produced a higher cytotoxic effect and a better vanadium cellular uptake. In this sense, we can discard a total decomposition of the complex leading to the release of the free ligands (being only one of them (phen) biologically active but with low cytotoxicity) at 24 h incubation. In addition, the metal complex did not show toxic effects against a non-tumorigenic cell line and could be stored and transported by albumin.

Supplementary Materials: The following are available online at https://www.mdpi.com/article/10.3390/inorganics10010004/s1, Figure S1. EPR spectrum of [VO(chrys)phenCl]. (A) Powder sample at 120 K (black) and 298 K (red). (B) Frozen DMSO solution EPR spectrum recorded at 120 K (black) together with simulation (red). EPR spectra of both powder and DMSO solution were recorded in a Bruker EMX-Plus spectrometer, equipped with a rectangular cavity. Experimental conditions: 100 kHz modulation field, 4 Gpp modulation amplitude and 2 mW microwave power. The spectra were baseline corrected using WinEPR Processing software (Bruker, Inc.) and simulations were performed with the Easy Spin 5.2.3. toolbox based on MATLAB assuming an axial spin-Hamiltonian. The spin Hamiltonian parameters obtained were $g_{\text{II}} = 1.941$; $A_{\text{II}} = 162.2 \times 10^{-4}$ cm^{-1}; $g_{\perp} = 1.977$; $A_{\perp} = 59.5 \times 10^{-4}$ cm^{-1}. Figure S2. Spectral variation of a dissolution of [VO(chrys)phenCl] in (A) DMSO, (B) DMSO/H$_2$O 1/99 (1×10^{-2} M), during 4 h. Figure S3. Cell viability assay at different of [VO(chrys)phenCl] concentrations after treatment for 24 h on HEK293 cells. The results are expressed as a percentage of the control level and represent the mean $\pm$ the standard error of the mean (SEM) from three separate experiments. * indicates significant values in comparison with the control level ($p < 0.05$). Figure S4. The fluorescence spectra of BSA at various temperatures for VOchrysphen (0, 4, 6, 8,10, 30 µM). λex = 280 nm, [BSA] = 6 µM. Figure S5. Cell viability assay of a mixture of sodium metavanadate, chrysin and phen (1:1:1), physiological pH at different concentrations after treatment for 24 h on A549 cells. The results are expressed as a percentage of the control level and represent the mean $\pm$ the standard error of the mean (SEM) from three separate experiments. * indicates significant values in comparison with the control level ($p < 0.05$). Figure S6. Electrospray ionization–mass spectrometry (ESI-MS) spectrum of [VO(chrys)phenCl].

Author Contributions: Conceptualization, P.A.M.W.; validation, E.G.F., L.G.N. and P.A.M.W.; formal analysis, A.A.D. and L.G.N.; investigation, A.A.D.; measurements, P.J.G. and M.R.; resources, P.A.M.W.; writing—original draft preparation, A.A.D., E.G.F. and P.J.G.; writing—review and editing, P.A.M.W.; visualization, P.A.M.W.; supervision, P.A.M.W. and L.G.N.; project administration, P.A.M.W.; funding acquisition, P.A.M.W. All authors have read and agreed to the published version of the manuscript.

Funding: This research was funded by ANPCyT, 2019-0945 and UNLP X871.

Institutional Review Board Statement: Not applicable.

Informed Consent Statement: Not applicable.

Data Availability Statement: The data presented in this study are available in the Supplementary Materials.

Acknowledgments: This work was supported by UNLP (X871), CICPBA, and ANPCyT (PICT 2018-0985), Argentina. LGN and EGF are members of the Research Career, CONICET, Argentina. PAMW is member of the Research Career, CICPBA. AAD is fellowship holder from CONICET.

Conflicts of Interest: The authors declare no conflict of interest. The funders had no role in the design of the study; in the collection, analyses, or interpretation of data; in the writing of the manuscript, or in the decision to publish the results.

References

1. Stompor-Gorący, M.; Bajek-Bil, A.; Machaczka, M. Chrysin: Perspectives on Contemporary Status and Future Possibilities as Pro-Health Agent. *Nutrients* **2021**, *13*, 2038. [CrossRef]
2. Crans, D.C.; Henry, L.R.; Cardiff, G.; Posner, B.I. Developing vanadium as an antidiabetic or anticancer drug: A clinical and historical perspective. In *Essential Metals in Medicine: Therapeutic Use and Toxicity of Metal Ions in the Clinic*; Peggy, L., Carver, Eds.; De Gruyter: Berlin, Germany, 2019; pp. 203–230. [CrossRef]
3. Siegel, R.L.; Miller, K.D.; Jemal, A. Cancer statistics, 2021. *CA Cancer J. Clin.* **2021**, *71*, 7–33. [CrossRef]
4. Sung, H.; Ferlay, J.; Siegel, R.L.; Laversanne, M.; Soerjomataram, I.; Jemal, A.; Bray, F. Global cancer statistics 2020: Globocan estimates of incidence and mortality worldwide for 36 cancers in 185 countries. *CA Cancer J. Clin.* **2021**, *71*, 209–249. [CrossRef] [PubMed]
5. Naso, L.G.; Ferrer, E.G.; Lezama, L.; Rojo, T.; Etcheverry, S.B.; Williams, P.A.M. Role of oxidative stress in the antitumoral action of a new vanadyl(IV) complex with the flavonoid chrysin in two osteoblast cell lines: Relationship with the radical scavenger activity. *J. Biol. Inorg. Chem.* **2010**, *15*, 889–902. [CrossRef]
6. Naso, L.G.; Valcarcel, M.; Villace, P.; Roura-Ferrer, M.; Salado, C.; Ferrer, E.G.; Williams, P.A.M. Specific antitumor activities of natural and oxovanadium(IV) complexed flavonoids in human breast cancer cells. *New J. Chem.* **2014**, *38*, 2414–2421. [CrossRef]
7. Naso, L.G.; Martínez Medina, J.J.; Okulik, N.B.; Ferrer, E.G.; Williams, P.A.M. Study on the cytotoxic, antimetastatic and albumin binding properties of the oxidovanadium(IV) chrysin complex. Structural elucidation by computational methodologies. *Chem. Biol. Interact.* **2022**, *351*, 109750. [CrossRef] [PubMed]
8. Naso, L.G.; Martínez Medina, J.J.; D'Alessandro, F.; Rey, M.; Rizzi, A.; Piro, O.E.; Echeverría, G.A.; Ferrer, E.G.; Williams, P.A.M. Ternary Copper(II) complex of 5-hydroxytryptophan and 1,10-phenanthroline with several pharmacological properties and an adequate safety profile. *J. Inorg. Biochem.* **2020**, *204*, 110933. [CrossRef]
9. Halevas, E.; Mavroidi, B.; Antonoglou, O.; Hatzidimitriou, A.; Sagnou, M.; Pantazaki, A.A.; Litsardakis, G.; Pelecanou, M. Structurally characterized gallium-chrysin complexes with anticancer potential. *Dalton Trans.* **2020**, *49*, 2734–2746. [CrossRef] [PubMed]
10. Halevas, E.; Mitrakas, A.; Mavroidi, B.; Athanasiou, D.; Gkika, P.; Antoniou, K.; Samaras, G.; Lialiaris, E.; Hatzidimitriou, A.; Pantazaki, A.; et al. Structurally characterized copper-chrysin complexes display genotoxic and cytotoxic activity in human cells. *Inorg. Chim. Acta* **2021**, *515*, 120062. [CrossRef]
11. Zahirović, A.; Kahrović, E.; Cindrić, M.; Kraljević Pavelić, S.; Hukić, M.; Harej, A.; Turkušić, E. Heteroleptic ruthenium bioflavonoid complexes: From synthesis to in vitro biological activity. *J. Coord. Chem.* **2017**, *24*, 4030–4053. [CrossRef]
12. Islas, M.S.; Martínez Medina, J.J.; Piro, O.E.; Echeverría, G.A.; Ferrer, E.G.; Williams, P.A.M. Comparisons of the spectroscopic and microbiological activities among coumarin-3-carboxylate, o-phenanthroline and zinc(II) complexes. *Spectrochim. Acta A* **2018**, *198*, 212–221. [CrossRef]
13. Rizzi, A.C.; Neuman, N.I.; González, P.J.; Brondino, C.D. EPR as a Tool for Study of Isolated and Coupled Paramagnetic Centers in Coordination Compounds and Macromolecules of Biological Interest. *Eur. J. Inorg. Chem.* **2016**, *2*, 192–207. [CrossRef]
14. Ferrer, E.G.; Salinas, M.V.; Correa, M.J.; Naso, L.; Barrio, D.A.; Etcheverry, S.B.; Lezama, L.; Rojo, T.; Williams, P.A.M. Synthesis, characterization, antitumoral and osteogenic activities of Quercetin vanadyl(IV) complexes. *J. Biol. Inorg. Chem.* **2006**, *11*, 791–801. [CrossRef] [PubMed]
15. Martínez Medina, J.J.; Naso, L.G.; Pérez, A.L.; Rizzi, A.; Okulik, N.B.; Ferrer, E.G.; Williams, P.A.M. Apigenin oxidovanadium(IV) cation interactions. Synthesis, spectral, bovine serum albumin binding, antioxidant and anticancer studies. *J. Photochem. Photobiol. A* **2017**, *344*, 84–100. [CrossRef]
16. Islas, M.S.; Naso, L.G.; Lezama, L.; Valcarcel, M.; Salado, C.; Roura-Ferrer, M.; Ferrer, E.G.; Williams, P.A.M. Insights into the mechanisms underlying the antitumor activity of an oxidovanadium(IV) compound with the antioxidant naringenin. Albumin binding studies. *J. Inorg. Biochem.* **2015**, *149*, 12–24. [CrossRef]
17. Chasteen, N.D. Vanadyl(IV) EPR Spin Probe. Inorganic and Biochemical Aspects. In *Biological Magnetic Resonance*; Berliner, L.J., Reuben, J., Eds.; Plenum Press: New York, NY, USA, 1981; Volume 3, pp. 53–119.
18. Benítez, J.; Becco, L.; Correia, I.; Leal, S.M.; Guiset, H.; Costa Pessoa, J.; Lorenzo, J.; Tanco, S.; Escobar, P.; Moreno, V.; et al. Vanadium polypyridyl compounds as potential antiparasitic and antitumoral agents: New achievements. *J. Inorg. Biochem.* **2011**, *105*, 303–312. [CrossRef]
19. Kivelson, D.; Lee, S.J. ESR Studies and the Electronic Structure of Vanadyl Ion Complexes. *Chem. Phys.* **1964**, *41*, 1896–1903. [CrossRef]

20. Chand, P.; Murali Krishna, R.; Lakshamana Rao, J.; Lakshaman, S.V.J. EPR and optical studies of vanadyl complexes in two host-crystals of Tutton salts of thallium. *Rad. Eff. Def. Solids* **1993**, *127*, 245–254. [CrossRef]
21. Liu, K.T.; Yu, J.T.; Lou, S.H.; Lee, C.H.; Huang, Y.; Lii, K.H. Electron paramagnetic resonance study of V^{4+}-doped $KTiOPO_4$ single crystals. *J. Phys. Chem. Solids* **1994**, *55*, 1221–1226. [CrossRef]
22. Bandyopadhyay, A.K. Optical and ESR investigation of borate glasses containing single and mixed transition metal oxides. *J. Mater. Sci.* **1981**, *16*, 189–203. [CrossRef]
23. Nunes, P.; Correia, I.; Cavaco, I.; Marques, F.; Pinheiro, T.; Avecilla, F.; Pessoa, J.C. Therapeutic potential of vanadium complexes with 1,10-phenanthroline ligands, quo vadis? Fate of complexes in cell media and cancer cells. *J. Inorg. Biochem.* **2021**, *217*, 111350. [CrossRef]
24. Holko, P.; Ligeza, J.; Kisielewska, J.; Kordowiak, A.M.; Klein, A. The Effect of Vanadyl Sulphate (VOSO$_4$) on Autocrine Growth of Human Epithelial Cancer Cell Lines Pol. *J. Pathol.* **2008**, *59*, 3–8.
25. Wu, B.L.; Wu, Z.W.; Yang, F.; Shen, X.F.; Wang, L.; Chen, B.; Li, F.; Wang, M.K. Flavonoids from the seeds of Oroxylum indicum and their anti-inflammatory and cytotoxic activities. *Phytochem. Lett.* **2019**, *32*, 66–69. [CrossRef]
26. Kumar, N.; Afjei, R.; Massoud, T.F.; Paulmurugan, R. Comparison of cell-based assays to quantify treatment effects of anticancer drugs identifies a new application for Bodipy-L-cystine to measure apoptosis. *Sci. Rep.* **2018**, *8*, 16363. [CrossRef]
27. Guerrero-Palomo, G.; Rendón-Huerta, E.P.; Montaño, L.F.; Fortoul, T.I. Vanadium compounds and cellular death mechanisms in the A549 cell line: The relevance of the compound valence. *J. Appl. Toxicol.* **2019**, *39*, 540–552. [CrossRef] [PubMed]
28. Zitka, O.; Skalickova, S.; Gumulec, J.; Masarik, M.; Adam, V.; Hubalek, J.; Trnkova, L.; Kruseova, J.; Eckschlager, T.; Kizek, R. Redox status expressed as GSH:GSSG ratio as a marker for oxidative stress in paediatric tumour patients. *Onc. Lett.* **2012**, *4*, 1247–1253. [CrossRef]
29. Matsuyama, S.; Reed, J.C. Mitochondria-dependent apoptosis and cellular pH regulation. *Cell Death Differ.* **2000**, *7*, 1155–1165. [CrossRef]
30. Perry, S.W.; Norman, J.P.; Barbieri, J.; Brown, E.B.; Gelbard, H.A. Mitochondrial membrane potential probes and the proton gradient: A practical usage guide. *Biotechniques* **2011**, *50*, 98–115. [CrossRef] [PubMed]
31. Correia, I.; Chorna, L.; Cavaco, I.; Roy, S.; Kuznetsov, M.; Ribeiro, N.; Justino, G.; Santos-Silva, T.; Santos, M.; Santos, H.; et al. Interaction of [V^{IV}O(acac)$_2$] with Human Serum Transferrin and Albumin. *Chem. Asian J.* **2017**, *12*, 2062–2084. [CrossRef]
32. Chadborn, N.; Bryant, J.; Bain, A.J.; O'Shea, P. Ligand-dependent conformational equilibria of serum albumin revealed by tryptophan fluorescence quenching. *Biophys. J.* **1999**, *76*, 2198–2207. [CrossRef]
33. Lakowicz, J.R. *Principles of Fluorescence Spectroscopy*; Springer Science & Business Media: New York, NY, USA, 2013.
34. Ross, P.D.; Subramanian, S. Thermodynamics of protein association reactions: Forces contributing to stability. *Biochemistry* **1981**, *20*, 3096–3102. [CrossRef]
35. Costa Pessoa, J.; Correia, I. Misinterpretations in Evaluating Interactions of Vanadium Complexes with Proteins and Other Biological Targets. *Inorganics* **2021**, *9*, 17. [CrossRef]
36. Le, M.; Rathje, O.; Levina, A.; Lay, P. High cytotoxicity of vanadium(IV) complexes with 1,10-phenanthroline and related ligands is due to decomposition in cell culture medium. *J. Biol. Inorg. Chem.* **2017**, *22*, 663–672. [CrossRef]
37. Shan, F.; Shao, Z.; Jiang, S.; Cheng, Z. Erlotinib induces the human non-small-cell lung cancer cells apoptosis via activating ROS-dependent JNK pathways. *Cancer Med.* **2016**, *11*, 3166–3175. [CrossRef]
38. Onishi, M. *Photometric Determination of Traces of Metals*, 4th ed.; Wiley: NewYork, NY, USA, 1989.
39. Stoll, S.; Schweiger, A. EasySpin, a comprehensive software package for spectral simulation and analysis in EPR. *J. Magn. Reson.* **2006**, *178*, 42–55. [CrossRef]
40. Ling, L.; Tan, K.; Lin, H.; Chiu, G. The role of reactive oxygen species and autophagy in safingol-induced cell death. *Cell Death Dis.* **2011**, *2*, e129. [CrossRef] [PubMed]
41. Hissin, P.J.; Hilf, R.A. Fluorometric Method for Determination of Oxidized and Reduced Glutathione in Tissues. *Anal. Biochem.* **1976**, *74*, 214–226. [CrossRef]
42. Bradford, M.A. Rapid and Sensitive Method for the Quantitation of Microgram Quantities of Protein Utilizing the Principle of Protein-Dye Binding. *Anal. Biochem.* **1976**, *72*, 248–254. [CrossRef]
43. Zamzami, N.; Métivier, D.; Kroemer, G. Quantitation of Mitochondrial Transmembrane Potential in Cells and in Isolated Mitochondria. *Methods Enzymol.* **2000**, *322*, 208–213. [CrossRef] [PubMed]
44. Levina, A.; Pires Vieira, A.; Wijetunga, A.; Kaur, R.; Koehn, J.; Crans, D.; Lay, P. A Short-Lived but Highly Cytotoxic Vanadium(V) Complex as a Potential Drug Lead for Brain Cancer Treatment by Intratumoral Injections. *Angew. Chem.* **2020**, *59*, 15834–15838. [CrossRef]

 *inorganics*

MDPI

Article

The Effect of Vanadium Inhalation on the Tumor Progression of Urethane-Induced Lung Adenomas in a Mice Model

Nelly López-Valdez, Marcela Rojas-Lemus and Teresa I. Fortoul *

Departamento de Biología Celular y Tisular, Facultad de Medicina, Universidad Nacional Autónoma de México (UNAM), México City 04510, Mexico; nellylopezvaldez@gmail.com (N.L.-V.); marcelarojaslemus@hotmail.com (M.R.-L.)
* Correspondence: fortoul@unam.mx; Tel.: +52-55-5623-2182

Abstract: Lung cancer has the highest death rates. Aerosol drug delivery has been used for other lung diseases. The use of inhaled vanadium (V) as an option for lung cancer treatment is explored. Four groups of mice were studied: (1) Saline inhalation alone, (2) Single intraperitoneal (i.p.) dose of urethane, (3) V nebulization twice a week (Wk) for 8 Wk, and (4) A single dose of urethane and V nebulization for 8 Wk. Mice were sacrificed at the end of the experiment. Number and size of tumors, PCNA (proliferating cell nuclear antigen) and TUNEL (terminal deoxynucleotidyl tranferase dUTP nick-end labeling) immunohistochemistry were evaluated and compared within groups. Results: The size and number of tumors decreased in mice exposed to V-urethane and the TUNEL increased in this group; differences in the PCNA were not observed. Conclusions: Aerosol V delivery increased apoptosis and possibly the growth arrest of the tumors with no respiratory clinical changes in the mice.

Keywords: urethane; vanadium; aerosol delivery; lung cancer; apoptosis; antineoplastic

Citation: López-Valdez, N.; Rojas-Lemus, M.; Fortoul, T.I. The Effect of Vanadium Inhalation on the Tumor Progression of Urethane-Induced Lung Adenomas in a Mice Model. *Inorganics* **2021**, *9*, 78. https://doi.org/10.3390/inorganics9110078

Academic Editor: Dinorah Gambino

Received: 9 October 2021
Accepted: 27 October 2021
Published: 29 October 2021

Publisher's Note: MDPI stays neutral with regard to jurisdictional claims in published maps and institutional affiliations.

1. Introduction

Lung cancer is a worldwide health problem, and this tumor has the highest estimated death rates because of the delay in the diagnosis and in the beginning of its treatment [1]. Usually, the patients have been mistreated because they do not have a previous history of smoking, which is the reason why lung cancer is not the first suspected diagnosis, even though its frequency among this group of patients has steadily increased. Some of the risk factors reported among non-smokers are environmental and occupational exposures, sex hormones, and wood smoke exposure [2,3]. The main histological subtype observed in recent years is adenocarcinoma [4].

Lung cancer treatment depends on the size of the tumor, the histologic type, and the clinical variables. Surgery, radiotherapy, and chemotherapy and or its combinations are the treatment options. The 5-year relative survival rate is about 17% for all patients and all of the stages of this pathology; if the disease is detected in early stages, the survival rate increases to 54%. Nonetheless, only 15% of the cases are diagnosed at early stages [5]. In addition, the resistance to chemotherapy and the cost of the new options, oriented to specific molecular targets, is expensive [4,6,7]. On the other hand, resistance to these treatments has also increased [8]. Additionally, because of these described events, it is important to keep looking for new treatment options for lung cancer.

The urethane model for lung tumors is a chemical carcinogenesis method that has been used to study tumor progression and treatments [9–11].

Metal compounds have been tested as possible antineoplastic agents, such as platinum compounds. Vanadium (V) is a transitional element with controversial effects [12]. Vanadium compounds have emerged as possible options for therapeutic uses because of the induction of reactive oxygen species (ROS), the activation of apoptotic cell death mechanisms, autophagy, and the inhibition of cell proliferation [13–15]. As antineoplastic agents, organic compounds have been studied in vitro in a variety of cancer types such as pancreatic ductal carcinoma [16].

Recently other optional routes for drug delivery have been proposed for the treatment of diverse pathologies [17] such as lung cancer [11]. Hamzawy et al. reported intratracheal administration of temozolomide in lung tumors induced by urethane [18] and Roger-Parra intranasally delivered anti-collagen-V for lung cancer treatment with promising results [11].

Aerosol drug delivery has been a technique used for the treatment of a variety of diseases and recently for experimental antineoplastic therapies [11]. This route increases the bioavailability and decreases the time of action because the doses decrease the risk of side effects, and the drug reaches the lung tissue without the metabolic changes that occur in the liver [19,20]. These benefits have been observed in asthma, COPD, and some lung infections [21,22]. In this study, we explore the administration of V by inhalation as a possible alternative route for the treatment of lung cancer in the urethane-induced lung tumor mice model.

2. Results

During the whole experiment, no signs or changes in the patterns of food intake and water consumption were observed. Body weight was statistically different when comparing the beginning (T0) with the end of the experiment (8th week) in each group. Body weight at 8th week was not statistically different among group I (control), group II (urethane), and group IV (urethane-V). In group III (vanadium), weight was significantly higher compared with the control, urethane, and urethane-V mice groups (Figure 1).

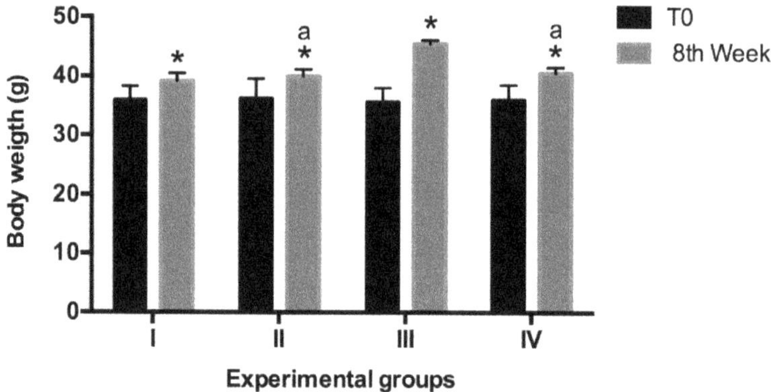

Figure 1. Weight of the mice per group. The weight at the beginning (Time 0) compared to the end of the experiment (8th week) was different. In group III (vanadium) the weight gain was higher than in the other groups. N = 10, values are expressed as the mean of body weight $\pm$ SEM in grams (ANOVA $p \leq 0.05$ Tukey's post hoc).* statistically significant differences vs. T0; a: statistically significant differences versus group III 8th week.

Except in group III (vanadium) where an increase in the weight of the mice was recorded, no other changes in the physical appearance or motor behavior of the animals were observed.

2.1. Lung Histology

Panoramic photomicrographs show the presence of pulmonary adenomas in mice treated with urethane, i.e., groups II and IV (Figure 2B,D, respectively). Groups I and III did not develop lung adenomas (Figure 2A,C, respectively).

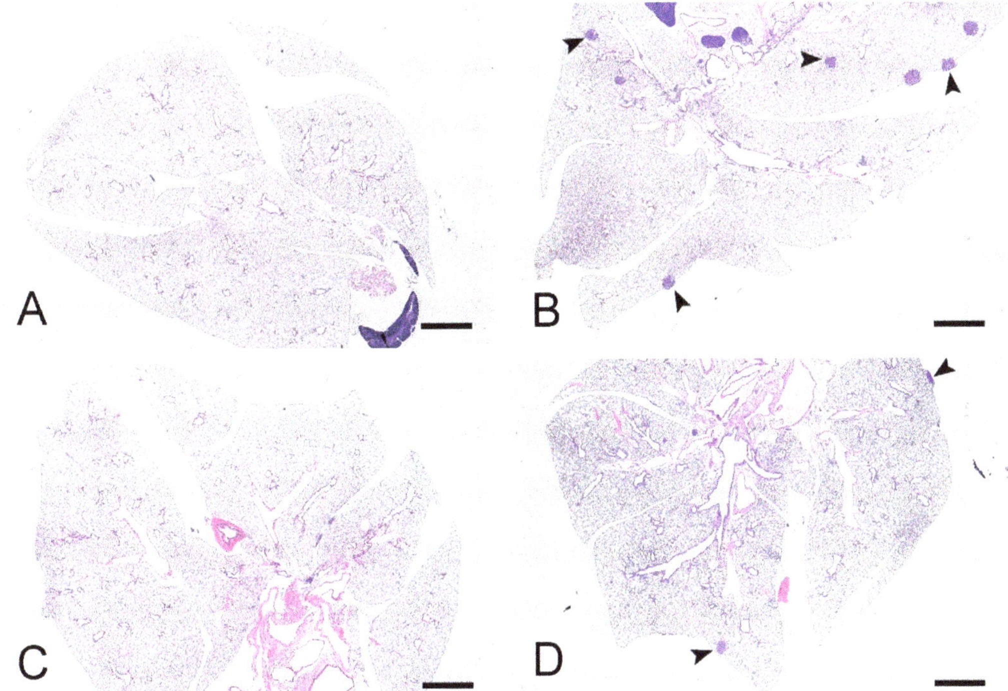

Figure 2. Adenoma development in mice treated with urethane. No tumors are observed in the control (**A**) and in vanadium (**C**) groups. Arrow heads (➤) indicate the adenomas in urethane (**B**) and urethane–vanadium (**D**) groups. Hematoxylin–Eosin. Bar 2 mm.

Detailed changes in the lung tissue observed in the experimental groups are shown in Figure 3. In group I (Figure 3A), bronchioles and alveolar walls with a well-preserved structure were observed; inflammatory foci as well as focal hyperplasia of the bronchiolar and alveolar epithelium, as well as solid adenomas, are clearly identified in group II (Figure 3B). In the V-exposed group (group III; Figure 3C), perivascular and peribronchiolar lymphocytic inflammatory infiltrate were observed, whereas no adenomas were identified; in the urethane-V group (group IV), focal bronchiolar epithelial hyperplasia, small and scanty adenomas, and lymphocytic infiltrate were the main observed features (Figure 3D).

The mean number of tumors in group II was 9 ± 1.13, whereas in group IV it was 2 ± 0.51; when both groups were compared, a statistically significant difference was observed (Figure 4).

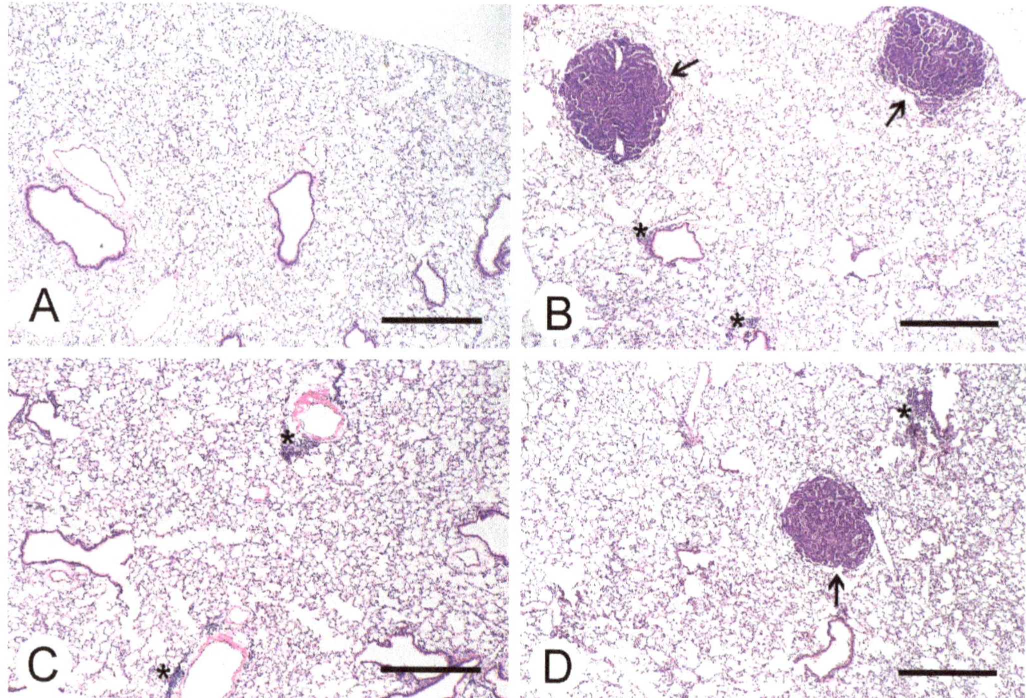

Figure 3. Details of the changes observed in the lungs. Control group (**A**) with no changes in the lung's parenchyma, whereas in the V-exposed group (**C**) epithelial hyperplasia and inflammatory infiltrate (*) are observed. In urethane (**B**) and urethane-V (**D**) groups, the adenomas (↑) are observed as well as the hyperplastic epithelium (*). Hematoxylin–Eosin stain. Bar 0.5 mm.

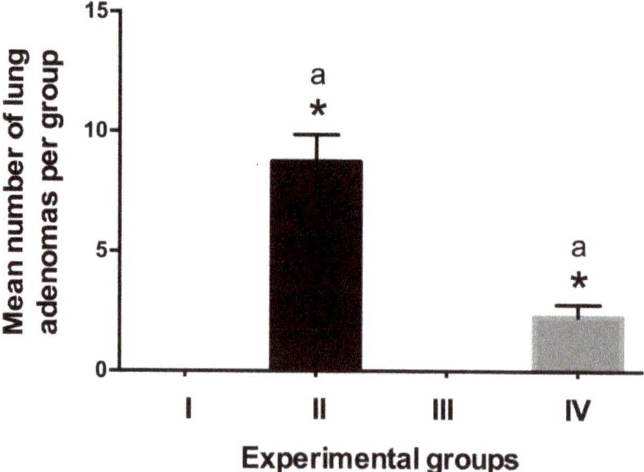

Figure 4. Quantitation of developed adenomas per group. No tumors were observed in group I (control) and group III (vanadium). In group II (urethane), the tumors were larger than in group IV (urethane-V group). N = 10, values are expressed as the mean number of adenomas ±SEM (ANOVA $p \leq 0.05$ Tukey's post hoc). * Statistically significant differences versus group I; a: statistically significant differences between group II and IV.

The area occupied by tumors in group II was 0.6 ± 0.07 mm^2, whereas in group IV it was 0.39 ± 0.05 mm^2, a difference which was statistically significant (Figure 5). No qualitative differences were observed in the amount and the spread of the infiltrate in groups II and IV.

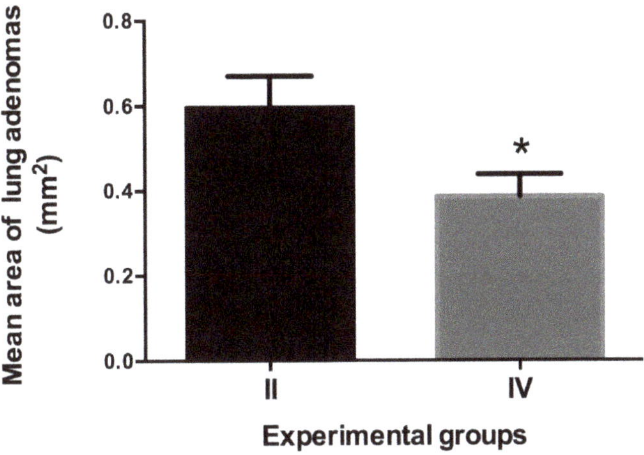

Figure 5. Area (mm^2) occupied by lung adenomas. In group II (urethane) the adenomas were larger compared with those observed in group IV (urethane-V). N = 10, values expressed as the mean area $\pm$ SEM (two-tailed Student's *t*-test with Welch's correction, $p \leq 0.05$). * Statistically significant differences between group II versus group IV.

2.2. Proliferative Index

PCNA positive nuclei stain in the lung parenchyma was observed in the four groups. In group I, the positive cells were observed mainly in some bronchiolar cells as well as in group IV. In groups II and IV, the nuclei stain was in the tumor cells (Figure 6) and no statistical difference was observed in the proliferative index calculated only in the adenomas in groups II (PI 24.18%) and IV (PI 25.93%) (Figure 7).

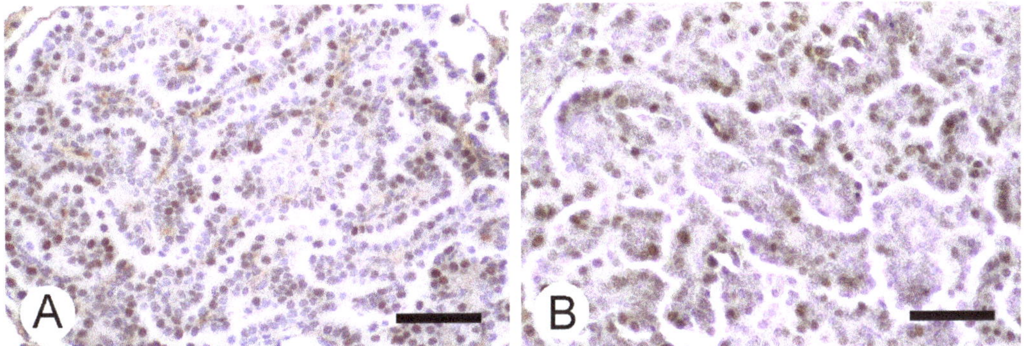

Figure 6. Immune stain for PCNA in lung adenomas. Positive PCNA nuclei (ochre color) were observed in urethane (**A**) and urethane-V (**B**) with no statistical difference between them. Hematoxylin counterstain. Bar 50 µm.

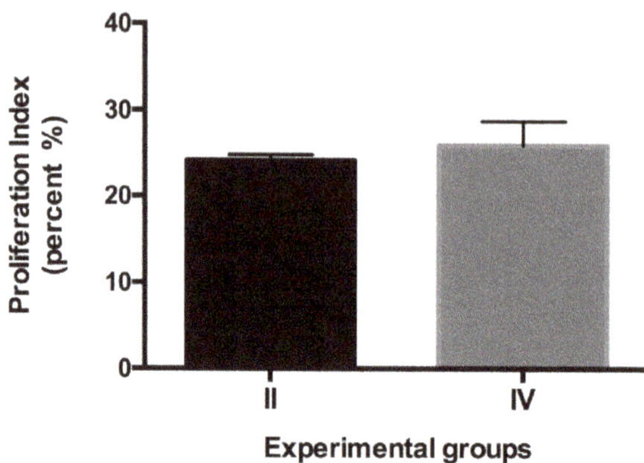

Figure 7. Proliferation index. The index indicated that between groups II and IV, no statistical difference was observed. N = 10, values expressed as mean percentage of PCNA positive nuclei ± SEM (two-tailed Student's *t*-test with Welch's correction). No statistically significant differences were observed.

2.3. TUNEL Assay

Positive TUNEL nuclei stain in the lung parenchyma was observed in the four groups. In group I, the positive cells were scanty inflammatory cells in the parenchyma, whereas in the V-exposed group (III), the stain was in the inflammatory cells and scarce in the bronchiolar epithelia. In groups II and IV, the nuclei stain was in the tumor cells, the inflammatory foci, and in the bronchiolar epithelium (Figure 8).

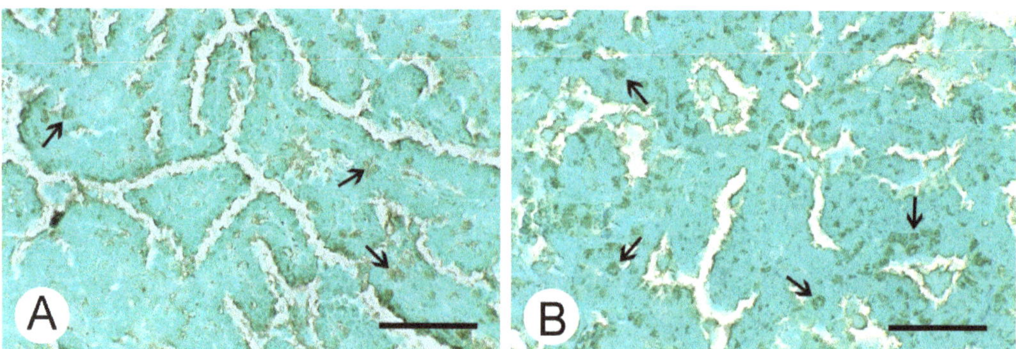

Figure 8. TUNEL immuno-essay. Positive nuclei TUNEL stain (ochre color) were observed in urethane group (**A**) and urethane-V group (**B**) (↑). Light green counterstain. Bar 50 μm

A clear statistical difference was observed in the apoptotic index calculated only in the adenomas in groups II (AI 5.1%) and IV (AI 10%) (Figure 9).

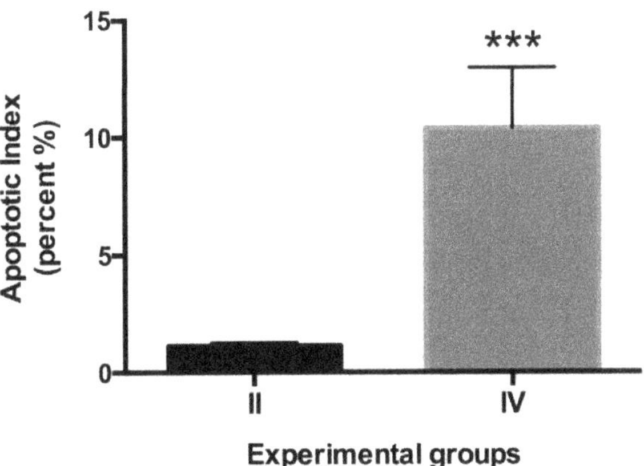

Figure 9. Apoptotic index. The index was higher in group IV (urethane-V) compared with group II (urethane). N = 10, values are expressed as mean percentage of TUNEL positive nuclei $\pm$ SEM (two-tailed Student's *t*-test with Welch's correction $p \leq 0.05$). *** Statistically significant differences between group II versus group IV.

3. Discussion

In the present study, we report that V aerosol delivery interfered with the development of lung adenomas induced by urethane. The decrease in the area and the number of tumors was the result of increasing apoptosis of the tumor cells evaluated by TUNEL; however, no effect was observed in the PCNA proliferation marker. No respiratory clinical compromise was observed in the mice exposed to aerosolized V, and only a difference in weight gain was notorious in the V-exposed groups.

Vanadium as an antineoplastic agent has been previously explored. Köpf-Maier et al. found that vanadocene dichloride (VDC) reduces cell proliferation in leukemia tumor cells [23]. In female Sprague Dawley rats, Thompson et al., in MNU-1 (1-methyl-1-nitrosurea)-induced mammary carcinogenesis, show that supplementation with vanadium sulfate reduced the incidence and the average amount of neoplasms [24]. Other studies from Köpf-Maier's group report that the antitumor activity of VDC on the Fluid Erlich ascites tumor is because of its heterochromatin accumulation [25], mitotic aberration induction, transitory mitosis suppression, and reversible cell accumulation in the late S and G2 phases [23]. Bishayee et al. [26] suggest that the antineoplastic action of VDC might be the result of the effect on the antioxidant status in the liver and the modulation of drug metabolism enzymes of phases I and II. Sankar-Ray et al. suggested the suppression of cell proliferation, induction of apoptosis, and DNA cross-links reduction as other possible antineoplastic mechanisms. In our experimental model, an increase in apoptotic cells was observed, but not in PCNA proliferation biomarkers [27].

These results suggest that apoptosis could be the mechanisms by which V is acting on the adenomas. Recently, Rozzo et al, [15] reported the effect of V compounds in the melanoma A375 cell line in which apoptosis is observed, as well as the arrest of cell cycle in two different phases, probably by different mechanisms. The findings reported here suggest that V in aerosol delivery could be acting by inducing apoptosis and possible the tumors' cell cycle arrest [28–30].

Lu et. al. demonstrated that some synthetic V complexes showed pro-apoptotic activities in MGC803 (human gastric cancer cell line) cells related to the increase in proteins such as Bax, caspases 3 and 9, as well as the decrease in Bcl2 [31]. On the other hand, Xi et al. reported the induction of apoptotic cell death in A549 and BEAS-2B lung cancer lines associated with the overexpression of caspase 3 induced by the exposure to V nanoparticles [32].

The generation of reactive oxygen species (ROS) [28] and their effect on the neoplastic cells such as: DNA damage, oxidative alterations of other cellular organelles leading to apoptosis, and different types of cell death mechanisms, could explain the antineoplastic effects of V compounds [5].

Aerosol Delivery

In a variety of respiratory diseases, the aerosol delivery of drug treatments has been used with good results [33]. The best examples are inhaled steroids for asthma, which reduce the symptoms and the systemic effects of steroids [21]. Hamzamy et al. reported the intratracheal administration of temozolomide in gold nanoparticles or liposomes as antineoplastic carriers for the treatment of urethane-induced lung adenomas in mice [18]. Gagnadoux et al., with gemcitabine delivered by the same route, reported the potential use of aerosol delivery for lung cancer treatment [20]. The aerosol delivery of chemotherapy for lung cancer treatment in patients with non-small cell lung carcinoma (NSCLC) was reported. 5-fluorouracil (5-FU) was delivered by an ultrasonic nebulizer in two different situations: one in patients prior to surgical resection, in which the authors demonstrated that the concentration of the 5-FU was 5 to 15 times higher in the tumor than in the normal lung tissue, and two, conducted in patients with unresectable tumor in which the 5-FU was also administered by aerosol delivery, two to three times a week, reporting less pulmonary or systemic side effects. Additionally, the main agent used in the therapeutic schemes for lung cancer, cisplatin, was delivered by inhalation with less systemic side effects and with promising preclinical results [34]. With the urethane model, Abdelaziz et al. reported the reduction in lung tumors in BALB/c mice by the inhalation of lactoferrin/Chondroitin-Functionalized Monoolein Nanocomposites, supporting the use of inhalation as a possible route for lung cancer treatment, stressing that by inhalation route the agent employed in the treatment reaches higher concentrations in the tumors [35].

In our study, with whole body exposure, the regression of the adenomas was almost complete, observing some areas of peribronchiolar inflammation with no clinical effects observed in the mice, only a weight increase in V-exposed groups, which could be explained by the anabolic activity reported for V [36].

Vanadium has been reported as a possible chemotherapeutic agent for different types of neoplasms. Some antineoplastic drugs have been administered by aerosol delivery with promising results [20,32]. Here, we propose V as a potential antineoplastic agent for lung cancer by aerosol delivery that will reduce the number of tumor cells and possibly the systemic side effects reported for V. Other V compounds need further analysis to find the dose and the best protocol for the administration for this element, which opens another possible treatment for NSCLC.

4. Materials and Methods

4.1. Animals

Forty CD-1 adult male mice weighing 33–35 g were housed in hanging plastic cages (10 animals per cage), kept in an animal facility (with an average temperature of 21 °C, 57% humidity and controlled lighting −12:12 h light/dark regime), and fed with Rodent Laboratory chow (PMI nutrition international, Brentwood, MO, USA and Agribrands Purina, Cuautitlan, Mexico) and filtered water ad libitum. Mice were obtained from the vivarium at the School of Medicine, UNAM, and managed according to the Mexican official norm NOM-062-ZOO-1999 for the production, care, and use of laboratory animals. The project was reviewed and approved by the Research and Ethical Committee from the School of Medicine (#04-2005).

4.2. Experimental Protocol

Adult male mice were randomly assigned into four groups of 10 mice each: group I (negative control) inhaled saline 0.9% during the exposures; group II (urethane alone as positive control), received a single dose (ip) of urethane 1mg/g (ethyl carbamate, 99% purity,

Sigma Aldrich, St. Louis, MO, USA) in accordance with previous studies using the urethane-induced lung tumorigenesis model [37]; group III (vanadium) inhaled V_2O_5 (0.02 mol/L) (99.99%, Sigma, St. Louis, MO, USA) in saline 1h twice a week (Tuesday and Thursday) for the 8 weeks of exposure time; and group IV (urethane and V) was as in groups II (urethane alone) and III (vanadium). At the end of the 8-week exposure, mice were anesthetized with (ip) lethal dose of pentobarbital sodium (PiSa Pharmaceutical, Guadalajara, Jalisco, México) 0.3 mg/mL and perfused via aorta with saline followed by 4% paraformaldehyde, whereas the lungs were fixed intratracheally with 4% paraformaldehyde at Total Lung Capacity (TLC) [38].

4.3. Vanadium Exposure and Cardiothoracic Block Dissection

Mice of the V-exposed groups (III and IV) inhaled 1h twice a week in an acrylic box chamber measuring 45 cm × 21 cm × 35 cm (3.3 L total volume), that could house 20 mice per session. An ultra-nebulizer DeVillbiss Ultraneb 99 (DeVillbiss Healthcare, Somerset, PA, USA) system was used to nebulize the vanadium solution at a flow rate of 10 L/min; according to the manufacturer's provided information, about 80% of the aerosolized particles reaching the mice would be expected to have a mass median aerodynamic diameter (MMAD) of 0.5–5 µm. The concentration of vanadium in the chamber was quantified as follows: a filter was placed at the external outlet of the nebulizer during the entire exposure period and samples were collected at a flow rate of 10 L/min. The filter was removed and weighed after each exposure; the V on each filter was quantified as follows: six-filters per inhalation exposures were evaluated. The source of the fog was located at the top of the chamber to ensure a homogeneous exposure. Mice behavior was always observed to detect any changes. As it has been reported in earlier studies, the final concentration in the chamber was 1.56 mg V/m3 [36] and it is in the range of the World Health Organization (0.01–60 mg/m^3) for V concentrations detected in occupational exposures [39]. The V concentration in the blood was analyzed by mass spectrometry of induction-coupled plasma (ICP-MS) using a Bruker equipment, model Aurora M90, with coupled autosampler (Bruker Corp., Billerica, MA, USA). The concentration of the metal was 436 ppb or nanograms of V per of dry weight tissue (ng/g).

4.4. Tissue Sampling and Preparation

Experimental and control groups were sacrificed at the end of the exposures (8 Wk). Animals were anesthetized with sodium pentobarbital aqueous solution of (PiSa Pharmaceutical, Guadalajara, Jalisco, México) 0.3 mg/mL (ip) and perfused via aorta with saline, followed by 4% paraformaldehyde (pH 7.4) in phosphate buffer. The cardiopulmonary block was removed, and then the lungs were dissected and processed for paraffin wax embedding; 5 µm thickness tissue sections were obtained and stained with hematoxylin-eosin, and proved for immunohistochemical evaluation with anti-PCNA antibody and TUNEL assay. Changes were assessed with a light microscope BX51 (Olympus, Miami, FL, USA). Samples were photographed with a digital camera attached to the microscope (Media Cybernetics Inc., Bethesda, MD, USA). The number and area of the tumors were measured with Motic Images 2.0 software (Motic, Kowloon Bay, Kowloon, Hong Kong).

4.5. Immunohistochemistry for PCNA

Tissue sections were placed in poly-L-lysine (SIGMA, St Louis, MO, USA) coated slides. Antigen retrieval was achieved by incubation in a citrate buffer (pH 7.4) at 103425 Pa for 3 min, after which the slides were washed in phosphate-buffered saline (PBS). Endogenous peroxidase was blocked with 3% H2O2 (J.T. Baker, Phillipsburg, NJ, USA) for 10 min. The sections were rinsed several times with PBS-Albumin, washed for 10 min in PBS, (MP Biomedicals Inc, Kuwait city, Kuwait) and incubated for 1 h at 37 °C in rabbit monoclonal anti-PCNA (Abcam, Cambridge, MA, USA), and diluted 1:100 in PBST- (PBS with 0.1% Tween 20 and Albumin). The sections were washed in PBS and incubated for 30 min at 37 °C with the biotinylated universal link secondary antibody (Dako, Carpinteria, CA,

USA), rinsed several times in PBS, and incubated for 30 min at 37 °C in HRP streptavidin complex (Dako, Carpinteria, CA, USA). Immunoreactivity was visualized by incubation in 0.05% 3,3′-diaminobenzidine tetrahydrochloride (Zymed Laboratories Inc, San Francisco, CA, USA). Samples skipping primary antibody were also included as negative controls. Immunoreactivity to PCNA-exposed lungs was measured in tissue sections to calculate the proliferation index in the adenomas; the total adenomas were evaluated from the slides of the lung of each animal and the number of positive nuclei were also counted; positive nuclei were considered when an ochre color stain was observed when the developer for the reaction was diaminobenzidine. The evaluation was completed with an image analyzer, using the software Image-Pro-Plus version 6.0. (Media Cybernetics Inc., Silver Spring, MD, USA) coupled to a digital camera (Evolution MP Color, Media Cybernetics Inc., Silver Spring, MD, USA) on a light microscope Olympus BX51 (Olympus America Inc., Melville, NY, USA).

Proliferative index (PI) was determined by a relation between the number of positive nuclei per adenoma by the total number of cells in each tumor by 100.

4.6. TUNEL Assay

The enzymatic TUNEL assay (TdT-mediated dUPT-biotin nick end labeling) (Dead-End™ Colorimetric TUNEL System, Promega Corp., Madison, WI, USA) was used to evaluate the apoptotic index. The assay identifies DNA strand breaks by marking the free 3′-OH terminal, with biotinylated desoxyuridine with the enzymatic reaction with terminal deoxynucleotidyl transferase (tdT). The biotin signaling is detected by the streptavidin-marked nuclei with the HRP enzyme bound by the biotinylated nucleotides which are visualized with HRP,3′3′-diaminobenzidine (DAB). The apoptotic nuclei are observed in ochre color in the light microscope. The assay was performed according to the providers' indications. The slides were counterstained with light green to increase the visibility of the apoptotic nucleus.

The apoptotic index (AI) was calculated in 40X photomicrographs from the different groups. Five fields randomly were selected. TUNEL positive nuclei were counted within the tumors and the total nuclei from each tumor. The index was calculated with the formula: Total amount of marked nuclei/ total nuclei X 100. The total nuclei were counted in Feulgen-stained slides [40]. Apoptotic index was compared between Group II and Group IV.

4.7. Statistical Analysis

In the different experimental groups comparison of the number of the tumors was carried out with an analysis of variance (ANOVA) with Tukey's post hoc test, whereas the analysis of the tumors area, the PI and AI was carried out with Student's *t*-test with Welch's correction (GraphPad Prism Software V 6.0, La Jolla, CA, USA). Differences were considered when $p < 0.05$.

5. Conclusions

Inhaled vanadium decreases the number and size of urethane-induced lung adenomas in mice by inducing apoptosis of the tumor cells, and with no observed clinical effects.

A limitation of this study was that the effect of V was only analyzed at a single point. It would be interesting to evaluate the effects of V overtime, to establish a wider picture of its beneficial and side effects.

Further studies about the mechanisms that lead to apoptotic cell death overtime in this model would be of interest, as well as if other types of cell death could be involved in this process.

In addition, to study if V interferes with urethane's metabolism throughout the whole experiment, we would add more information about the effects observed in this study.

Author Contributions: Conceptualization, N.L.-V.; methodology, N.L.-V. and M.R.-L.; formal analysis, N.L.-V.; investigation, N.L.-V. and M.R.-L.; resources, N.L.-V. and T.I.F.; writing—original draft preparation, N.L.-V.; writing—review and editing, N.L.-V. and T.I.F.; project administration, T.I.F. All authors have read and agreed to the published version of the manuscript.

Funding: This research received no external funding.

Acknowledgments: The authors thank Raquel Guerrero-Alquicira for the tissue processing and to Armando Zepeda-Rodríguez and Francisco Pasos-Nájera for the artwork with the figures, all from the Departamento de Biología Celular y Tisular, Facultad de Medicina, UNAM. Alejandra Núñez-Fortoul edited English of the final version of the manuscript.

Conflicts of Interest: The authors declare no conflict of interest.

References

1. Siegel, R.L.; Miller, K.D.; Jemal, A. Cancer Statistics, 2017. *CA Cancer J. Clin.* **2017**, *67*, 7–30. [CrossRef]
2. McCarthy, W.J.; Meza, R.; Jeon, J.; Moolgavkar, S.H. Chapter 6: Lung cancer in never smokers: Epidemiology and risk prediction models. *Risk Anal.* **2012**, *32* (Suppl. 1), S69–S84. [CrossRef]
3. Green, L.S.; Fortoul, T.I.; Ponciano, G.; Robles, C.; Rivero, O. Bronchogenic cancer in patients under 40 years old. The experience of a Latin American country. *Chest* **1993**, *104*, 1477–1481. [CrossRef]
4. Jamal-Hanjani, M.; Wilson, G.A.; McGranahan, N.; Birkbak, N.J.; Watkins, T.B.K.; Veeriah, S.; Shafi, S.; Johnson, D.H.; Mitter, R.; Rosenthal, R.; et al. Tracking the Evolution of Non-Small-Cell Lung Cancer. *N. Engl. J. Med.* **2017**, *376*, 2109–2121. [CrossRef] [PubMed]
5. Guerrero-Palomo, G.; Rendon-Huerta, E.P.; Montano, L.F.; Fortoul, T.I. Vanadium compounds and cellular death mechanisms in the A549 cell line: The relevance of the compound valence. *J. Appl. Toxicol.* **2019**, *39*, 540–552. [CrossRef] [PubMed]
6. Blumenthal, G.M.; Karuri, S.W.; Zhang, H.; Zhang, L.; Khozin, S.; Kazandjian, D.; Tang, S.; Sridhara, R.; Keegan, P.; Pazdur, R. Overall response rate, progression-free survival, and overall survival with targeted and standard therapies in advanced non-small-cell lung cancer: US Food and Drug Administration trial-level and patient-level analyses. *J. Clin. Oncol.* **2015**, *33*, 1008–1014. [CrossRef] [PubMed]
7. Jamal-Hanjani, M.; Quezada, S.A.; Larkin, J.; Swanton, C. Translational implications of tumor heterogeneity. *Clin. Cancer Res.* **2015**, *21*, 1258–1266. [CrossRef] [PubMed]
8. Suda, K.; Rozeboom, L.; Rivard, C.J.; Yu, H.; Ellison, K.; Melnick, M.A.C.; Hinz, T.K.; Chan, D.; Heasley, L.E.; Politi, K.; et al. Therapy-induced E-cadherin downregulation alters expression of programmed death ligand-1 in lung cancer cells. *Lung Cancer* **2017**, *109*, 1–8. [CrossRef]
9. Kemp, C.J. Animal Models of Chemical Carcinogenesis: Driving Breakthroughs in Cancer Research for 100 Years. *Cold Spring Harb. Protoc.* **2015**, *2015*, 865–874. [CrossRef]
10. Gurley, K.E.; Moser, R.D.; Kemp, C.J. Induction of Lung Tumors in Mice with Urethane. *Cold Spring Harb. Protoc.* **2015**, *2015*, pdb–prot077446. [CrossRef] [PubMed]
11. Parra, E.R.; Alveno, R.A.; Faustino, C.B.; Correa, P.Y.; Vargas, C.M.; de Morais, J.; Rangel, M.P.; Velosa, A.P.; Fabro, A.T.; Teodoro, W.R.; et al. Intranasal Administration of Type V Collagen Reduces Lung Carcinogenesis through Increasing Endothelial and Epithelial Apoptosis in a Urethane-Induced Lung Tumor Model. *Arch. Immunol. Ther. Exp. (Warsz)* **2016**, *64*, 321–329. [CrossRef] [PubMed]
12. Levine, A.J. The Evolution of Tumor Formation in Humans and Mice with Inherited Mutations in the p53 Gene. *Curr. Top. Microbiol. Immunol.* **2017**, *407*, 205–221. [CrossRef] [PubMed]
13. Petanidis, S.; Kioseoglou, E.; Hadzopoulou-Cladaras, M.; Salifoglou, A. Novel ternary vanadium-betaine-peroxido species suppresses H-ras and matrix metalloproteinase-2 expression by increasing reactive oxygen species-mediated apoptosis in cancer cells. *Cancer Lett.* **2013**, *335*, 387–396. [CrossRef] [PubMed]
14. Rehder, D. The role of vanadium in biology. *Metallomics* **2015**, *7*, 730–742. [CrossRef] [PubMed]
15. Rozzo, C.; Sanna, D.; Garribba, E.; Serra, M.; Cantara, A.; Palmieri, G.; Pisano, M. Antitumoral effect of vanadium compounds in malignant melanoma cell lines. *J. Inorg. Biochem.* **2017**, *174*, 14–24. [CrossRef]
16. Kowalski, S.; Hac, S.; Wyrzykowski, D.; Zauszkiewicz-Pawlak, A.; Inkielewicz-Stepniak, I. Selective cytotoxicity of vanadium complexes on human pancreatic ductal adenocarcinoma cell line by inducing necroptosis, apoptosis and mitotic catastrophe process. *Oncotarget* **2017**, *8*, 60324–60341. [CrossRef]
17. Paredes Aller, S.; Quittner, A.L.; Salathe, M.A.; Schmid, A. Assessing effects of inhaled antibiotics in adults with non-cystic fibrosis bronchiectasis—Experiences from recent clinical trials. *Expert Rev. Respir. Med.* **2018**, *12*, 769–782. [CrossRef]
18. Hamzawy, M.A.; Abo-Youssef, A.M.; Salem, H.F.; Mohammed, S.A. Antitumor activity of intratracheal inhalation of temozolomide (TMZ) loaded into gold nanoparticles and/or liposomes against urethane-induced lung cancer in BALB/c mice. *Drug Deliv.* **2017**, *24*, 599–607. [CrossRef]
19. Carafa, M.; Marianecci, C.; Donatella, P.; D'i Marzio, L.; Celia, M.; Fresta, F.; Alhaique, F. Novel concept in pulmonary delivery. In *Chronic Obstructive Pulmonary Disease-Current Concept and Practice*; Ong, K.C., Ed.; IntechOpen Limited: London, UK, 2012.

20. Gagnadoux, F.; Pape, A.L.; Lemarie, E.; Lerondel, S.; Valo, I.; Leblond, V.; Racineux, J.L.; Urban, T. Aerosol delivery of chemotherapy in an orthotopic model of lung cancer. *Eur. Respir. J.* **2005**, *26*, 657–661. [CrossRef]
21. Barnes, P.J. Inhaled Corticosteroids. *Pharmaceuticals* **2010**, *3*, 514–540. [CrossRef]
22. Borghardt, J.M.; Kloft, C.; Sharma, A. Inhaled Therapy in Respiratory Disease: The Complex Interplay of Pulmonary Kinetic Processes. *Can. Respir. J.* **2018**, *2018*, 2732017. [CrossRef] [PubMed]
23. Kopf-Maier, P.; Wagner, W.; Hesse, B.; Kopf, H. Tumor inhibition by metallocenes: Activity against leukemias and detection of the systemic effect. *Eur. J. Cancer* **1981**, *17*, 665–669. [CrossRef]
24. Thompson, H.J.; Chasteen, N.D.; Meeker, L.D. Dietary vanadyl(IV) sulfate inhibits chemically-induced mammary carcinogenesis. *Carcinogenesis* **1984**, *5*, 849–851. [CrossRef] [PubMed]
25. Kopf-Maier, P.; Wagner, W.; Liss, E. Induction of cell arrest at G1/S and in G2 after treatment of Ehrlich ascites tumor cells with metallocene dichlorides and cis-platinum in vitro. *J. Cancer Res. Clin. Oncol.* **1983**, *106*, 44–52. [CrossRef] [PubMed]
26. Bishayee, A.; Oinam, S.; Basu, M.; Chatterjee, M. Vanadium chemoprevention of 7,12-dimethylbenz(a)anthracene-induced rat mammary carcinogenesis: Probable involvement of representative hepatic phase I and II xenobiotic metabolizing enzymes. *Breast Cancer Res. Treat* **2000**, *63*, 133–145. [CrossRef]
27. Sankar Ray, R.; Roy, S.; Ghosh, S.; Kumar, M.; Chatterjee, M. Suppression of cell proliferation, DNA protein cross-links, and induction of apoptosis by vanadium in chemical rat mammary carcinogenesis. *Biochim. Biophys. Acta* **2004**, *1675*, 165–173. [CrossRef] [PubMed]
28. Evangelou, A.M. Vanadium in cancer treatment. *Crit. Rev. Oncol. Hematol.* **2002**, *42*, 249–265. [CrossRef]
29. Mateos-Nava, R.A.; Rodriguez-Mercado, J.J.; Altamirano-Lozano, M.A. Premature chromatid separation and altered proliferation of human leukocytes treated with vanadium (III) oxide. *Drug Chem. Toxicol.* **2017**, *40*, 457–462. [CrossRef] [PubMed]
30. Leon, I.E.; Porro, V.; Di Virgilio, A.L.; Naso, L.G.; Williams, P.A.; Bollati-Fogolin, M.; Etcheverry, S.B. Antiproliferative and apoptosis-inducing activity of an oxidovanadium(IV) complex with the flavonoid silibinin against osteosarcoma cells. *J. Biol. Inorg. Chem.* **2014**, *19*, 59–74. [CrossRef]
31. Lu, L.P.; Suo, F.Z.; Feng, Y.L.; Song, L.-L.; Li, Y.; Li, Y.-J.; Wang, K.-T. Synthesis and biological evaluation of vanadium complexes as novel anti-tumor agents. *Eur. J. Med. Chem.* **2019**, *176*, 1–10. [CrossRef]
32. Xi, W.S.; Tang, H.; Liu, Y.Y.; Liu, C.Y.; Gao, Y.; Cao, A.; Liu, Y.F.; Chen, Z. Cytotoxicity of vanadium oxide nanoparticles and titanium dioxide-coated vanadium oxide nanoparticles to human lung cells. *J. Appl. Toxicol.* **2020**, *40*, 567–577. [CrossRef] [PubMed]
33. Chen, Y.; Zhao, Y.; Dai, C.L.; Liang, Z.; Run, X.; Iqbal, K.; Liu, F.; Gong, C.X. Intranasal insulin restores insulin signaling, increases synaptic proteins, and reduces Abeta level and microglia activation in the brains of 3xTg-AD mice. *Exp. Neurol.* **2014**, *261*, 610–619. [CrossRef]
34. Gagnadoux, F.; Hureaux, J.; Vecellio, L.; Urban, T.; Le Pape, A.; Valo, I.; Montharu, J.; Leblond, V.; Boisdron-Celle, M.; Lerondel, S.; et al. Aerosolized chemotherapy. *J. Aerosol. Med. Pulm. Drug Deliv.* **2008**, *21*, 61–70. [CrossRef]
35. Abdelaziz, H.M.; Elzoghby, A.O.; Helmy, M.W.; Abdelfattah, E.A.; Fang, J.Y.; Samaha, M.W.; Freag, M.S. Inhalable Lactoferrin/Chondroitin-Functionalized Monoolein Nanocomposites for Localized Lung Cancer Targeting. *ACS Biomater. Sci. Eng.* **2020**, *6*, 1030–1042. [CrossRef] [PubMed]
36. Fortoul, T.I.; Rojas-Lemus, M.; Rodriguez-Lara, V.; Gonzalez-Villalva, A.; Ustarroz-Cano, M.; Cano-Gutierrez, G.; Gonzalez-Rendon, S.E.; Montano, L.F.; Altamirano-Lozano, M. Overview of environmental and occupational vanadium exposure and associated health outcomes: An article based on a presentation at the 8th International Symposium on Vanadium Chemistry, Biological Chemistry, and Toxicology, Washington DC, August 15–18, 2012. *J. Immunotoxicol.* **2014**, *11*, 13–18. [CrossRef] [PubMed]
37. Roomi, M.W.; Roomi, N.W.; Kalinovsky, T.; Rath, M.; Niedzwiecki, A. Chemopreventive Effect of a Novel Nutrient Mixture on Lung Tumorigenesis Induced by Urethane in Male A/J Mice. *Tumori. J.* **2009**, *95*, 508–513. [CrossRef]
38. Fortoul, T.I.; Soto-Mota, A.; Rojas-Lemus, M.; Rodriguez-Lara, V.; Gonzalez-Villalva, A.; Montano, L.F.; Paez, A.; Colin-Barenque, L.; Lopez-Valdez, N.; Cano-Gutierrez, G.; et al. Myocardial connexin-43 and N-Cadherin decrease during vanadium inhalation. *Histol. Histopathol.* **2016**, *31*, 433–439. [CrossRef] [PubMed]
39. Costigan, M.C.; Dobson, R. *Vanadium Pentoxide and Other Inorganic Vanadium Compounds*; World Health Organization & International Programme on Chemical Safety 2001; World Health Organization: Geneva, Switzerland, 2001. Available online: https://apps.who.int/iris/handle/10665/42365 (accessed on 20 August 2021).
40. Chieco, P.; Derenzini, M. The Feulgen reaction 75 years on. *Histochem. Cell Biol.* **1999**, *111*, 345–358. [CrossRef]

Article

Vanadium(V) Complexes with Siderophore Vitamin E-Hydroxylamino-Triazine Ligands

Maria Loizou [1], Ioanna Hadjiadamou [1], Chryssoula Drouza [2,*], Anastasios D. Keramidas [1,*], Yannis V. Simos [3,*] and Dimitrios Peschos [3]

[1] Department of Chemistry, University of Cyprus, Nicosia 2901, Cyprus; loizou.maria@ucy.ac.cy (M.L.); hadjiadamou.ioanna@ucy.ac.cy (I.H.)

[2] Department of Agricultural Production, Biotechnology and Food Science, Cyprus University of Technology, Limassol 3036, Cyprus

[3] Department of Physiology, Faculty of Medicine, School of Health Sciences, University of Ioannina, 45110 Ioannina, Greece; dpeschos@uoi.gr

* Correspondence: chryssoula.drouza@cut.ac.cy (C.D.); akeramid@ucy.ac.cy (A.D.K.); isimos@uoi.gr (Y.V.S.); Tel.: +357-2289-2764 (A.D.K.); Fax: +357-2289-5454 (A.D.K.)

Abstract: Novel vitamin E chelate siderophore derivatives and their V^V and Fe^{III} complexes have been synthesised and the chemical and biological properties have been evaluated. In particular, the α- and δ-tocopherol derivatives with bis-methyldroxylamino triazine (α-tocTHMA) and (δ-tocDPA) as well their V^V complexes, $[V_2^VO_3(\alpha\text{-tocTHMA})_2]$ and $[V_2^{IV}O_3(\delta\text{-tocTHMA})_2]$, have been synthesised and characterised by infrared (IR), nuclear magnetic resonance (NMR), electron paramagnetic resonance (EPR) and ultra violet-visible (UV-Vis) spectroscopies. The dimeric vanadium complexes in solution are in equilibrium with their respefrctive monomers, $H_2O + [V_2^VO_2(\mu\text{-O})]^{4+} = 2\,[V^VO(OH)]^{2+}$. The two amphiphilic vanadium complexes exhibit enhanced hydrolytic stability. EPR shows that the complexes in lipophilic matrix are mild radical initiators. Evaluation of their biological activity shows that the compounds do not exhibit any significant cytotoxicity to cells.

Keywords: vanadium; vitamin E; EPR; tocopherol; ^{51}V NMR

Citation: Loizou, M.; Hadjiadamou, I.; Drouza, C.; Keramidas, A.D.; Simos, Y.V.; Peschos, D. Vanadium(V) Complexes with Siderophore Vitamin E-Hydroxylamino-Triazine Ligands. *Inorganics* 2021, 9, 73. https://doi.org/10.3390/inorganics9100073

Academic Editor: Dinorah Gambino

Received: 21 July 2021
Accepted: 27 September 2021
Published: 29 September 2021

Publisher's Note: MDPI stays neutral with regard to jurisdictional claims in published maps and institutional affiliations.

1. Introduction

The understanding of the physiological role of vanadium ions in biological systems as well as the biological activity of vanadium compounds have stimulated the interest of the scientific community towards the vanadium chemistry [1–8]. Pharmaceuticals based on vanadium complexes have attracted the interest of scientists due to the biological activity of vanadium molecules and their low toxicity [4,5,9–19]. In addition, vanadium compounds exert antitumor effects through activation of apoptotic pathways, cell cycle arrest and the generation of Reactive Oxygen Species (ROS), inducing lower toxicity than anticancer platinum-based molecules [4,20–25].

α-Tocopherol acts in biological organisms as a strong lipophilic antioxidant, without any other biological activity. However, the vitamin E (tocopheryl and tocotrienyl) derivatives, such as α-tocopheryl succinate, have anticancer properties [26–34]. The hydrophobic domain of the vitamers of vitamin E is responsible for docking the agents in circulating lipoproteins and biological membranes [35]. Conjugate molecules of vitamin E vitamers with pharmaceuticals, such as metal complexes, can be used to transfer the drug in the active site of vitamin E vitamers, inducing biological responses.

Recently, we reported the first study of the synthesis of complexes comprising tocopherol ligating to metals [36]. The ligands in this study are β-tocopherol molecules substituted with chelate groups in o-position derivatives (Scheme 1, $H_3\beta$-tocDEA), thus, enabling coordination of the metal ion from the phenolic oxygen. The $[V^VO(\beta\text{-tocDEA})]$ has been found to be cytotoxic to cancer cells. Some of the features of these amphiphilic vanadium complexes have been their ability to induce the formation of free radicals [37], and their higher stability in aqueous solutions than the respective counterparts deprived of their lipophilic part. The hydrolytic stability of the vanadium complexes is enhanced in amphiphilic media [36,38,39], therefore, the high hydrolytic stability of the amphiphilic vanadium complexes has been attributed to their amphiphilic nature; presumably through a more favourable solvation [40].

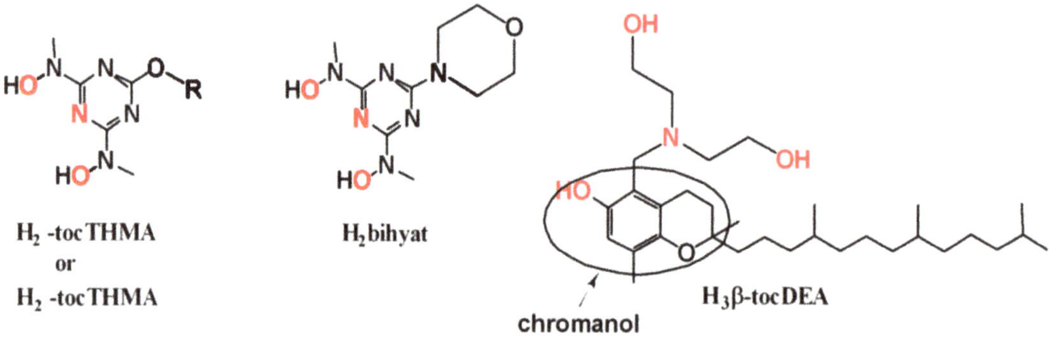

Scheme 1. Hydroxylamino-triazine ligands and $H_3\beta$-tocDEA. RO- is α- or δ-tocopherol. The donor atoms for metal ion coordination are in red colour.

Herein, we have attached a siderophore moiety on the phenoxy oxygen of the chromanol (Scheme 1), forming two new ligands, the 2,4-dichloro-6-(((R)-2,5,7,8-tetramethyl-2-((4R,8R)-4,8,12-trimethyltridecyl)chroman-6-yl)oxy)-1,3,5-triazine ($H_2\alpha$-tocTHMA) and 2,4-dichloro-6-(((R)-2,8-dimethyl-2-((4R,8R)-4,8,12-trimethyltridecyl)chroman-6-yl)oxy)-1,3,5-triazine($H_2\delta$-tocTHMA). The labile hydrogen atom of the hydroxy group has been replaced with the triazine moiety forming an inert ether bond and, thus, the new organic molecules will act as ligand owing lower antioxidant activity than free tocopherols; the formation of the tocopheryl radical requires deprotonation of chromanol group. In addition, the high lipophilicity of both the tocopherol derivatives and their complexes assures easy penetration in cell membranes [38]. As chelate group for V^V we have chosen the siderophore hydroxylamino-triazine (Scheme 1), targeting to enhance the hydrolytic stability of the V^V complexes as much as possible. This chelate group, for example in the ligand H_2bihyat (Scheme 1) forms very strong complexes with hard acids such as Fe^{III}, V^V, Mo^{VI} and U^{VI} [41–44], with Fe^{III} and U^{VI} to exert the higher affinity for this chelate coordination. The V^V complexes of this study exhibiting a chromanol hydroxy group unavailable for coordination, present no significant toxicity to cells. These results are in contrast to the high toxicity of the previous reported vanadium complexes, in which the vanadium ion was coordinated directly with the hydroxy group of the chromanol [36].

2. Experimental Section

2.1. Reagents

All reagents were purchased from Aldrich and Merck, (Kenilworth, NJ, USA). Vanadium complexes used for cell viability studies were dissolved in dimethyl sulfoxide (DMSO). DMSO was also used as vehicle control. Microanalyses for C, H and N were performed using a Euro-Vector EA3000 CHN elemental analyser (Milan, Italy). Infrared (IR) spectra were recorder on a Shimadzu Prestige 21, 7102 Riverwood Drive, Columbia, Maryland 21046, U.S.A. MALDI-TOF mass spectra were recorded on a Bruker Autoflex III Smartbeam (Billerica, MA, USA) instrument using α-Cyano-4-hydroxycinnamic acid (HCCA) as matrix.

2,2′-((2-hydroxyoctadecyl)azanediyl)bis(ethan-1-ol) (**C18DEA**), [V^VO(C18DEA)] were prepared according to reference [37].

2.2. Synthesis

*2,4-dichloro-6-(((R)-2,5,7,8-tetramethyl-2-((4R,8R)-4,8,12-trimethyltridecyl)chroman-6-yl)oxy)-1,3,5-triazine (**H₂α-tocTCL**).* Cyanuric chloride (1.86 g, 10.0 mmol) and *N,N*-diisopropylethylamine (1.45 g, 11.2 mmol) were dissolved in 40 mL of THF. The resulting colourless solution cooled at 0 °C. A THF solution (5 mL) of α-tocopherol (4.31 g, 10.0 mmol) was added to the above solution. The yellow solution was stirred for 24 h at room temperature. Then, the solution was evaporated, and the residue was extracted with chloroform/water. The organic phase was evaporated to give a yellow-orange oil as the product (3.50 g, 61%). ^{1}H-NMR, (300MHz, CDCl$_3$) δ ppm: 0.85–0.88 (m, 12H, -CH$_3$ methyl groups of the tocopherol), 1.12–1.15 (m, 6H), 1.26 (br s, 10H), 1.35–1.46 (m, 5H), 1.50–1.58 (m, 3H), 1.80–1.85 (m, C16-H), 1.95–1.99 (d, C8-H, C14-H), 2.12 (s, C10-H), 2.60–2.64 (t, C15-H). Elemental Analysis for C$_{32}$H$_{49}$C$_{12}$N$_3$O$_2$: Found: C, 66.20; H, 8.88; N, 7.13, Calcd.: C, 66.42; H, 8.54; N, 7.26.

*2,4-dichloro-6-(((R)-2,8-dimethyl-2-((4R,8R)-4,8,12-trimethyltridecyl)chroman-6-yl)oxy)-1,3,5-triazine (**H₂δ-tocTCL**).* Cyanuric chloride (1.95 g, 10.6 mmol) was dissolved in 40 mL of THF with the dropwise addition of *N,N*-diisopropylethylamine (1.50 g, 11.6 mmol) and the solution cooled at 0 °C. To the resulting solution, equivalent quantity of δ-tocopherol (4.26 g, 10.6 mmol) dissolved in 5 mL of THF was added. The solution was left to stir for 24 h at room temperature and the colour of the solution changed to yellow. The next day the solution was evaporated and extracted with chloroform/water. The chloroform extract was evaporated to give a yellow-orange oil as the product (3.00 g, 52%). ^{1}H-NMR, (300 MHz, CDCl$_3$) δ ppm: 0.87–0.90 (m, 12H, -CH$_3$ methyl groups of the tocopherol), 1.10–1.16 (m, 6H), 1.31 (br s, 10H), 1.36–1.45 (m, 5H), 1.52–1.61 (m, 3H), 1.80–1.87 (m, C16-H), 2.16 (s, C10-H), 2.72–2.83 (t, C15-H). Elemental analysis for C$_{30}$H$_{45}$C$_{12}$N$_3$O$_2$: Found: C, 65.31; H, 8.33; N, 7.28, Calcd.: C, 65.44; H, 8.24; N, 7.63.

*Synthesis of N,N′-(6-(((R)-2,5,7,8-tetramethyl-2-((4R,8R)-4,8,12-trimethyltridecyl)chroman-6-yl)oxy)-1,3,5-triazine-2,4-diyl)bis(N-methylhydroxylamine) (**H₂α-tocTHMA**).* **1st method:** α-tocTCL (3.00 g, 5.18 mmol) was dissolved in THF (120 mL) at 0 °C. A cooled (0 °C) solution of *N*-Methylhydroxylamine hydrochloride (1.80 g, 21.5 mmol) and sodium hydroxide (0.86 g, 2.2 mmol) in water (10 mL) was added dropwise in the above solution. The reaction mixture was refluxed for 24 h. Then, the solution was evaporated under vacuum to dry, and the residue was extracted with chloroform/water. The organic phase was evaporated under vacuum to dry resulting in H$_2$α-tocTHMA as an orange-brown oil. The yield was 2.1 g, 68%.

2nd method: α-tocTCL (3.00 g, 5.18 mmol) was dissolved in THF (120 mL) at 0 °C. A cooled (0 °C) solution of *N*-Methylhydroxylamine hydrochloride (1.80 g, 21.5 mmol) and sodium hydroxide (0.86 g, 2.2 mmol) in water (10 mL) was added dropwise to the above solution. The reaction mixture was kept under stirring at room temperature for 4 days. Then, it was evaporated under vacuum to dry, and the residue was dissolved in chloroform and filtrated to remove the insoluble in chloroform NaCl. The solution was evaporated under vacuum to dry yielding H$_2$α-tocTHMA, 2.0 g, 66% as an orange-brown oil. ^{1}H-NMR, (300 MHz, CDCl$_3$) δ ppm: 0.85–0.90 (m, 12H methyl groups of tocopherol), 1.14–1.17 (m, 6H), 1.25–1.27 (br s, 10H), 1.35–1.46 (m, 5H), 1.50–1.57 (m, 3H), 1.79–1.84 (m, C16-H), 1.96–2.00 (d, C8-H, C14-H), 2.11–2.13 (s, C10-H), 2.59–2.62 (t, C15-H), 3.3–2.5 (s, 6H methyl groups of *N*-methylhydroxylamine). Elemental Analysis for C$_{34}$H$_{57}$N$_5$O$_4$: Found: C, 68.29; H, 9.31; N, 11.59, Calcd.: C, 68.08; H, 9.58; N, 11.68. [MALDI-TOF(+)-MS]: calcd for (C$_{34}$H$_{57}$N$_5$O$_4$Na) {[M + Na]$^+$} m/z 622.43, found 623.13 (100%).

*Synthesis of N,N′-(6-(((R)-2,8-dimethyl-2-((4R,8R)-4,8,12-trimethyltridecyl)chroman-6-yl)oxy)-1,3,5-triazine-2,4-diyl)bis(N-methylhydroxylamine) (**H₂α-tocTHMA**).* H$_2$δ-tocTHMA was synthesised following the same methodology as for the synthesis of H$_2$α-tocTHMA. The yields were 82% and 53% for the 1st and the 2nd synthetic methods respectively. ^{1}H-NMR, (300 MHz, CDCl$_3$) δ ppm: 0.87–0.90 (m, 12H methyl groups of tocopherol), 1.08–1.19 (m, 6H), 1.26–1.29 (br s, 10H), 1.36–1.48 (m, 5H), 1.52–1.57 (m, 3H), 1.79–1.81 (t, C16-H), 2.14–

2.17 (d, C10-H), 2.70–2.75 (t, C15-H), 3.3–2.5 (s, 6H methyl groups of *N*-methylhydroxylamine), 6.40–6.75 (2 dd, H-aromatic C7, C13). Elemental analysis for $C_{32}H_{53}N_5O_4$: Found: C, 67.11; H, 9.18; N, 12.07, Calcd.: C, 67.22; H, 9.34; N, 12.25. [MALDI-TOF(+)-MS]: calcd for $(C_{32}H_{53}N_5O_4Na)$ {[M + Na]$^+$} *m/z* 594.37, found 595.18 (100%).

*Synthesis of [V^V_2 $O_2(\mu$-$O)(\alpha$-tocTHMA)$_2$], **1**. **1st Method:** $V^{IV}OSO_4 \cdot 5H_2O$ (0.13 g, 0.49 mmol) was stirred in 35 mL MeOH at 35 °C, under nitrogen. Then, $H_2\alpha$-tocTHMA (0.29 g, 0.49 mmol) dissolved in the minimum amount of methanol, was added to the above methanolic solution resulting in a deep brown solution. The solution was stirred for 24 h at room temperature. Then, it was filtered to remove any precipitation, and the filtrate was kept at room temperature for 5 days. During that time a black solid of **1** was formed, which was filtered and dried under vacuum. The yield was 65 mg, 20%.

2nd Method: α-tocTHMA (1.2 g, 2.0 mmol) dissolved in the minimum amount of methanol was added to a stirring methanol solution (40 mL) of [$V^{IV}O(acac)_2$] (0.53 g, 2.0 mmol) under nitrogen. The dark brown solution was stirring at room temperature for 10 min. Then, the solution was filtered, and the filtrate was left unstirred for 5 days, under air, at room temperature. During that time a black solid of **1** was formed which was filtered and dried under vacuum. The yield was 0.35 g, 26%. FTIR (ATR, cm^{-1}): 2924 (C-H), 1580 (C = N$_{triazine}$), 1524 (ar C–C), 1094 (N–O), 966 (V=O), 798 (V-O-V). Elemental analysis for $C_{68}H_{110}N_{10}O_{11}V_2$: Found: C, 60.47.11; H, 8.32; N, 10.10, Calcd.: C, 60.70; H, 8.24; N, 10.41. [MALDI-TOF(+)-MS]: calcd for $(C_{34}H_{55}N_5Na_2O_6V)$ {[M-O-M ($-$M+2Na)]$^+$} *m/z* 726.34, found 727.19 (100%).

*Synthesis of [V^V_2 $O_2(\mu$-$O)(\delta$-tocTHMA)$_2$], **2**. Similar with **1** synthetic methodologies were used for the synthesis of **2**. The yields were 24% and 19% for the 1st and the 2nd synthetic methods respectively. FTIR (ATR, cm^{-1}): 2928 (C-H), 1578 (C = N$_{triazine}$), 1526 (ar C–C), 1070 (N–O), 964 (V=O), 798 (V-O-V). Elemental analysis for $C_{64}H_{102}N_{10}O_{11}V_2$: Found: C, 59.45; H, 7.83; N, 10.47, Calcd.: C, 59.61; H, 7.97; N, 10.86. [MALDI-TOF(+)-MS]: calcd for $(C_{32}H_{52}N_5Na_2O_6V)$ {[M-O-M ($-$M + 2Na)]$^+$} *m/z* 698.31, found 699.30 (100%).

*Synthesis of [$Fe^{III}(\alpha$-tocTHMA)(Hα-tocTHMA)], **3**. Ferric chloride (0.03 g, 0.20 mmol) was dissolved under Ar in 20 mL MeOH forming a yellow solution. $H_2\alpha$-tocTHMA (0.24 g, 0.40 mmol) dissolved in the minimum amount of methanol was added in the above yellow solution, resulting in a deep blue solution. The solution was filtered, and the filtrate cooled at -18 °C resulting in a black precipitate of **3** which was filtered and dried under vacuum. The yield was 120 mg, 24%. Elemental analysis for $C_{68}H_{111}FeN_{10}O_8$: Found: C, 65.11; H, 8.89; N, 10.95, Calcd.: C, 65.21; H, 8.93; N, 11.18. [MALDI-TOF(+)-MS]: calcd for $(C_{34}H_{55}N_5O_4Fe)$ {[M$-$L]$^+$} *m/z* 653.36, found 654.31 (100%).

*Synthesis of [$Fe^{III}(\delta$-tocTHMA)(Hδ-tocTHMA)], **4**. Complex **4** was synthesised using the same methodology as the one used for **3**. The yield was 105 mg, 22%. Elemental analysis for $C_{64}H_{103}FeN_{10}O_8$: Found: C, 64.09; H, 8.81; N, 11.69, Calcd.: C, 64.25; H, 8.68; N, 11.71. [MALDI-TOF(+)-MS]: calcd for $(C_{32}H_{51}N_5O_4Fe)$ {[M$-$L]$^+$} *m/z* 625.33, found 626.19 (100%).

2.3. Spectroscopic Studies

All NMR samples were prepared from the dissolution of the solids in CDCl$_3$ or 10% DMSO-d_6:90% D$_2$O at room temperature immediately before NMR spectrometric determinations. NMR spectra were recorded on a Bruker Avance 300 spectrometer at 300 MHz for ^{1}H, 75.4 MHz for ^{13}C and 78.9 MHz for ^{51}V NMR. A 30°-pulse width was applied for both the ^{1}H and ^{51}V NMR measurements, and the spectra were acquired with 3000 and 30,000 Hz spectral window, using 1 and 0.1 s relaxation delay respectively. The spectra were analysed using Topspin 4.0 and MultispecNMR 5.0 (https://sourceforge. net/projects/multispecnmr/, accessed on 1 March 2021). 2D [45] *gr*NOESY spectra were obtained by using standard pulse sequences of Bruker Topspin 3.0 software. These spectra were acquired using 256 increments (with 56 scans each) and mixing time 0.43 s.

UV-Vis measurements were recorded on a Photonics UV-Vis spectrophotometer Model 400, equipped with a CCD array, operating in the range 250 to 1000 nm. The spectra were analysed using MultispecUVVIS 5.0 (https://sourceforge.net/projects/multispecuvvis/, accessed on 1 March 2021).

2.4. Reactivity with DPPH$^\bullet$

The rate of DPPH$^\bullet$ consumption was measured by UV-vis spectroscopy at 520 nm for 30 min. Stock solutions of each compound (12 mM) were prepared in dry toluene at room temperature. The final concentrations of the compounds were 80–300 µM, while the concentration of DPPH$^\bullet$ was 100 µM. The samples were incubated at 25 °C for 4 min. The reaction was initiated by the addition of the DPPH$^\bullet$ solution. The samples were measured in triplicate. Second-order rate constants were calculated to determine the radical scavenging activity (RSC) of antioxidants. The decay of DPPH$^\bullet$ from the medium has been assumed to follow pseudo-first-order kinetics, under the conditions of the reaction $[DPPH^\bullet]_0$, $[AH]_0$. One of the reactants is in large excess compared to the other, so the concentration of the minor component decreased exponentially [46]. The $[DPPH^\bullet]$ concentration is calculated from Equation (1):

$$[DPPH^\bullet] = [DPPH^\bullet]_0 \, e^{-k_{obsd}\, t} \tag{1}$$

where $[DPPH^\bullet]$ is the radical concentration at time t, and $[DPPH^\bullet]_0$ is the radical concentration at time zero, and k_{obsd} is the pseudo-first-order rate constant. The pseudo-first-order rate constant k_{obsd} was linearly dependent on the concentration of antioxidants [AH], and from the slope of their plot, second-order rate constants (k_2) were calculated to evaluate the radical scavenging capacity of each compound.

2.5. Measurement of Oxidative Inducing Effect of Vanadium Compounds by EPR Spectroscopy

An ELEXSYS E500 Bruker EPR spectrometer operating at cw X-band, resonance frequency ~9.5 GHz and modulation frequency 100 MHz was used. The resonance frequency was accurately measured with solid DPPH (g = 2.0036). The EPR oxidative inducing effect experiments were conducted by monitoring the evolution of α-tocopheryl radicals versus time [37] at room temperature. The assays were prepared in a 5 mm quartz tube by adding 100 µL or 150 µL of a CHCl$_3$ stock solution (4.95 mM) of complex to 0.500 g of a commercial extra virgin olive oil. Radical initiators are the **1**, **2** and [VO(C18DEA)] whereas the addition step consists of the initial time of the reaction, time = 0 min. EPR spectra were recorded for 25- or 30-time domains, each one consisting of 50 scans. The spectra were processed using appropriate software, MultispecEPR 5.0 (https://sourceforge.net/projects/multispecepr/, accessed on 1 March 2021).

2.6. Cell Culture

The human tongue squamous cell carcinoma (Cal33, DSMZ$^\circledR$ACC 447), the human cell line derived from cervical cancer (HeLa, ATCC$^\circledR$CCL-2) and the embryonic mouse fibroblasts (NIH/3T3, CRL-1658™) were used in this study [47]. Cells were grown in monolayer cultures in high glucose Dulbecco's modified eagle medium (DMEM) supplemented with 10% (v/v) foetal bovine serum, 2 mM L-glutamine and 1% (v/v) penicillin-streptomycin (100×: 10,000 units/mL of penicillin and 10,000 µg/mL streptomycin) in a humidified incubator (5% CO$_2$, 95% air) at 37 °C.

2.7. Measurement of Cell Viability

Cells were plated in 96-well plates at density 5 $\times$ 10^3 cells/well and treated with the ligands or the complexes for 24 and 48 h. Cell viability was measured after incubation of each well with 50 µL of MTT (stock solution of 3 mg/mL) for 3 h and absorbance was determined at 570 nm (background absorbance measured at 690 nm) using a microplate spectrophotometer (Multiskan Spectrum, Therno Fisher Scientific, Waltham, MA, USA). All experiments were performed in triplicate.

Stock solutions of H$_2$α-tocTHMA and H$_2$δ-tocTHMA and **1–4** prepared in pure DMSO were diluted into the culture medium so that the final concentration of DMSO was less than 1%. The same amount of DMSO was added to the control sample. Stock solutions were kept at 4 °C.

3. Results and Discussion

3.1. Synthesis and Characterisation

The triazine tocopherol molecules $H_2\alpha$-tocTHMA and $H_2\delta$-tocTHMA were synthesized by two-step substitution reactions of cyanuric chloride. The synthetic process is summarised in Scheme 2.

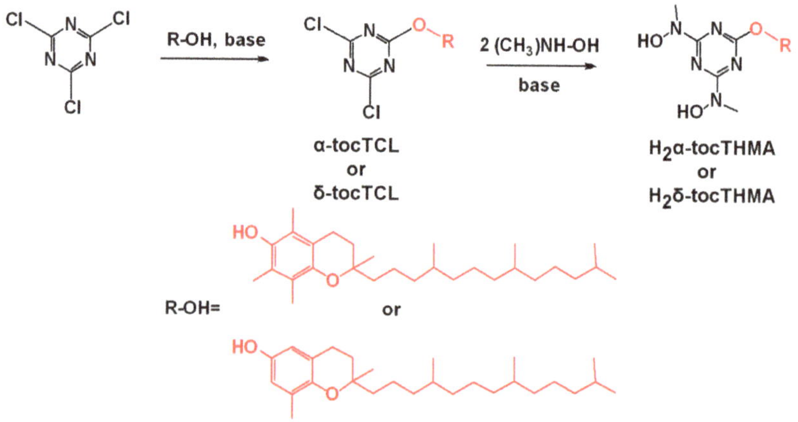

Scheme 2. Synthetic route for the organic compounds.

Reaction of equimolar quantities of $H_2\alpha$-tocTHMA or $H_2\delta$-tocTHMA with $[V^{IV}O(acac)_2]$ or $V^{IV}OSO_4$ results in the formation of V^V complexes **1** and **2** (Scheme 3). The V^{IV} is oxidised to V^V by the atmospheric O_2. X-band EPR spectroscopy of frozen $CHCl_3$ solutions of either **1** or **2** did not exhibit any signal supporting that all V^{IV} has been oxidised to V^V.

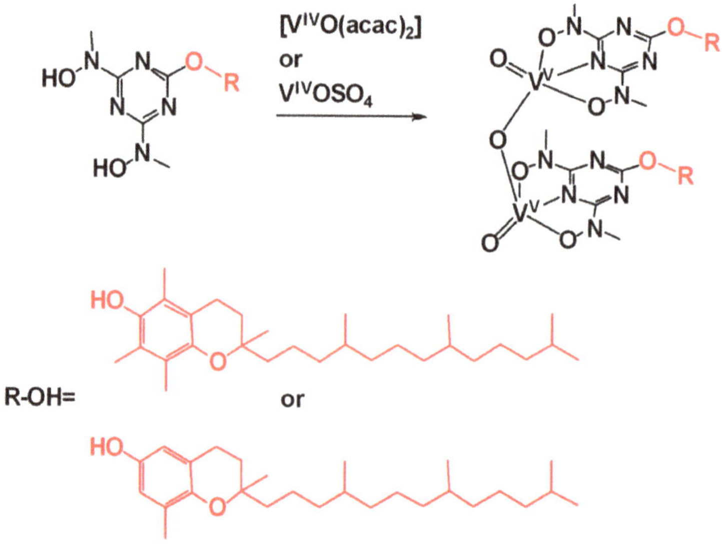

Scheme 3. Synthetic route for the vanadium complexes.

The complexes were characterised by elemental analysis, UV-vis, IR and NMR spectroscopies. The structures of the vanadium complexes are based on the data of the experimental analysis and the X-ray structures of the respective V^V-bihyat^{2-} complexes [42].

3.2. Complexes Characterisation by IR

The IR spectra of **1** and **2** are shown in Figure 1. Both **1** and **2** gave a strong peak at 966 cm^{-1} attributed to the stretching of the V=O bond. The peaks at 798 cm^{-1} are characteristic to V-O-V stretching vibrations [48,49], thus, confirming the dinuclear structure of the complex. The complexes also show two strong stretching N-O vibrations shifted around 80 cm^{-1} at higher energy compared to the free ligand suggesting coordination of the metal ion from the hydroxylamine-triazine chelating group.

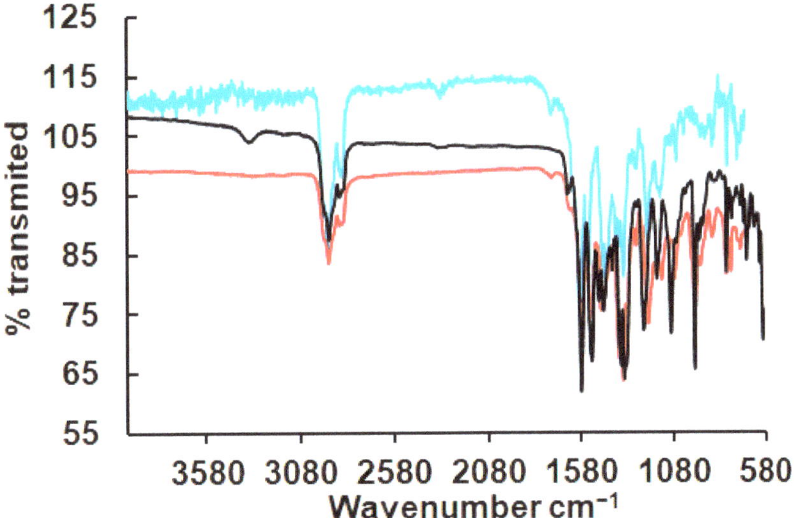

Figure 1. IR spectra of: **1** [v(V=O, 966 cm^{-1}; (C=N triazine), 1579, 1529 cm^{-1}; (Ph-O), 1242 cm^{-1}; (N-O), 1092 cm^{-1}]; 798 (V-O-V)] (black line), **2** [v(V=O, 966 cm^{-1}; (C=N triazine), 1579, 1529 cm^{-1}; (Ph-O), 1240 cm^{-1}; (N-O), 1084 cm^{-1}]; 798 (V-O-V) (red line), **3** [(C=N triazine), 1569, 1525 cm^{-1}; (Ph-O), 1244 cm^{-1}; (N-O), 1093 cm^{-1}] (blue line).

The IR spectra of the complexes **3** and **4** gave peaks at 2952, 2924 and 2896 cm^{-1} assigned to the C-H stretching of lipid chains. The C=N ring vibrations show peaks at 1569 and 1525 cm^{-1} whereas Ph-O and N-O stretching vibrations were detected at 1244 and 1093 cm^{-1} respectively. The N-O peaks are significantly shifted compared to the free ligand (~70 cm^{-1}) due to the ligation of the ligand to Fe^{III}.

3.3. Complex Characterisation by ^{51}V NMR, 2D {^{1}H} grNOESY

The ^{51}V spectra of each of the V^V complexes (**1, 2**) in CDCl$_3$ solutions gave two signals at −216 and −387 ppm (Figure 2). The intensity of the peaks is dependent on the concentration of the complexes in solution. At low concentration (i.e., 1 mM) the component at −216 ppm is the major, whereas at more concentrate solutions (i.e., 7 mM) the spectra of each of the complexes shows only the peak at −387 ppm. For more concentrated solutions three broad additional peaks of equal intensity appear at higher field (−402, −439 and −648 ppm), presumably originated from a higher nuclearity compound. The ^{51}V NMR spectra changes observed by the variation of the concentration are attributed to the equilibrium between the monomer (**1m**), dimer (**1**) and oligomers (Scheme 4). The quantities of **1m** and **1** are equal at concentration 2.5 mM, calculating a K$_{eq}$ = [**1m**^{2}]/[**1**] = 1.25 × 10^{-3} M. The conversion of the **1** to **1m** species takes approximately 2 min after the dissolution of **1** in $CDCl_3$, and it can be observed by the change of the colour of the solution from purple to blue. Dimerization of the hydroxylamine-triazine ligands through the formation of M-O-M bridge has been previously observed for U^{VI}-, Mo^{VI}- and V^{V}-bihyat^{2-} complexes [42,43]. The respective ^{51}V NMR spectrum of $CDCl_3$ solution of the structurally characterised by single crystal X-ray dinuclear complex $[V^{V}_2O_2(\mu_2\text{-O})(bihyat)_2]$ exhibits peaks at -192, -402 and -485 ppm. However, the ^{51}V NMR peaks of the $CDCl_3$ solution of $[V^{V}_2O_2(\mu_2\text{-O})(bihyat)_2]$ at -192 ppm had been mistakenly attributed to the decomposition of the compound.

Figure 2. ^{51}V NMR spectra of CD_3Cl solutions of **1**, (**A**) 1.00 mM (**B**) 7.00 mM.

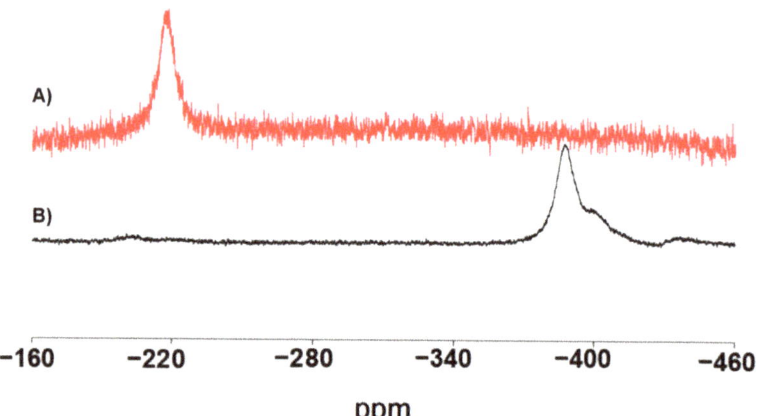

Scheme 4. Equilibrium between the monomer, **1m** and the dimer, **1**.

The 2D {^{1}H} grNOESY of **1** is shown in Figure 3. The two methyl groups show difference in chemical shifts due to the different chemical environment. 2D {^{1}H} grNOESY shows positive cross peaks between the protons of the two methyl groups assigned to the slow rotation methylhydroxylamine giving two peaks at 2.869 and 2.573 ppm in proton NMR. The rotation of tocopherol is performed around the ether bond between tocopherol and triazine moieties (Figure 3).

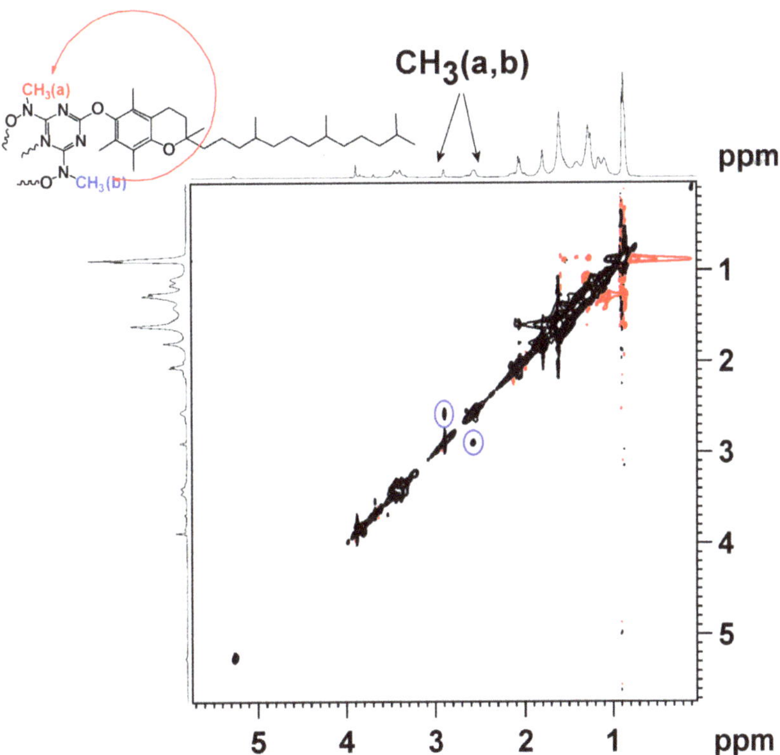

Figure 3. 2D {^{1}H} grNOESY spectrum of CDCl$_3$ solution of **1** (1.00 mM). Exchange of CH$_3$(a) and CH$_3$(b) cross peaks in blue circles.

3.4. Complexes Characterisation by UV-Vis

The UV-vis spectra of the CHCl$_3$ solutions of **1** and **2** are shown in Figure 4. Both gave strong peaks in the visible region [$\lambda(\varepsilon)$ of 1 = 493 nm (3600 M^{-1} cm^{-1}), 682 nm (1140 M^{-1} cm^{-1}), $\lambda(\varepsilon)$ of 2 = 484 nm (2300 M^{-1} cm^{-1}), 645 nm (860 M^{-1} cm^{-1})] assigned to the ligand to metal charge transfer transitions (LMCT). The concentration of the complexes in solutions were 0.500 mM, and according to ^{51}V NMR the species in the solution have the monomeric structure, **1m** and **2m** (Scheme 4). Although [V^{V}O$_2$(bihyat)]$^-$ has a similar structure with the monomers **1m** and **2m**, it does not exhibit any strong absorption peaks in the visible region. Thus, the strong colour of **1m** and **2m** is due to electron transitions from the chromanol ring to the metal. Chromanol ring can contribute electronically to the metal ion through the resonance of the triazine ring (Scheme 5) [42]. The shift of the UV-vis peaks of **1** to lower energy compared to those of **2** agrees with the higher electron density of α- than δ-tocopherol, supporting our hypothesis regarding the significance of the chromanol role on the LMCT effect.

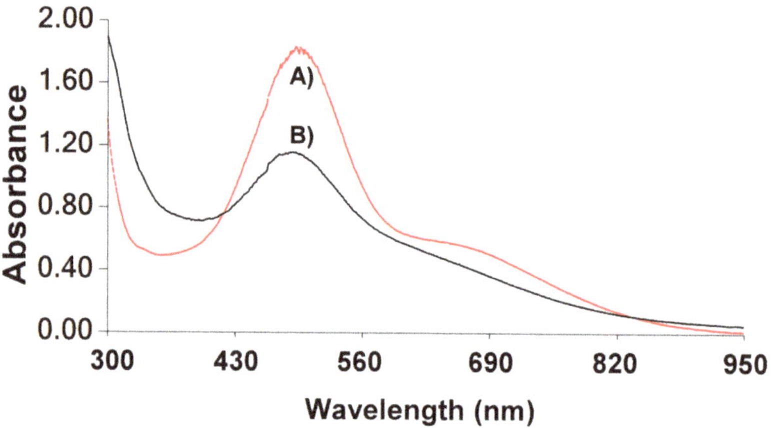

Figure 4. UV-vis spectra of CHCl$_3$ solutions of 0.500 mM of (**A**) **1m** [$\lambda(\varepsilon)$ = 493 nm (3600 M^{-1}cm^{-1}), 682 nm (1140 M^{-1} cm^{-1})] and (**B**) **2m** [$\lambda(\varepsilon)$ = 484 nm (2300 M^{-1} cm^{-1}), 645 nm (860 M^{-1} cm^{-1})].

Scheme 5. Triazine resonance structures.

The CHCl$_3$ solutions of Fe^{III} complexes **3** and **4** gave peaks at 560 nm (3200 M^{-1} cm^{-1}) and 535 nm (2200 M^{-1} cm^{-1}) respectively (Figure 5A). These spectra are similar to other hydroxylamine-triazin iron complexes, for example the Fe^{III}-bihyat compounds [41]. Complexes **3** and **4** exhibit the same pattern as the vanadate complexes; the α-tocopherol complex **3** absorbs at lower energy than the δ-tocopherol complex **4**. This is in line with the proposed electron transfer resonance mechanism proposed in Scheme 5.

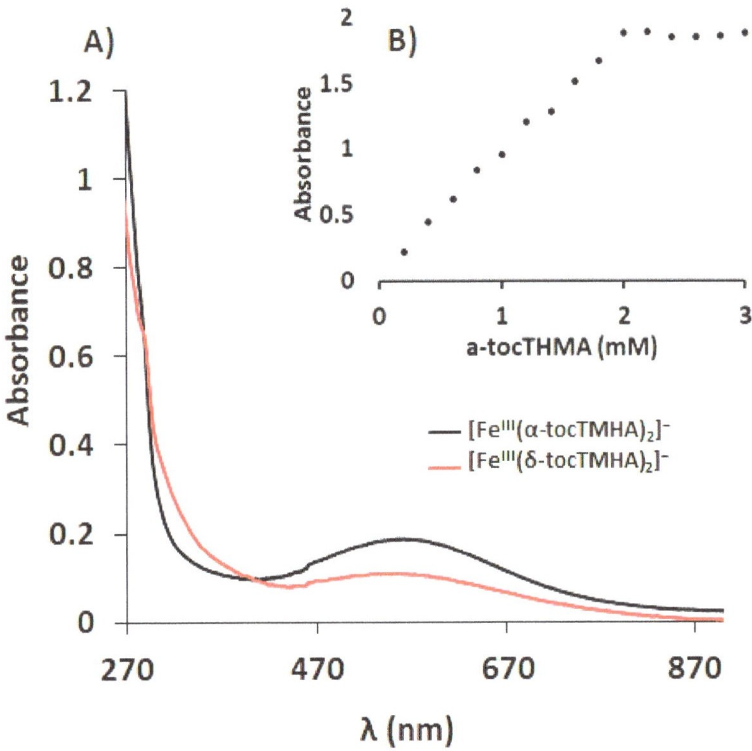

Figure 5. (**A**) UV-vis spectra of CHCl$_3$ solutions of 0.100 mM of **3** [black line, $\lambda(\varepsilon)$ = 560 nm (3200 M^{-1} cm^{-1})] and **4** [red line, $\lambda(\varepsilon)$ = 535 nm (2200 M^{-1} cm^{-1})]. (**B**) Absorbance at 560 nm vs. the concentration of *a*-tocTMHA added in the CHCl$_3$ solution of FeCl$_3$ (1.00 mM).

Addition of various quantities of either H$_2$α-tocTHMA or H$_2$δ-tocTHMA to a CHCl$_3$ solution of Fe^{III}Cl$_3$ gave the same spectra with **3** and **4** respectively. Titration of the CHCl$_3$ solution of Fe^{III}Cl$_3$ with either H$_2$α-tocTHMA (Figure 5B) or H$_2$δ-tocTHMA reveal that only the 1:2 Fe^{III}-Ligand complexes are formed in the solution.

3.5. Characterisation of the Complexes in 10% DMSO:90% D$_2$O Solutions by ^{51}V NMR

The ^{51}V NMR spectra of 10% DMSO:90% D$_2$O solutions of inorganic vanadate with either **1** or **2** at pD = 5.0–7.5 clearly shows very different chemical shifts for the peaks of vanadate oligomers from those of the complexes, undoubtedly assigning the peaks at −560 ppm to the new vanadium complexes (Figures 6 and 7). The addition of D$_2$O in the DMSO solutions (10% DMSO:90% D$_2$O, pD = 6.0–7.5) of **1** or **2** at concentrations 0.10 mM do not hydrolyse the complexes as evidenced by the ^{51}V NMR spectroscopy (the spectra show only one peak originated from the complex and there is no formation of any inorganic vanadate species (Figure 6)). In the ^{51}V NMR spectra, a shift from −387 ppm in CDCl$_3$ to −560 ppm in 10% DMSO:90% D$_2$O observed for **1** originated from the structural change of the complex from tetragonal pyramidal to dioxido octahedral geometry. A similar shift was observed upon changing from CDCl$_3$ to D$_2$O solutions for [VO$_2$(bihyat)]$^-$ as well [42]. The 10% DMSO:90% D$_2$O solutions of **1** or **2** were stable at these conditions for more than 72 h. The high hydrolytic stability of **1** and **2** is attributed to their amphiphilic nature [36,38,39]. The lipophilicity of **1** and **2** may enhance the hydrolytic stabilisation over the non-lipophilic vanadium complexes, exhibiting the same coordination environment, through a more favourable solvation [40].

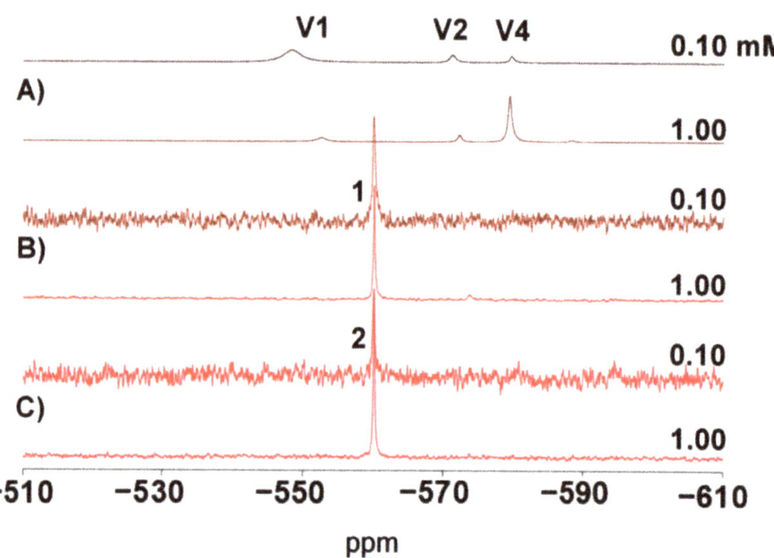

Figure 6. ^{51}V NMR spectra DMSO-d_6:D$_2$O (1:9) solutions of (**A**) NaVO$_3$, (**B**) **1**, (**C**) **2** at pD ~ 7. V1, V2, V4 are the vanadium oligomers.

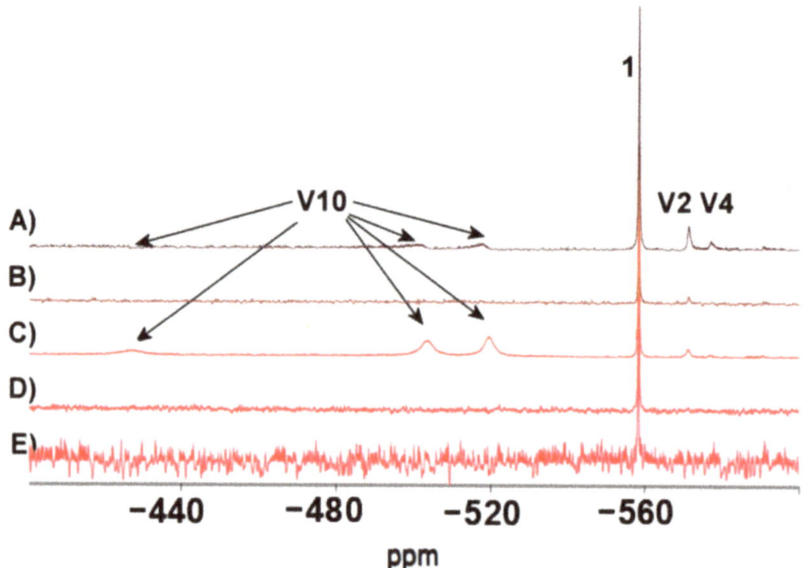

Figure 7. ^{51}V NMR spectra DMSO-d_6:D$_2$O (1:9) solutions of (**A**) 1.00 mM of **1** + 1.00 mM NaVO$_3$, pD = 5.9 (**B**) 1.00 mM of **1** pD = 5.9 (**C**) 1.00 mM of **1** + 1.00 mM NaVO$_3$, pD = 5.0 (**D**) 1.00 mM of **1**, pD = 5.0. (**E**) 0.10 mM of **1**, pD = 7.0. V10, V2, V4 are the vanadium oligomers.

3.6. Characterisation of the Complexes in 10% DMSO:90% D$_2$O Solutions by UV-Vis Spectroscopy

The UV-vis spectra of the 10%DMSO: 90%D$_2$O solutions of **1** and **2** were similar to the spectra of the complexes in CHCl$_3$ (Figure 8). The only difference between the spectra in the two different solvents is the lower intensity of the peaks in 10%DMSO: 90%D$_2$O than CHCl$_3$. However, complexes **1** and **2** show significant different absorption coefficients compared to those of $[VO_2(bihyat)]^-$ in various solvents [42]. The extinction coefficients of **1** and **2** are lower in protic polar solvents than in the non-polar ones in the same manner as $[VO_2(bihyat)]^-$. The absorbance values from the UV spectra of **1**, **2** solutions appear to obey Beer's law, even at low concentrations at 50 µM, suggesting that the complexes are hydrolytically stable in those solutions (Figure 9), in agreement with ^{51}V NMR spectroscopy. The spectra of the 10% DMSO:90% D_2O solutions of the iron complexes **3** and **4** are also similar with their spectra in $CHCl_3$ solutions.

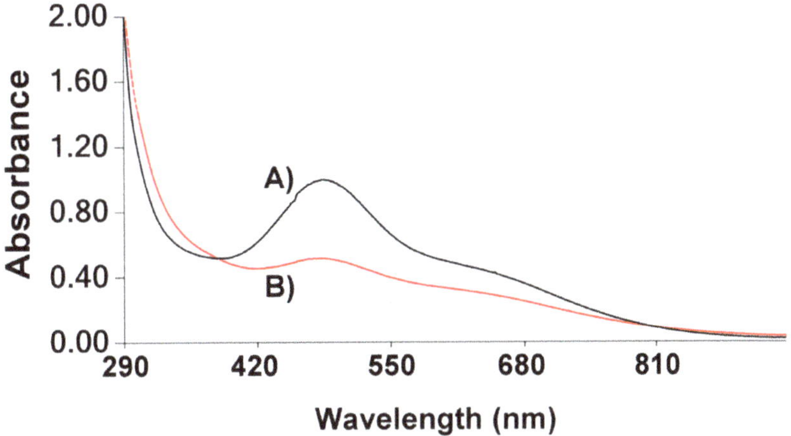

Figure 8. UV-vis spectra of DMSO-d_6:D_2O (1:9) solutions of 0.500 mM of (**A**) **1** [λ(ε) = 490 nm (2000 $M^{-1}cm^{-1}$), 675 nm (900 $M^{-1}cm^{-1}$)] and (**B**) **2** [λ(ε) = 490 nm (1100 $M^{-1}cm^{-1}$), 675 nm (660 $M^{-1}cm^{-1}$)].

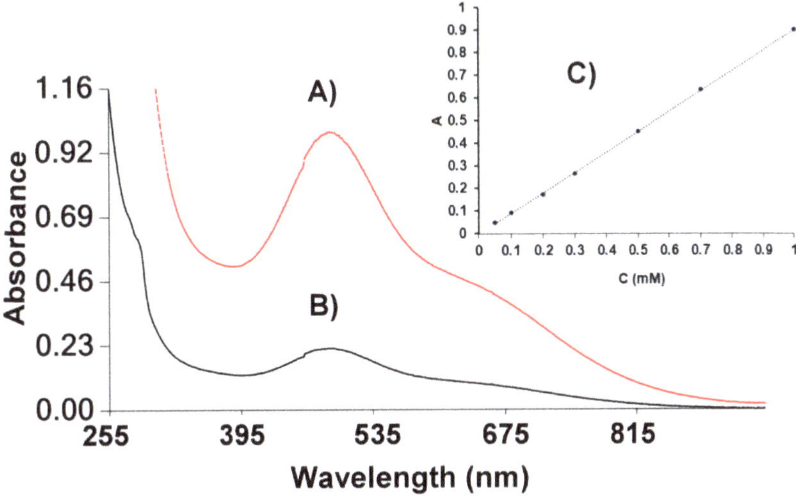

Figure 9. UV-vis spectra of DMSO-d_6:D_2O (1:9) solutions of **1** (**A**) 0.500 mM and (**B**) 0.100 mM, (**C**) graph of the absorption at 490 nm vs. concentration.

3.7. Reactivity with DPPH$^•$

The radical scavenging activity (RSC) values of the organic compounds and the complexes towards scavenging the DPPH$^•$ radical are shown in Table 1. α-tocTHMA and δ-tocTHMA exhibit very low antioxidant activity, much lower than free α-tocopherol. The reason for this low activity is the replacement of the labile phenoxy proton of α-tocopherol with an inert ether bond of α-tocTHMA and δ-tocTHMA. Large number of vanadium complexes exhibit radical scavenging activity [50]. However, complexes **1–4** either did not show any or very little decrease of the peak intensity at 520 nm, resulting in the conclusion that they do not have any antioxidant activity.

Table 1. Rate constants (k_2) for the RSC of the compounds under study.

Compounds	k_2 (M^{-1}s^{-1}) Toluene
α-tocTMHA	7.0 ± 0.16
δ-tocTMHA	12.5 ± 0.08
1m	-0.88 ± 0.02
2m	-0.88 ± 0.02
3	5.0 ± 0.09
4	-0.75 ± 0.02
[V^{IV}O(acac)$_2$]	6.8 ± 0.3
α-tocopherol [33]	560 ± 80

3.8. Oxidative Inducing Effect of Vanadium Compounds by EPR Spectroscopy

The ability of the new vanadium compounds to produce radicals was examined by monitoring the generation of α-tocopheryl radicals in olive oil by cw X-band EPR using 2D intensity vs. time experiments (Figure 10). The ability of complexes **1** and **2** were compared with that of the [V^{V}O(C18DEA)] used in a previous study [36,37]. [V^{V}O(C18DEA)] has been studied for its activity towards the production of radicals in olive oil, therefore, it is used in this work as a reference. Based on previous studies, it has been reported that V^V and/or V^{IV} coordinated catalytic sites are able to activate phenolics in the lipophilic matrix of oil mediated by dioxygen activation; in this oxidative environment free radicals are trapped by α-tocopherol to give α-tocopheryl radical. The generation of α-tocopheryl radicals is monitored by X-band cw-EPR vs. time. The graph of the signal intensity vs. time is a very useful quantification tool to determine the ability of the complexes to initiate radicals. Experiments were run for two different quantities of each radical initiator for the study, (0.490 μmole or 0.720 μmole). The intensity of the EPR peaks, at the same time period after addition of the radical initiator in olive oil is higher for [V^{V}O(C18DEA)] than **1**, meaning that [V^{V}O(C18DEA)] produces more α-tocopheryl radicals than **1**. The intensity of the EPR signal is lower at higher concentrations of the radical initiator due to the faster oxidation of the polar antioxidants that regenerate α-tocopherol in olive oil; the mechanism has been previously investigated [37]. Apparently, [V^{V}O(C18DEA)], and consequently [V^{V}O(β-tocDEA)] which is stronger initiator than [V^{V}O(C18DEA)] [36], are by far much more potent radical initiators than **1**.

[V^{V}O(C18DEA)] and [V^{V}O(β-toc)DEA] vanadium complexes have been reported to have high cytotoxicity [36]. If the cytotoxicity of the complexes is related to their oxidative power measured by EPR, then **1** is expected to be less cytotoxic than [V^{V}O(C18DEA)] or [V^{V}O(β-toc)DEA].

3.9. Cytotoxic Activity

None of the complexes exerted cytotoxic activity against the three cell lines, a fact that differentiated them significantly from the ligands. As seen in Figure 11, exposure of Cal33 cells for 24 h to increasing concentrations of the ligands and the complexes had no severe effect on the ability of the cells to proliferate (Figure 11A,C). Prolongation of the incubation time revealed that complexes 1 and 4 exerted a no-dose-dependent cytotoxicity across the different doses (1 to 100 μM) leading to a 40% reduction of cell population. On the contrary, the cytotoxic activity of the ligands H$_2\alpha$-tocTHMA and H$_2\delta$-tocTHMA as well as the complexes **2** and **3** was depicted mainly at doses higher than 25 μM. Order of cytotoxic activity (100 µM) was found as following: $H_2\delta$-tocTHMA < $H_2\alpha$-tocTHMA < **4** < **1** < **3** < **2** (24 h), $H_2\delta$-tocTHMA = $H_2\alpha$-tocTHMA < **2** < **3** < **1** < **4** (48 h).

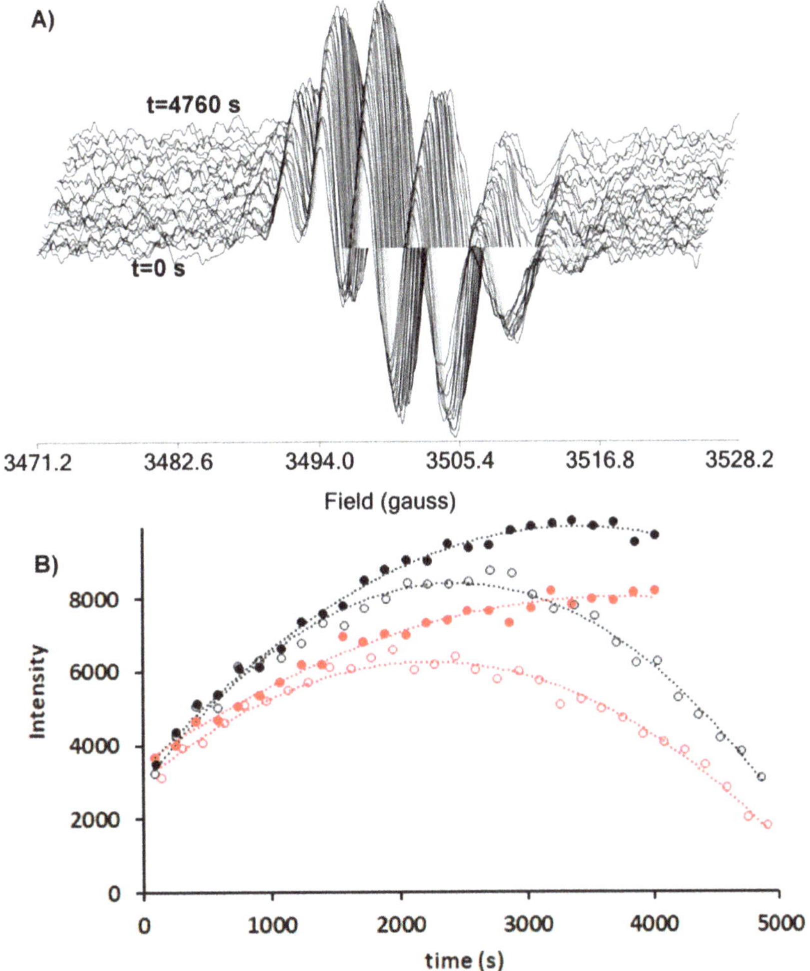

Figure 10. (**A**) X-band EPR spectra of virgin olive oil (0.500 g) vs. time after addition of **1** (0.720 µmole) at RT. The period between two continues spectra is 150 s. The total number of spectra is 30. (**B**) Intensity of the α-tocopheryl radical signal in X-band EPR spectra vs. time after the addition of the radical initiator in extra virgin olive oil (0.500 g), t = 0 s. Complex **1** (red colour), [VO(C18DEA)] (black colour). Different concentrations of radical initiator with respect to each V^V catalytic site per molecule: filled circles (0.490 µmole), open circles (0.720 µmole). The fitting curves have been generated from quadratic equations.

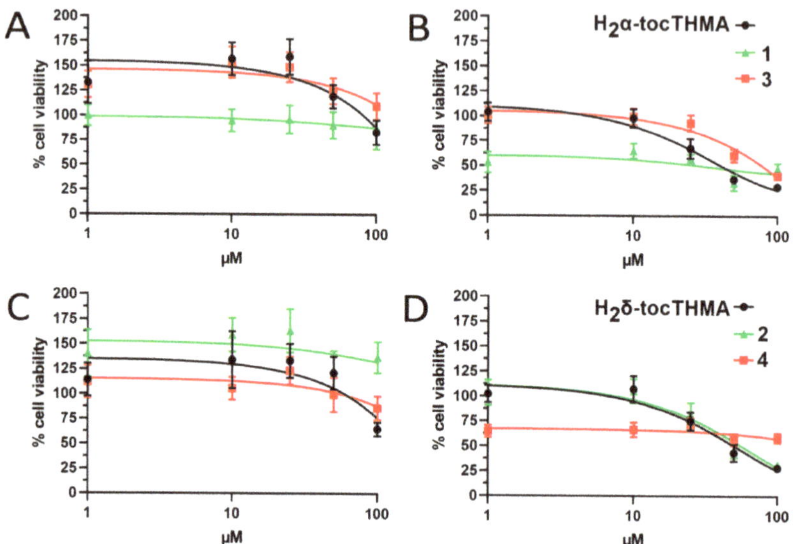

Figure 11. Cytotoxicity of H$_2\alpha$-tocTHMA (**A**,**B**) and H$_2\delta$-tocTHMA (**C**,**D**) complexes against Cal33 cells after exposure for 24 (**A**,**C**) and 48 h (**B**,**D**), respectively.

A similar cytotoxic profile was also seen against Hela cells (Figure 12). Twenty-four hours of exposure to the ligands and the complexes exerted a mild effect on cell viability even at the highest dose. At 48 h a slightly greater reduction in cell viability was recorded for both ligands and the complexes. Order of cytotoxic activity (100 µM): H$_2\alpha$-tocTHMA < 3 < 2 < H$_2\delta$-tocTHMA < 1 < 4 (24 h), H$_2\alpha$-tocTHMA < 1 < 3 = H$_2\delta$-tocTHMA < 2 < 4 (48 h).

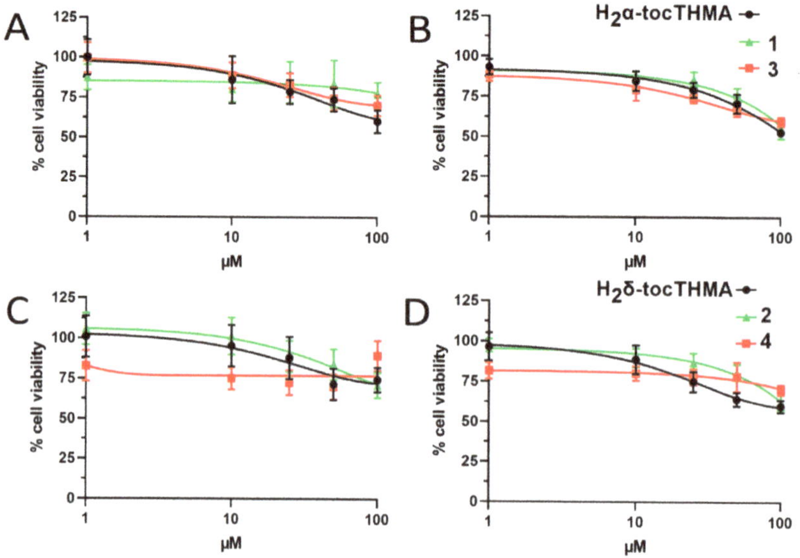

Figure 12. Cytotoxicity of H$_2\alpha$-tocTHMA (**A**,**B**) and H$_2\delta$-tocTHMA (**C**,**D**) complexes against HeLa cells after exposure for 24 (**A**,**C**) and 48 h (**B**,**D**), respectively.

Against embryonic mouse fibroblasts (NIH/3T3), the complexes exerted minimal toxicity after 24 h of incubation (Figure 13A,C). $H_2\alpha$-tocTHMA presented a strong cytotoxic effect after 48 h, a pattern similar to that seen against Cal33 cells. On the contrary, complexes **1** and **3** had no effect on cell viability and remained relatively non-toxic (Figure 13B). Complex **4** maintains the same cytotoxic profile, as seen in HeLa and Cal33 cells, exerting a mild reduction in cell viability across the range of doses (1–100 µM) whereas $H_2\delta$-tocTHMA and **2** were cytotoxic at concentrations higher than 25 µM (Figure 13D). Order of cytotoxic activity (100 µM) was found as follows: **2** > $H_2\delta$-tocTHMA > **1** = $H_2\alpha$-tocTHMA > **3** > **4** (24 h), $H_2\delta$-tocTHMA = **2** > $H_2\alpha$-tocTHMA > **1** > **4** > **3** (48 h).

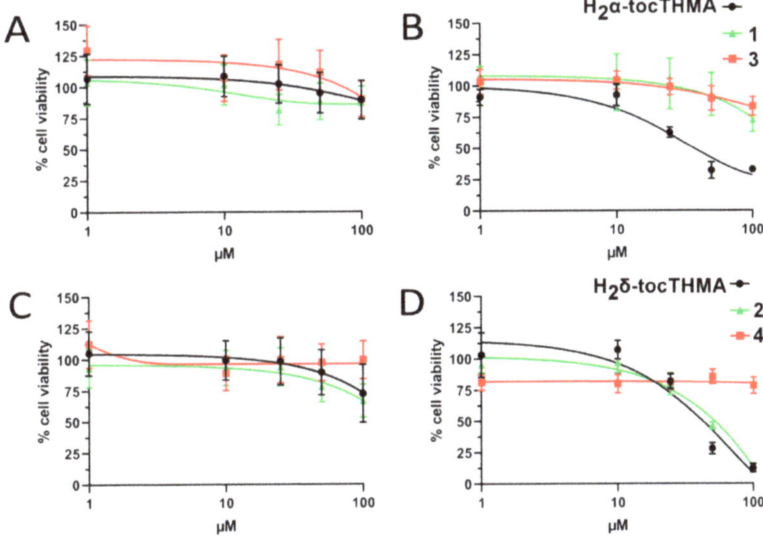

Figure 13. Cytotoxicity of $H_2\alpha$-tocTHMA (**A,B**) and $H_2\delta$-tocTHMA (**C,D**) complexes against NIH/3T3 cells after exposure for 24 (**A,C**) and 48 h (**B,D**), respectively.

4. Conclusions

Stepwise substitution reactions of cyanuric chloride with α- or δ-tocopherol and then with N-methylhydroxylamine resulted in the synthesis of the amphiphilic $H_2\alpha$-tocTHMA and $H_2\delta$-tocTHMA ligands. Reaction of the ligands with either the V^{IV} starting materials $[V^{IV}O(acac)_2]$ or $V^{IV}OSO_4$ afforded the complexes **1** and **2**. Reaction of $Fe^{III}Cl_3$ with $H_2\alpha$-tocTHMA or $H_2\delta$-tocTHMA resulted in the formation of complexes **3** and **4** respectively. The new compounds have been characterised by NMR, UV/Vis and infrared spectroscopies. The RSC activities for all compounds have been determined by the DPPH• assay and the results showed than none of the molecules, ligands or complexes, exhibits antioxidant activity. On the contrary, EPR spectroscopy showed that **1** and **2** are radical initiators. All complexes exhibit high hydrolytic stability even at low concentrations similar to those used in cell viability studies. All complexes, **1–4**, do not exerted significant cytotoxic activity against NIH/3T3, Cal33 and HeLa cell lines. The low cytotoxic activity is attributed to the low antioxidant-prooxidant activity of the tocopherol—triazine conjugate molecules. This is in line with the fact that **1** and **2** are moderate radical initiators. Previous studies support that the tocopherol–metal complexes with the hydroxy group of chromanol accessible to metal ion coordination, are stronger radical initiators than **1** and **2**, and they exert high cytotoxic activity. However, we cannot exclude the structural differences of the chelate moieties that might induce various biological responses. Currently, the vitamin E metal complexes are rare, and more work is required, including synthesis of new compounds with specific structural features, in order to understand the mechanism of their reactivity.

Author Contributions: Conceptualization, A.D.K. and C.D.; data curation, M.L., I.H. and Y.V.S.; methodology, A.D.K., C.D., Y.V.S. and D.P.; writing—original draft, A.D.K., C.D. and Y.V.S.; writing—review and editing, A.D.K. and C.D. All authors have read and agreed to the published version of the manuscript.

Funding: This research received no external funding.

Institutional Review Board Statement: Not applicable.

Informed Consent Statement: Not applicable.

Data Availability Statement: Data is contained within the article.

Conflicts of Interest: The authors declare no conflict of interest.

References

1. Michibata, H.; Sakurai, H. *Vanadium in Biological Systems*; Chasteen, N.D., Ed.; Kluer Academic Publishers: Dordrecht, The Netherlands, 1990; pp. 153–171.
2. Crans, D.C.; Keramidas, A.D.; Hoover-Litty, H.; Anderson, O.P.; Miller, M.M.; Lemoine, L.M.; Pleasic-Williams, S.; Vandenberg, M.; Rossomando, A.J.; Sweet, L.J. Synthesis, structure, and biological activity of a new insulinomimetic peroxovanadium compound: Bisperoxovanadium imidazole monoanion. *J. Am. Chem. Soc.* **1997**, *119*, 5447–5448. [CrossRef]
3. Rehder, D. The coordination chemistry of vanadium as related to its biological functions. *Coord. Chem. Rev.* **1999**, *182*, 297–322. [CrossRef]
4. Kioseoglou, E.; Petanidis, S.; Gabriel, C.; Salifoglou, A. The chemistry and biology of vanadium compounds in cancer therapeutics. *Coord. Chem. Rev.* **2015**, *301–302*, 87–105. [CrossRef]
5. Levina, A.; Lay, P.A. Stabilities and Biological Activities of Vanadium Drugs: What is the Nature of the Active Species? *Chem. Asian J.* **2017**, *12*, 1692–1699. [CrossRef] [PubMed]
6. Harwood, C.S. Iron-Only and Vanadium Nitrogenases: Fail-Safe Enzymes or Something More? *Annu. Rev. Microbiol.* **2020**, *74*, 247–266. [CrossRef] [PubMed]
7. Amante, C.; De Sousa-Coelho, A.L.; Aureliano, M. Vanadium and melanoma: A systematic review. *Metals* **2021**, *11*, 828. [CrossRef]
8. Sciortino, G.; Maréchal, J.D.; Garribba, E. Integrated experimental/computational approaches to characterize the systems formed by vanadium with proteins and enzymes. *Inorg. Chem. Front.* **2021**, *8*, 1951–1974. [CrossRef]
9. Costa Pessoa, J.; Garribba, E.; Santos, M.F.A.; Santos, S. Vanadium and proteins: Uptake, transport, structure, activity and function. *Coord. Chem. Rev.* **2015**, *301–302*, 49–86. [CrossRef]
10. Pessoa, J.C.; Etcheverry, S.; Gambino, D. Vanadium compounds in medicine. *Coord. Chem. Rev.* **2015**, *301–302*, 24–48. [CrossRef]
11. Rehder, D. The future of/for vanadium. *Dalton Trans.* **2013**, *42*, 11749–11761. [CrossRef]
12. Crans, D.C.; Trujillo, A.M.; Pharazyn, P.S.; Cohen, M.D. How environment affects drug activity: Localization, compartmentalization and reactions of a vanadium insulin-enhancing compound, dipicolinatooxovanadium(V). *Coord. Chem. Rev.* **2011**, *255*, 2178–2192. [CrossRef]
13. McLauchlan, C.C.; Peters, B.J.; Willsky, G.R.; Crans, D.C. Vanadium-phosphatase complexes: Phosphatase inhibitors favor the trigonal bipyramidal transition state geometries. *Coord. Chem. Rev.* **2015**, *301–302*, 163–199. [CrossRef]
14. Willsky, G.R.; Chi, L.H.; Godzala, M.; Kostyniak, P.J.; Smee, J.J.; Trujillo, A.M.; Alfano, J.A.; Ding, W.; Hu, Z.; Crans, D.C. Antidiabetic effects of a series of vanadium dipicolinate complexes in rats with streptozotocin-induced diabetes. *Coord. Chem. Rev.* **2011**, *255*, 2258–2269. [CrossRef]
15. Crans, D.C.; Tarlton, M.L.; McLauchlan, C.C. Trigonal bipyramidal or square pyramidal coordination geometry? Investigating the most potent geometry for vanadium phosphatase inhibitors. *Eur. J. Inorg. Chem.* **2014**, *2014*, 4450–4468. [CrossRef]
16. Crans, D.C.; Schoeberl, S.; Gaidamauskas, E.; Baruah, B.; Roess, D.A. Antidiabetic vanadium compound and membrane interfaces: Interface-facilitated metal complex hydrolysis. *J. Biol. Inorg. Chem.* **2011**, *16*, 961–972. [CrossRef]
17. Sánchez-Lombardo, I.; Alvarez, S.; McLauchlan, C.C.; Crans, D.C. Evaluating transition state structures of vanadium-phosphatase protein complexes using shape analysis. *J. Inorg. Biochem.* **2015**, *147*, 153–164. [CrossRef]
18. Crans, D.C. Antidiabetic, Chemical, and Physical Properties of Organic Vanadates as Presumed Transition-State Inhibitors for Phosphatases. *J. Org. Chem.* **2015**, *80*, 11899–11915. [CrossRef]
19. Crans, D.C.; Henry, L.; Cardiff, G.; Posner, B.I. Developing Vanadium as an Antidiabetic or Anticancer Drug: A Clinical and Historical Perspective. *Met. Ions Life Sci.* **2019**, *19*, 203–230. [CrossRef]
20. Kostova, I. Titanium and vanadium complexes as anticancer agents. *Anti-Cancer Agents Med. Chem.* **2009**, *9*, 827–842. [CrossRef] [PubMed]
21. Bishayee, A.; Waghray, A.; Patel, M.A.; Chatterjee, M. Vanadium in the detection, prevention and treatment of cancer: The in vivo evidence. *Cancer Lett.* **2010**, *294*, 1–12. [CrossRef]
22. Evangelou, A.M. Vanadium in cancer treatment. *Crit. Rev. Oncol. Hematol.* **2002**, *42*, 249–265. [CrossRef]
23. León, I.E.; Cadavid-Vargas, J.F.; Di Virgilio, A.L.; Etcheverry, S. Vanadium, ruthenium and copper compounds: A new class of non-platinum Metallodrugs with anticancer activity. *Curr. Med. Chem.* **2016**, *24*, 112–148. [CrossRef]

24. Crans, D.C.; Yang, L.; Haase, A.; Yang, X. Health Benefits of Vanadium and Its Potential as an Anticancer Agent. *Met. Ions Life Sci.* **2018**, *18*, 251–259. [CrossRef]

25. Doucette, K.A.; Hassell, K.N.; Crans, D.C. Selective speciation improves efficacy and lowers toxicity of platinum anticancer and vanadium antidiabetic drugs. *J. Inorg. Biochem.* **2016**, *165*, 56–70. [CrossRef] [PubMed]

26. Yan, B.; Stantic, M.; Zobalova, R.; Bezawork-Geleta, A.; Stapelberg, M.; Stursa, J.; Prokopova, K.; Dong, L.; Neuzil, J. Mitochondrially targeted vitamin E succinate efficiently kills breast tumour-initiating cells in a complex II-dependent manner. *BMC Cancer* **2015**, *15*, 401. [CrossRef] [PubMed]

27. Dong, L.F.; Grant, G.; Massa, H.; Zobalova, R.; Akporiaye, E.; Neuzil, J. α-Tocopheryloxyacetic acid is superior to α-tocopheryl succinate in suppressing HER2-high breast carcinomas due to its higher stability. *Int. J. Cancer* **2012**, *131*, 1052–1058. [CrossRef] [PubMed]

28. Dong, L.-F.; Neuzil, J. Vitamin E Analogues as Prototypic Mitochondria-Targeting Anti-cancer Agents. In *Mitochondria: The Anti-Cancer Target for the Third Millennium*; Neuzil, J., Pervaiz, S., Fulda, S., Eds.; Springer: Dordrecht, The Netherlands, 2014; pp. 151–181.

29. Kovarova, J.; Bajzikova, M.; Vondrusova, M.; Stursa, J.; Goodwin, J.; Nguyen, M.; Zobalova, R.; Pesdar, E.A.; Truksa, J.; Tomasetti, M.; et al. Mitochondrial targeting of α-tocopheryl succinate enhances its anti-mesothelioma efficacy. *Redox Rep.* **2014**, *19*, 16–25. [CrossRef]

30. Yap, W.N.; Chang, P.N.; Han, H.Y.; Lee, D.T.W.; Ling, M.T.; Wong, Y.C.; Yap, Y.L. γ-Tocotrienol suppresses prostate cancer cell proliferation and invasion through multiple-signalling pathways. *Br. J. Cancer* **2008**, *99*, 1832–1841. [CrossRef]

31. Zingg, J.M. Molecular and cellular activities of vitamin E analogues. *Mini-Rev. Med. Chem.* **2007**, *7*, 545–560. [CrossRef]

32. Dunn, B.K.; Richmond, E.S.; Minasian, L.M.; Ryan, A.M.; Ford, L.G. A nutrient approach to prostate cancer prevention: The selenium and vitamin e cancer prevention trial (SELECT). *Nutr. Cancer* **2010**, *62*, 896–918. [CrossRef]

33. Vraka, P.S.; Drouza, C.; Rikkou, M.P.; Odysseos, A.D.; Keramidas, A.D. Synthesis and study of the cancer cell growth inhibitory properties of α-, γ-tocopheryl and γ-tocotrienyl 2-phenylselenyl succinates. *Bioorg. Med. Chem.* **2006**, *14*, 2684–2696. [CrossRef]

34. Vraka, P.S.; Rikkou, M.N.; Drouza, C.; Keramidas, A.D.; Odysseos, A.D. Modulation of apoptotic signals with esterification of selenium and vitamin E in binary compounds. *FEBS J.* **2006**, *273*, 115.

35. Pussinen, P.J.; Karten, B.; Wintersperger, A.; Reicher, H.; McLean, M.; Malle, E.; Sattler, W. The human breast carcinoma cell line HBL-100 acquires exogenous cholesterol from high-density lipoprotein via CLA-1 (CD-36 and LIMPII analogous 1)-mediated selective cholesteryl ester uptake. *Biochem. J.* **2000**, *349*, 559–566. [CrossRef]

36. Hadjiadamou, I.; Vlasiou, M.; Spanou, S.; Simos, Y.; Papanastasiou, G.; Kontargiris, E.; Dhima, I.; Ragos, V.; Karkabounas, S.; Drouza, C.; et al. Synthesis of vitamin E and aliphatic lipid vanadium(IV) and (V) complexes, and their cytotoxic properties. *J. Inorg. Biochem.* **2020**, *208*, 111074. [CrossRef]

37. Drouza, C.; Dieronitou, A.; Hadjiadamou, I.; Stylianou, M. Investigation of Phenols Activity in Early Stage Oxidation of Edible Oils by Electron Paramagnetic Resonance and19F NMR Spectroscopies Using Novel Lipid Vanadium Complexes As Radical Initiators. *J. Agric. Food. Chem.* **2017**, *65*, 4942–4951. [CrossRef]

38. Crans, D.C.; Koehn, J.T.; Petry, S.M.; Glover, C.M.; Wijetunga, A.; Kaur, R.; Levina, A.; Lay, P.A. Hydrophobicity may enhance membrane affinity and anti-cancer effects of Schiff base vanadium(v) catecholate complexes. *Dalton Trans.* **2019**, *48*, 6383–6395. [CrossRef]

39. Lemons, B.G.; Richens, D.T.; Anderson, A.; Sedgwick, M.; Crans, D.C.; Johnson, M.D. Stabilization of a vanadium(v)-catechol complex by compartmentalization and reduced solvation inside reverse micelles. *New J. Chem.* **2013**, *37*, 75–81. [CrossRef]

40. Crans, D.C.; Keramidas, A.D.; Mahroof-Tahir, M.; Anderson, O.P.; Miller, M.M. Factors Affecting Solution Properties of Vanadium(V) Compounds: X-ray Structure of β-cis-NH4[VO2(EDDA)]. *Inorg. Chem.* **1996**, *35*, 3599–3606. [CrossRef]

41. Ekeltchik, I.; Gun, J.; Lev, O.; Shelkov, R.; Melman, A. Bis(hydroxyamino)triazines: Versatile and high-affinity tridentate hydroxylamine ligands for selective iron(III) chelation. *Dalton Trans.* **2006**, *6*, 1285–1293. [CrossRef] [PubMed]

42. Nikolakis, V.A.; Tsalavoutis, J.T.; Stylianou, M.; Evgeniou, E.; Jakusch, T.; Melman, A.; Sigalas, M.P.; Kiss, T.; Keramidas, A.D.; Kabanos, T.A. Vanadium(V) compounds with the bis-(hydroxyamino)-1,3,5-triazine ligand, H2bihyat: Synthetic, structural, and physical studies of [V 2 VO3(bihyat)2] and of the enhanced hydrolytic stability species cis-[VVO2(bihyat)]. *Inorg. Chem.* **2008**, *47*, 11698–11710. [CrossRef] [PubMed]

43. Stylianou, M.; Nikolakis, V.A.; Chilas, G.I.; Jakusch, T.; Vaimakis, T.; Kiss, T.; Sigalas, M.P.; Keramidas, A.D.; Kabanos, T.A. Molybdenum(VI) coordination chemistry of the N,N-disubstituted bis(hydroxylamido)-1,3,5-triazine ligand, H2bihyat. Water-assisted activation of the MoVI=O bond and reversible dimerization of cis-[MoVIO2(bihyat)] to [MoVI2O4(bihyat)2(H2O)2]. *Inorg. Chem.* **2012**, *51*, 13138–13147. [CrossRef]

44. Hadjithoma, S.; Papanikolaou, M.G.; Leontidis, E.; Kabanos, T.A.; Keramidas, A.D. Bis(hydroxylamino)triazines: High Selectivity and Hydrolytic Stability of Hydroxylamine-Based Ligands for Uranyl Compared to Vanadium(V) and Iron(III). *Inorg. Chem.* **2018**, *57*, 7631–7643. [CrossRef]

45. Chatterjee, P.B.; Abtab, S.M.T.; Bhattacharya, K.; Endo, A.; Shotton, E.J.; Teat, S.J.; Chaudhury, M. Hetero-bimetallic complexes involving vanadium(V) and rhenium(VII) centers, connected by unsupported μ-oxido bridge: Synthesis, characterization, and redox study. *Inorg. Chem.* **2008**, *47*, 8830–8838. [CrossRef]

46. Espín, J.C.; Soler-Rivas, C.; Wichers, H.J.; García-Viguera, C. Anthocyanin-based natural colorants: A new source of antiradical activity for foodstuff. *J. Agric. Food. Chem.* **2000**, *48*, 1588–1592. [CrossRef]

47. Avdikos, A.; Karkabounas, S.; Metsios, A.; Kostoula, O.; Havelas, K.; Binolis, J.; Verginadis, I.; Hatziaivazis, G.; Simos, I.; Evangelou, A. Anticancer effects on leiomyosarcoma-bearing Wistar rats after electromagnetic radiation of resonant radiofrequencies. *Hell. J. Nucl. Med.* **2007**, *10*, 95–101.
48. Drouza, C.; Keramidas, A.D. Solid state and aqueous solution characterization of rectangular tetranuclear V(IV/V)-p-semiquinonate/hydroquinonate complexes exhibiting a proton induced electron transfer. *Inorg. Chem.* **2008**, *47*, 7211–7224. [CrossRef]
49. Drouza, C.; Stylianou, M.; Papaphilippou, P.; Keramidas, A.D. Structural and electron paramagnetic resonance (EPR) characterization of novel vanadium(V/IV) complexes with hydroquinonate-iminodiacetate ligands exhibiting "noninnocent" activity. *Pure Appl. Chem.* **2013**, *85*, 329–342. [CrossRef]
50. Etcheverry, S.B.; Williams, P.A.M. Medicinal chemistry of copper and vanadium bioactive compounds. In *New Developments in Medicinal Chemistry*; Nova Science Publishers, Inc.: Hauppauge, NY, USA, 2009; pp. 105–129.

Article

2-Aminopyrimidinium Decavanadate: Experimental and Theoretical Characterization, Molecular Docking, and Potential Antineoplastic Activity

Amalia García-García [1,2], Lisset Noriega [3], Francisco J. Meléndez-Bustamante [3], María Eugenia Castro [1], Brenda L. Sánchez-Gaytán [1], Duane Choquesillo-Lazarte [4], Enrique González-Vergara [1,*] and Antonio Rodríguez-Diéguez [2,*]

[1] Centro de Química del Instituto de Ciencias, Benemérita Universidad Autónoma de Puebla, 18 sur y Av. San Claudio, Col. San Manuel, Puebla 72570, CP, Mexico; amaliagarcia@correo.ugr.es (A.G.-G.); mareug.castro@correo.buap.mx (M.E.C.); brenda.sanchez@viep.com.mx (B.L.S.-G.)

[2] Departamento de Química Inorgánica, Facultad de Ciencias, Universidad de Granada, Av. Fuentenueva S/N, 18071 Granada, Spain

[3] Facultad de Ciencias Químicas, Benemérita Universidad Autónoma de Puebla, 18 sur y Av. San Claudio, Col. San Manuel, Puebla 72570, CP, Mexico; lisset.noriegad@alumno.buap.mx (L.N.); francisco.melendez@correo.buap.mx (F.J.M.-B.)

[4] Laboratorio de Estudios Cristalográficos, IACT, CSIC-UGR, Av. Las Palmeras n°4, 18100 Granada, Spain; duane.choquesillo@csic.es

* Correspondence: enrique.gonzalez@correo.buap.mx (E.G.-V.); antonio5@ugr.es (A.R.-D.); Tel.: +52-222-363-0623 (E.G.-V.); +34-958-248-524 (A.R.-D.)

Citation: García-García, A.; Noriega, L.; Meléndez-Bustamante, F.J.; Castro, M.E.; Sánchez-Gaytán, B.L.; Choquesillo-Lazarte, D.; González-Vergara, E.; Rodríguez-Diéguez, A. 2-Aminopyrimidinium Decavanadate: Experimental and Theoretical Characterization, Molecular Docking, and Potential Antineoplastic Activity. *Inorganics* **2021**, *9*, 67. https://doi.org/10.3390/inorganics9090067

Academic Editor: Dinorah Gambino

Received: 31 July 2021
Accepted: 26 August 2021
Published: 30 August 2021

Publisher's Note: MDPI stays neutral with regard to jurisdictional claims in published maps and institutional affiliations.

Abstract: The interest in decavanadate anions has increased in recent decades, since these clusters show interesting applications as varied as sensors, batteries, catalysts, or new drugs in medicine. Due to the capacity of the interaction of decavanadate with a variety of biological molecules because of its high negative charge and oxygen-rich surface, this cluster is being widely studied both in vitro and in vivo as a treatment for several global health problems such as diabetes mellitus, cancer, and Alzheimer's disease. Here, we report a new decavanadate compound with organic molecules synthesized in an aqueous solution and structurally characterized by elemental analysis, infrared spectroscopy, thermogravimetric analysis, and single-crystal X-ray diffraction. The decavanadate anion was combined with 2-aminopyrimidine to form the compound [2-ampymH]6[V10O28]·5H2O (**1**). In the crystal lattice, organic molecules are stacked by π–π interactions, with a centroid-to-centroid distance similar to that shown in DNA or RNA molecules. Furthermore, computational DFT calculations of Compound **1** corroborate the hydrogen bond interaction between pyrimidine molecules and decavanadate anions, as well as the π–π stacking interactions between the central pyrimidine molecules. Finally, docking studies with test RNA molecules indicate that they could serve as other potential targets for the anticancer activity of decavanadate anion.

Keywords: decavanadate; 2-aminopyrimidinium; experimental and theoretical characterization; DFT; docking RNA/DNA

1. Introduction

Polyoxidometalates (POMs) are defined as clusters made from early transition-metals, typically d^0 species V(V), Nb(V), Ta(V), Mo(VI), and W(VI), bridged by oxide anions. This class of compounds is highly interesting in molecular structural variety, reactivity, and applications in analytical chemistry, catalysis, medicine, and materials research [1]. POMs have great potential in biological applications, since every aspect that involves the interaction of POM with biological target macromolecules could be modified to improve their beneficial effects on a biological system. Thus, interesting POMs with anticancer [2] and antibiotic activities [3], among others, have been obtained to date.

In this context, the colorless aqueous solution of vanadate(V) turns orange as it acidifies. This phenomenon is associated with condensation reactions carried out by vanadate ions depending on the acidity range of the solution. Thus, at pH $\approx$ 6, the orange solution indicates that the decavanadate anion, $[H_n V_{10}O_{28}]^{(6-n)-}$ with n = 0–4, has been formed [4,5]. These clusters have attracted much interest due to their potential applications in a wide range of uses such as sensors [6], batteries [7–9], catalysts [10–12], or metallodrugs [13–15].

In particular, decavanadates have recently attracted attention due to their medicinal and biochemical behavior, since they have an important role in biological systems by having the ability to interact with proteins, enzymes, and cell membranes [16].

More than forty years ago, vanadate was first found as an impurity in commercial ATP obtained from horse skeletal muscle and was initially identified as a muscle inhibition factor by inhibiting the activity of the sodium pump [17]. Subsequently, the first enzyme that decavanadate was able to inhibit, the rabbit skeletal muscle adenylate kinase, was reported. After that, many enzymes have been found that can be inhibited by this decameric species, such as hexokinase, phosphofructokinase, inositol phosphate metabolism enzymes, or nicotinamide adenine dinucleotide (NADH)-vanadate reductase [18].

Due to the high negative charge of decavanadate, this species can interact with a multitude of molecules such as proteins, counterions, or lipid structures, affecting many biological processes such as muscle contraction, calcium homeostasis, necrosis, actin polymerization, oxidative stress markers, or glucose uptake, among others [19]. As a result, different compounds based on decavanadate and cationic organic ligands have been published in recent years, which can decrease glycemia [20–22], induce neuronal and cognitive restoration mechanisms to treat metabolic syndrome [13], affect the growth of protozoan parasites and bacteria [23–26], or show antitumor activity [27,28]. For example, the compound $Mg(H_2O)_6(C_4N_2H_7)_4V_{10}O_{28}\cdot4H_2O$ demonstrated dose-dependent antiproliferative activity on human cancer cells U87, IGR39, and MDA-MB-231 [29].

Previous works with Adenine and Cytosine have shown a hydrogen bond interaction with the decavanadate anion [6,30,31]. Thus, to obtain new bioactive compounds based on the decavanadate cluster, one organic ligand with potential biological activity was chosen. Here, we report the deployment of 2-aminopyrimidine, which is susceptible to protonation and can interact with the decavanadate anion. Recently, structures formed by decavanadate and ligands with nitrogenous groups with promising antidiabetic and anticancer properties have been published [22,29,32–34]. In addition, within the family of N-heterocyclic compounds, pyrimidines and their derivatives are an important class of compounds in medicinal chemistry [35–37].

In this way, a new member of a family of compounds based on decavanadate was obtained. Decavanadate anion interacts with a 2-aminopyrimidine ligand to afford a crystalline compound with the formula $[2\text{-ampymH}]_6[V_{10}O_{28}]\cdot5H_2O$ (**1**). The structural characterization of the compound was carried out by elemental analysis, infrared spectroscopy, thermogravimetric analysis, and single-crystal X-ray diffraction. In addition, the compound was studied using Density Functional Theory (DFT) computational methods. The frontier molecular orbitals and global reactivity indexes were analyzed for showing interesting characteristics of the donor-acceptor interactions. The insights about the compounds' reactivity were corroborated by analyzing the non-covalent interactions using the AIM approach.

On the other hand, since Sciortino et al. [38] recently published the interaction of decavanadate with G-actin protein with docking calculations, in this work, docking studies using small RNA and DNA molecules were used to test the hypothesis that the attributed anticancer activity of decavanadate could be due to interaction with these critical molecules. Structurally, the compound has a set of hydrogen bonds and π–π interactions resembling those found in DNA/RNA molecules, opening the field of POM to RNA interactions as potential target molecules for cancer treatment.

2. Results

2.1. Structural Description of [2-ampymH]$_6$[V$_{10}$O$_{28}$]·5H$_2$O (1)

The decavanadate anion is a very useful cluster in coordination chemistry as a building block for various structures. Since they have a high negative charge and oxygen-rich surfaces, interactions between this anion and metals or organic ligands are easily formed by coordination or hydrogen bonds [39]. Based on this cluster, the crystal structure of a new compound with a cationic ligand was determined in the present study. Table 1 presents the corresponding crystal data, and Tables S1–S7 in the supplementary section contain additional crystallographic information.

Table 1. Crystal data and structure refinement for Compound 1.

Compound	1
Empirical formula	C$_{24}$H$_{46}$N$_{28}$O$_{33}$V$_{10}$
Formula mass (g·mol^{-1})	1764.20
CCDC	2099300
Crystal system	Triclinic
Space group	$P\bar{1}$
a (Å)	9.783(5)
b (Å)	11.309(5)
c (Å)	12.853(5)
α (°)	110.166(5)
β (°)	95.645(5)
λ (°)	97.551(5)
Volume (Å^3)	1307.4(10)
Z	2
Density (calcd) (g·cm^{-3})	2.063
μ(Mo/CuKα) (mm^{-1})	1.815
Temperature (K)	300(2)
GoF on F^2 [a]	1.065
R$_1$ [1 > 2σ(I)] [b]	0.0386
R$_1$ [all data] [b]	0.0621
wR$_2$ [1 > 2σ(I)] [c]	0.0821
wR$_2$ [all data] [c]	0.0958

[a] $S = [\Sigma w(F_o^2 - F_c^2)^2/(N_{obs} - N_{param})]^{1/2}$; [b] $R_1 = \Sigma\|F_o| - |F_c\|/\Sigma|F_o|$;
[c] $wR_2 = \{\Sigma[w(F_o^2 - F_c^2)^2]/\Sigma[w(F_o^2)^2]\}^{1/2}$; $w = 1/[\sigma^2(F_o^2) + (aP)^2 + bP]$ where $P = (max(F_o^2,0) + 2F_c^2)/3$.

Compound **1** crystallizes in the $P\bar{1}$ space group of the triclinic system and consists of a three-dimensional supramolecular structure, where units of [V$_{10}$O$_{28}$]$^{6-}$ interact with the cationic ligand 2-aminopyrimidinium, [2-ampymH]$^+$ and crystallization water molecules through hydrogen bonds. The asymmetric unit contains half of a decavanadate anion arranged in an inversion center, three independent molecules of [2-ampymH]$^+$ (named as A, B, and C) protonated in the N3 position, and two and a half crystallization water molecules (two of them with 100% occupancy and one with occupancy set at 50%). Therefore, the (6-) charge of the decavanadate cluster is stabilized by six [2-ampymH]$^+$ cations, which are arranged on both sides of the cluster interacting by hydrogen bonds. Thus, the formula established by monocrystal X-ray diffraction was [2-ampymH]$_6$[V$_{10}$O$_{28}$]·5H$_2$O (Figure 1).

The protonation effect of the 2-ampym ligand mainly results in a variation in the internal angles N2-C1-N3 and, to a lesser extent, in the angles C1-N2-C2 and C1-N3-C4. Table 2 shows these angles compared with the values given by J. Scheinbeim et al. for a non-protonated 2-ampym molecule [40]. Protonation also affects the distances between the atoms within the cation. Specifically, there is a shortening of the N1-C1 bond, from 1.342 Å for the ligand to 1.317 Å on average for the cation. In addition, in the 2-ampym ligand, the N2-C2 and N3-C4 bonds are symmetrical, with a value of 1.331 Å, while, N3-C4 bond is greater for the protonated form than the N2-C2 one (1.351 Å versus 1.323 Å on average, respectively).

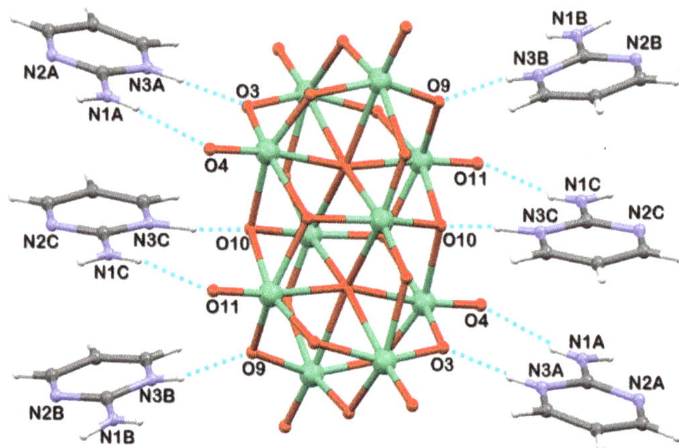

Figure 1. Structure of [2-ampymH]$_6$[V$_{10}$O$_{28}$]·5H$_2$O (**1**). The hydrogen bonds among [2-ampymH]$^+$ cations and decavanadate anion are shown (blue dashed line). Water molecules have been omitted for clarity. Vanadium atoms are represented in green.

Table 2. Difference between the angles ($\Delta°$) of the neutral 2-ampym molecule and the cations A, B, and C of the protonated ligand in Compound **1**.

Angle	[2-AmpymH]$^+$ (°)	2-Ampym (°)	$\Delta°$
N2A-C1A-N3A	121.3	125.2	3.9
N2B-C1B-N3B	120.9		4.3
N2C-C1C-N3C	123.2		2
C1A-N2A-C2A	117.3	115.7	1.6
C1B-N2B-C2B	117.3		1.6
C1C-N2C-C2C	116.6		0.9
C1A-N3A-C4A	120.7	116.2	4.5
C1B-N3B-C4B	120.9		4.7
C1C-N3C-C4C	118.4		2.2

The decavanadate cluster comprises ten edge-sharing VO$_6$ octahedra, containing ten vanadium atoms and twenty-eight oxygen atoms. In total, the cluster possesses sixty V-O bonds which can be differentiated by their bond distances. First, two oxygen atoms are located inside the polyanion (O$_c$) bonded to six vanadium atoms, each one with the largest bond distance in the range of 2.106–2.340 Å. Four other oxygen atoms are arranged on the surface (O$_{b1}$) and coordinate to three vanadium atoms each, with bond lengths of V-O$_{b1}$ in the range of 1.900–2.084 Å. Fourteen oxygen atoms at the corners (O$_{b2}$) coordinate to two vanadium atoms each, where V-O$_{b2}$ bond length distances are ranged from 1.684 to 2.069 Å. Lastly, eight terminal oxygen atoms (O$_t$) are coordinated to only one vanadium each, and the V = O$_t$ distance is the smallest one with values in the range of 1.595–1.611 Å. These bond distances agree with other [V$_{10}$O$_{28}$]$^{6-}$ anions reported in the literature [41–44]. Bond lengths are listed in Table S3.

The supramolecular structure is dominated by hydrogen bonds of type N-H···N, N-H···O, and O-H···O, in addition to π-stacking interactions between pyrimidine rings. On the one hand, as shown in Figure 2, the decavanadate anion interacts by hydrogen bonds of type N-H···O with six [2-ampymH]$^+$ cations: two hydrogen bonds are formed thanks to N1 and N3 atoms of molecules A and C, giving rise to a $R_2^2(8)$ ring motif, according to the nomenclature of Etter et al. [45], while molecule B only forms one hydrogen bond through the N3 atom. Furthermore, [2-ampymH]$^+$ cations interact with each other by hydrogen bonds, where N2 atoms of the rings act as acceptors and N1 atoms of the amino groups are the donors, forming a $R_2^2(8)$ ring motif. The distances and angles of the hydrogen bonds are shown in Table 3.

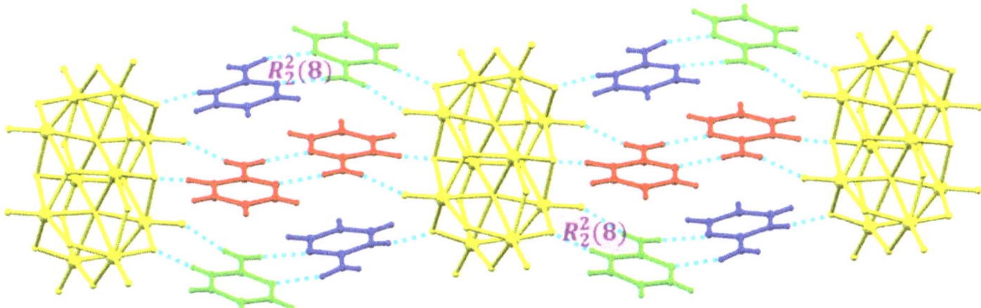

Figure 2. Partial supramolecular structure of Compound **1** based on hydrogen bonds (blue dashed line) among the [2-ampymH]$^+$ cations and decavanadate anions. The structure is colored by symmetric equivalence. C cations (in red) form hydrogen bonds with each other, while A and B cations (in green and blue, respectively) interact between them.

Table 3. Distances (Å) and angles (°) of the hydrogen bonds of Compound **1**.

D-H⋯A ^{b}	D-H	H⋯A	D-H⋯A	Angle (°)
C2A-H2A⋯O12	0.93	2.57	3.176(4)	122.9
C3A-H3A⋯O5$^{(i)}$	0.93	2.28	3.133(4)	152.6
N1A-H1AA⋯N2B$^{(ii)}$	0.83	2.17	2.997(4)	171.8
N1A-H1AB⋯O4$^{(iii)}$	0.79	2.15	2.933(4)	171.2
N3A-H3AA⋯O3$^{(iii)}$	0.85	1.80	2.650(3)	174.1
N1B-H1BA⋯N2A$^{(ii)}$	0.93	2.14	3.043(4)	165.6
N1B-H1BB⋯O1W	0.70	2.34	3.024(6)	168.1
N3B-H3BA⋯O9	0.78	1.87	2.625(3)	163.2
N1C-H1CA⋯N2C$^{(iv)}$	0.81	2.14	2.954(4)	179.6
N1C-H1CB⋯O11$^{(v)}$	0.81	2.10	2.896(3)	168.8
N3C-H3CA⋯O10	1.09	1.49	2.585(3)	174.9
O1W-H1WA⋯O3W$^{(ii)}$	0.85	2.36	3.162(13)	157.2
O1W-H1WB⋯O1	0.85	2.29	3.083(4)	154.4
O2W-H2WA⋯O11$^{(vi)}$	0.85	2.79	3.352(7)	125.5
O2W-H2WB⋯O11$^{(ii)}$	0.85	2.09	2.729(8)	131.2
O3W-H3WA⋯O6	0.85	2.30	3.049(9)	147.7
O3W-H3WA⋯O1W$^{(ii)}$	0.85	2.64	3.162(13)	121.3
O3W-H3WB⋯O4$^{(i)}$	0.85	2.30	3.143(11)	174.7

Symmetry operations: (i) = x, −y, 1 − z; (ii) = 2 − x, 1 − y, 2 − z; (iii) = 1 − x, −y, 1 − z; (iv) = 1 − x, −y, −z; (v) = 2 − x, 1 − y, 1 − z; (vi) = +x, −1 + y, +z; ^{b} D: donor; H: hydrogen; A: acceptor.

On the other hand, due to the arrangement of [2-ampymH]$^+$ cations, π–π stacking interactions are formed between pyrimidine rings, by forming a chain of six rings that repeats infinitely in space (Figure 3), with an average distance between centroids of 3.617 Å, which falls within the range of a π–π interaction. Some of the rings are stacked in an aligned way, while others form a π–π stacking shifted parallel about 22°, also producing a shorter distance between centroids because π-σ attraction forces predominate in this arrangement [46]. A similar structure has been published by S. Sedghiniya et al. [6], in which decavanadate anion interacts by hydrogen bonds with cations of Adenine, one of the five nitrogenous bases that form DNA and RNA. Similarly, in this structure, the cations interact with each other by hydrogen bonds and π–π stacking interactions, with a centroid-to-centroid distance of 3.5 Å and a displacement angle of approximately 24°.

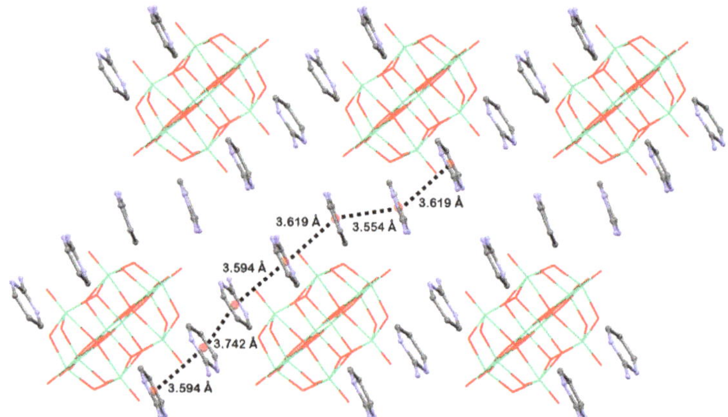

Figure 3. Partial supramolecular structure of Compound **1** where π–π stacking interactions between the aromatic rings of pyrimidine and the distance between centroids are shown (black dashed line). Vanadium atoms are represented in green.

This arrangement is clearly reminiscent of DNA structure, a polymer formed by base pairs that interact with hydrogen bonds and are stacked by π–π stacking interactions with a distance between centroids of 3.4 Å. Similarly, in Compound **1**, [2-ampymH]$^+$ molecules interact by hydrogen bonds and are stacked by π–π stacking interactions with a distance between centroids of approximately 3.6 Å (Figure 4).

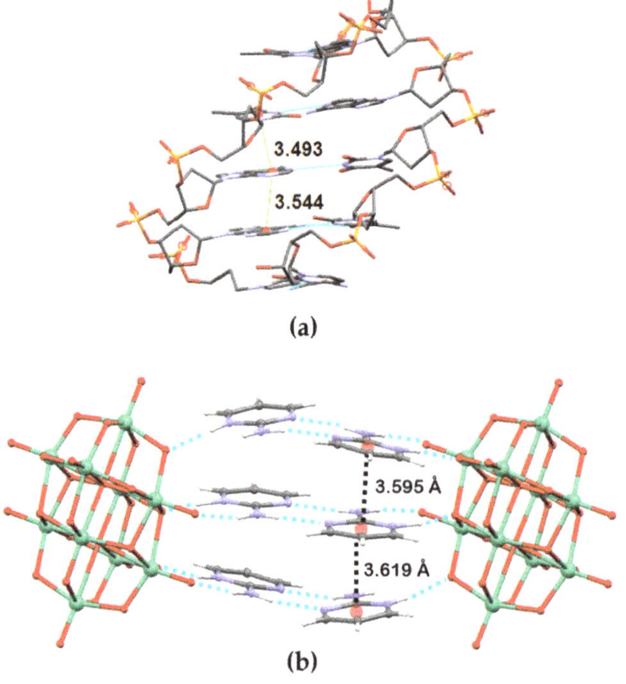

(a)

(b)

Figure 4. (**a**) Small fragment of DNA (1BNA) and (**b**) Compound **1** show a similar structure, with distances between centroids in the same range. Vanadium atoms are represented in green.

Lastly, this structure contains water molecules of crystallization (Figure 5). Two of the water molecules of the unit cell (O1W and O3W) are arranged between the decavanadate and the cations in general positions, while the third one (O2W) is located at the apex, making its contribution to the unit cell 50%. These water molecules form hydrogen bonds within the structure, helping to stabilize it. All distances of hydrogen bonds along with their angles are found in Table 3.

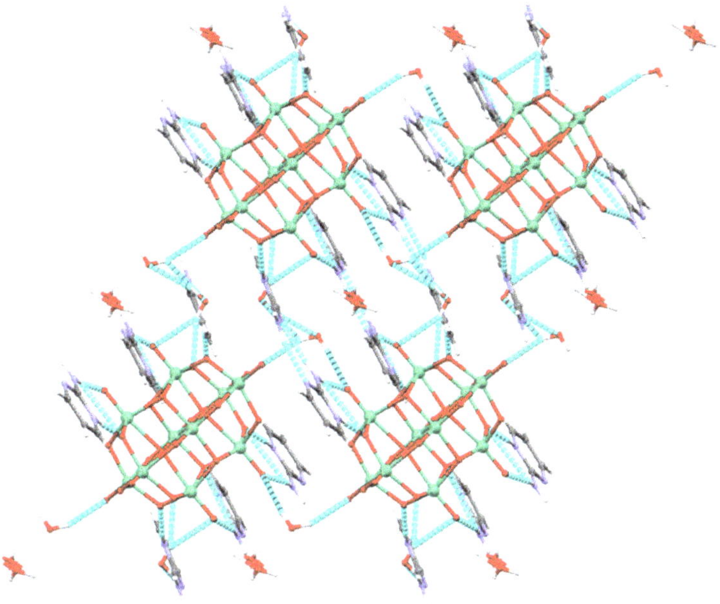

Figure 5. View of the supramolecular structure along the crystallographic *a*-axis, where all hydrogen bonds forming in Compound **1** can be observed (blue dashed line). Vanadium atoms are represented in green.

2.2. Infrared (IR) Spectroscopy

At the top of the 2-ampym ligand spectrum (Figure 6) were the characteristic bands at 3331 and 3165 cm^{-1} that belong to the stretching vibration modes of the N-H and C-H bonds, respectively. In addition, the bending vibration modes appeared at 1645 and 1128 cm^{-1}, respectively. Lower wavenumbers show the characteristic bands of the aromatic pyrimidine ring with peaks at 1556, 1469, 1356, 794, and 555 cm^{-1}, belonging to the C-N, C=C, C-C, and CCC bonds, respectively [47–49]. As for the spectrum of Compound **1**, it was possible to see both characteristic bands of the ligand 2-aminopyrimidine and the bands belonging to the different vibration modes of the V-O bond of decavanadate anion. Although the band belonging to the vibration of the O-H bond was not distinguished precisely, there was a wideband between 3400 and 2900 cm^{-1}, which could be assigned to water molecules in the structure. In addition, there was a slight displacement of the bands belonging to the V=O and N-H groups, so the interaction between both molecules was weak, mainly due to hydrogen bonds. Main IR bands are collected in Table 4.

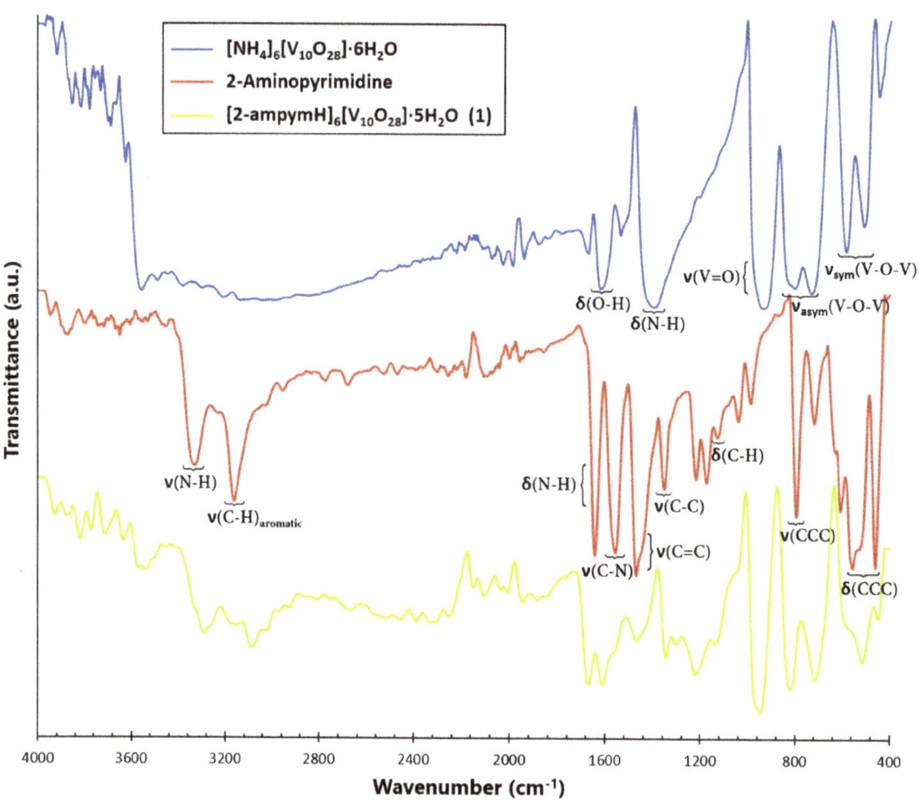

Figure 6. IR spectra of ammonium decavanadate, 2-aminopyrimidine ligand, and Compound **1** in the range 4000–400 cm^{-1} at room temperature.

Table 4. IR bands (cm^{-1}) of ammonium decavanadate, 2-aminopyrimidine ligand (2-ampym), and Compound **1** in the range of 4000–400 cm^{-1}.

Vibrational Mode	(NH$_4$)$_6$V$_{10}$O$_{28}$·6H$_2$O	2-Ampym	Compound 1
ν(N-H)	-	3331	3290
ν(C-H)$_{aromatic}$	-	3165	3082
δ(O-H)	1622	-	-
δ(N-H)	1394	1645	1670
ν(C-N)	-	1556	1610
ν(C=C)	-	1469	1462
ν(C-C)	-	1356	1346
δ(C-H)	-	1128	1132
ν(V=O)	927	-	941
ν_{asym}(V-O-V)	825	-	814
	802		711
	731		
ν(CCC)	-	794	
ν_{sym}(V-O-V)	580	-	580
	505		511
δ(C-C-C)	-	555	
		459	

2.3. Thermal Study

The thermal decomposition behavior of [2-ampymH]$_6$[V$_{10}$O$_{28}$]·5H$_2$O (**1**) in solid-state was studied by TG analysis in the range of 35–950 °C (Figure S1). The thermogram of Compound **1** revealed a mass loss of around 5% in the range of 35–180°C, which corresponded to five water molecules of crystallization (% mass, calc. (found) for 5 × H$_2$O: 5.5% (5.4%)). Then, a continuous mass loss occurred in the range from 180 to 450 °C, which may correspond to the thermal degradation of six 2-aminopyridinium cations (% mass, calc. (found) for 6 × C$_4$H$_6$N$_3$: 35.5% (36.3%)). The total mass loss of Compound **1** was 41.85%, which was in good agreement with the calculated value of 41%. No further losses in mass were observed above 460 °C, which can be attributed to the remaining [V$_{10}$O$_{28}$]$^{6-}$ fragment that may be thermally stable up to 600 °C, as previously reported [50,51]. Nevertheless, some authors point out that further thermal treatment of decavanadate anion could result in the formation of other vanadium oxides, such as V$_2$O$_5$ or other mixed-valence oxides [52–54].

2.4. Theoretical Calculations

Figure 7 shows the molecular structure and isosurfaces of the Highest Occupied Molecular Orbital (HOMO) and Lowest Unoccupied Molecular Orbital (LUMO) molecular orbitals of Compound **1** plotted with an isovalue of 0.02 a.u. The frontier orbitals HOMO and LUMO can be related to donating or accepting electrons, respectively. The higher energy of the HOMO orbital indicates a more pronounced behavior as an electron donor, while the lower energy of the LUMO orbital indicates a higher electron affinity. The HOMO orbital, with energy of −7.4671 eV, is located on the decavanadate anion, mainly on the O-bridge atoms, as shown in Figure 7b, while the LUMO orbital, with energy of −5.7381 eV, is located on the organic counterions of pyrimidine, mainly on the C and N atoms of the rings, as shown in Figure 7c.

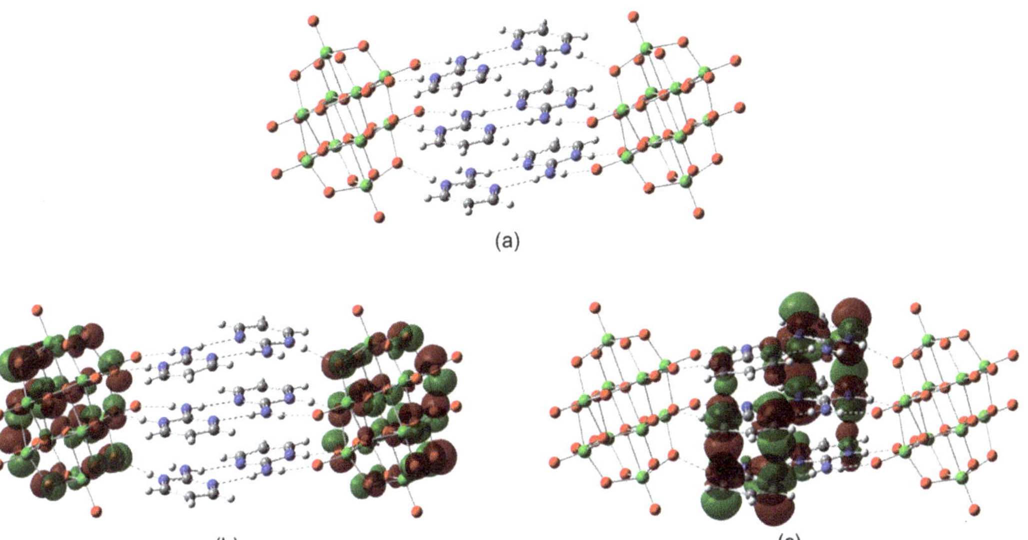

Figure 7. (a) Molecular structure, (b) HOMO isosurface, and (c) LUMO isosurface of Compound **1** calculated at the B3LYP/Def2SVP-LANL2DZ theory level using ECP = LANL2DZ for the V atom.

Figure 8 shows the molecular electrostatic potential (MEP) map of Compound **1**. The MEP map was mapped with an isovalue of 0.04 × 10^{-2} a.u. in the range of −3.45 × 10^{-2} to 3.45 × 10^{-2} and a.u.$^{-3}$. The red zones indicate areas with high charge density (nucleophilic zone), while the blue zones indicate the presence of positive charge (electrophilic zones). It was observed that the areas with the highest charge density were located on the decavanadate anion, while the positively charged areas were located in the pyrimidine molecules. The non-covalent interactions between the O atoms of decavanadate anion and the H atoms of pyrimidine molecules were found in intermediate electron density zones, represented in yellow-green colors.

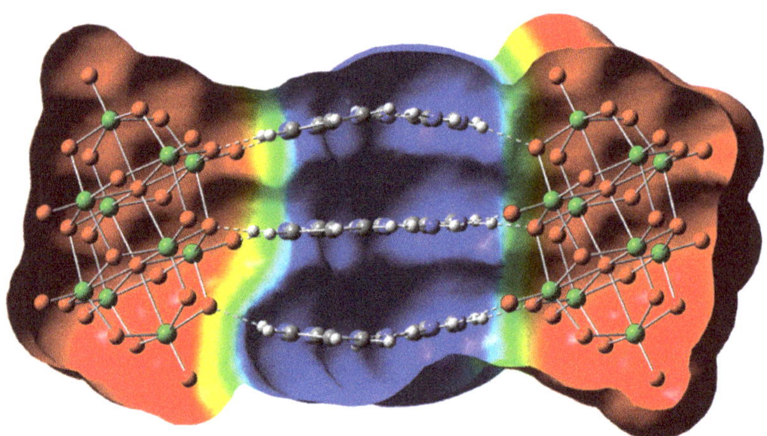

Figure 8. Molecular electrostatic potential map of Compound **1**, calculated at the theory level B3LYP/Def2SVP-LANL2DZ using ECP = LANL2DZ for the V atom. Vanadium atoms are represented in green. The red zones indicate areas with high charge density (nucleophilic zone), while the blue zones indicate the presence of positive charge (electrophilic zones).

The main non-covalent bonds between the decavanadate cluster and pyrimidine rings were characterized by topological electron density parameters, $\rho(r)$, the Laplacian of density, $\nabla^2\rho(r)$, the energy of interaction, $E_{H\cdots Y}$, and the interatomic distance, D_{int}. The results are shown in Table 5. In addition, Figure 9 shows the molecular graphs of Compound **1**.

Table 5. Topological parameters (in a.u.), interaction energies $E_{H\cdots Y}$ (in kcal mol^{-1}), and interatomic distances D_{int} (in Å).

BCP	$\rho(r)$	$\nabla^2\rho(r)$	$E_{H\cdots Y}$	D_{int}
O3···H3AA	0.0377	0.1440	11.26	1.8016
O4···H1AB	0.0166	0.0636	4.51	2.1456
O9···H3BA	0.0322	0.1345	9.88	1.8737
O10···H3CA	0.0788	0.1830	23.02	1.4946
O11···H1CB	0.0185	0.0705	5.11	2.0981
N2A···H1BA	0.0205	0.0684	4.80	2.1356
N2B···H1AA	0.0188	0.0691	4.54	2.1746
N2C···H1CA	0.0203	0.0758	5.05	2.9542

From the results, it can be seen that the positive values of $\nabla^2\rho(r)$ indicates that non-covalent interactions are hydrogen bonds. The interaction energy is calculated using the equation $E_{H\cdots Y} = \frac{1}{2}|V(r)|$. The maximum value was found for the interaction O10···H3CA with a value of 0.0788 a.u. and the highest interaction energy of 23.02 kcal mol^{-1}. In addition, the interaction O3···H3AA is strong with an energy of 11.26 kcal mol^{-1}. The interaction energies between pyrimidine molecules are in the range of 4.54–5.05 kcal mol^{-1} (N2A···H1BA, N2B···H1AA, and N2C···H1CA). Figure 7a,b shows the π–π stacking interactions between pyrimidine molecules. Many rings and cage critical points were also observed, which indicate the formation of a stable ring and cage-like structures that form between pyrimidine molecules and decavanadate, as well as between the central pyrimidine rings.

Lastly, the Hirshfeld surface of Compound **1,** as shown in Figure 10a, was mapped with the normalized contact distance, d_{norm}. It was observed that the red spots on the Hirshfeld surface were due to close intermolecular interactions of $O \cdots H$, between oxygen atoms of $[V_{10}O_{28}]^{6-}$ ion and hydrogen atoms of 2-aminopyrimidine. In Figure 10b, in the fingerprint plot, d_i indicates the distance from the surface to the nearest nucleus inside the surface, and d_e is the distance from the surface to the nearest nucleus outside the surface. The interaction $O \cdots H$ had the most significant contribution to the overall surface with 96.5%, blue-green region of the surface. Other interactions with minor contributions were $O \cdots N$ (1.4%) and $O \cdots O$ (1.3%), which corresponded to the small grey zone of the whole surface.

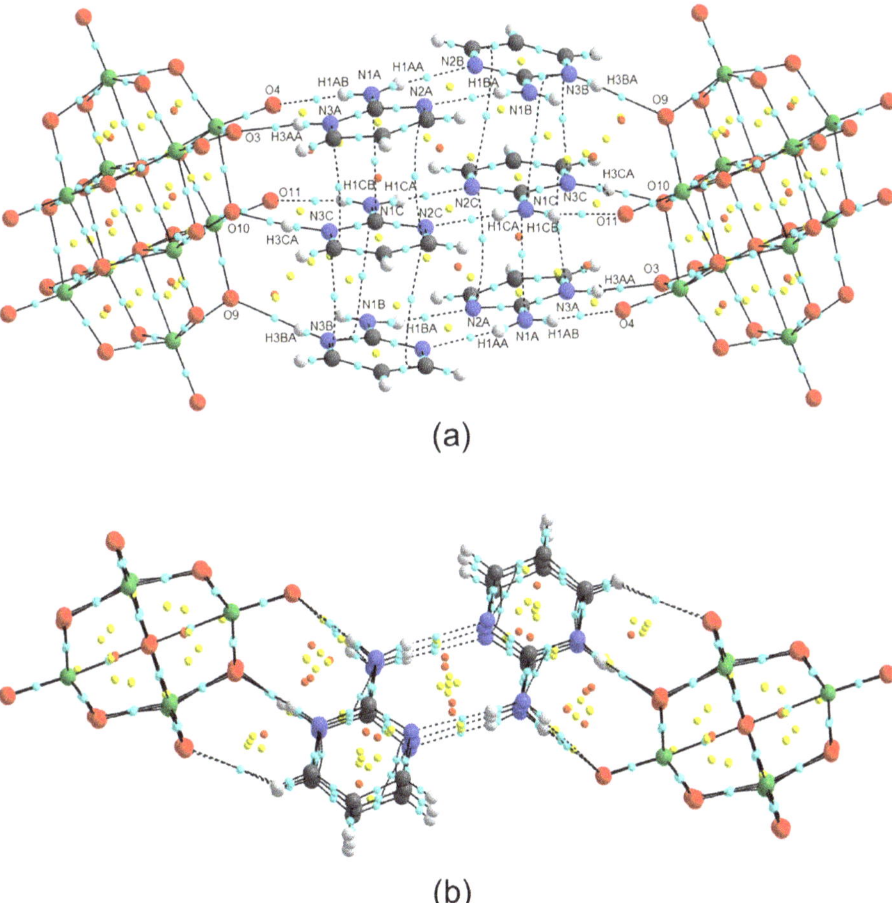

Figure 9. Molecular graphs of Compound 1: (a) showing the main H-bridge bonds between the decavanadate anion and pyrimidine molecules, (b) the interactions and the π-stacking arrangement between the pyrimidine rings as seen from a perpendicular orientation. Vanadium atoms are represented in green. Blue dots represent bond critical points (BCPs), yellow dots represent ring critical points (RCPs), and orange dots represent cage critical points (CCPs).

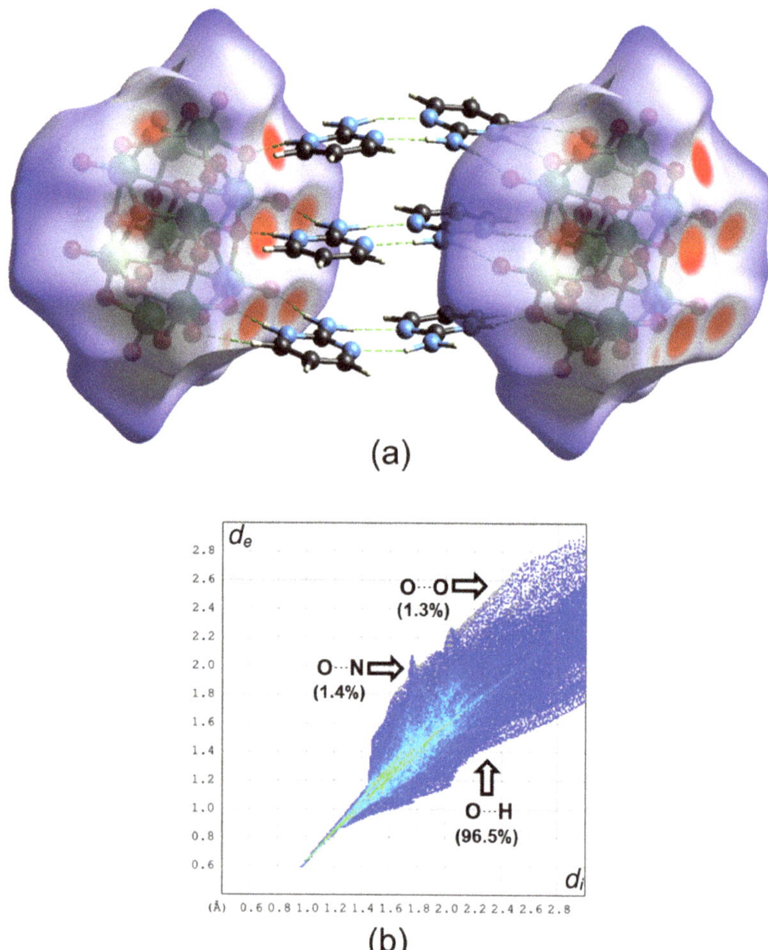

Figure 10. (**a**) Hirshfeld surface mapped with d_{norm} parameter for Compound **1** and (**b**) fingerprint plot of non-covalent interactions. In (**a**) red spots represent close intermolecular interactions and green dash lines represent hydrogen bonds. In (**b**) blue-green region represent the contribution to the total area of the surface.

2.5. Docking Analysis

Molecular docking analysis was performed to test DNA/RNA interactions with Compound **1** using DNA and several microRNAs (miRNA), such as lncRNA, miR-21, and let-7 miRNA (Figure 11). The docked binding energies and interactions with the miRNAs structures were between −8.91 and −8.39 Kcal/mol (Table 6). The interactions involve different positions between decavanadate and miRNAs that comprise hydrogen bonds and hydrophobic interactions mainly with the nitrogen bases Guanine and Adenine followed by Cytosine and Uracil (Figure 11a–c).

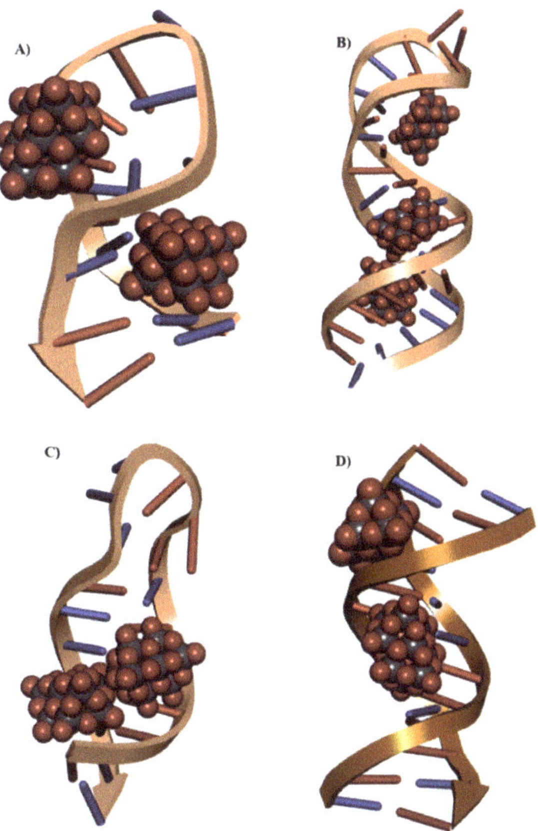

Figure 11. Docked structures of top molecular poses for: (**A**) decavanadate-pre-miRNA, (**B**) decavanadate-let-7 miRNA, (**C**) decavanadate-lncRNA, and (**D**) decavanadate-DNA. Vanadium atoms are represented as grey spheres. Red spheres represent oxygen atoms. Blue bars represent purine bases and red bars pyrimidine bases.

Table 6. Docking results. Binding energies for the decavanadate anion best molecular poses with miRNAs and DNA.

Compound	Target	Binding Energies (Kcal/mol)	H Bonds	Interactions
Decavanadate	2MNC (pre-miR-21)	−8.91	2	C8, G10, C9, C21, A20
		−8.67	3	G13, A14, C21, G13, A14, C17
	2JXV (let-7 miRNA)	−8.66	1	G8, U24, U25, U9, C26, A7I6
		−8.58	5	G11, G12, G19, A20, G15
	6PK9 (lncRNA)	−8.51	3	A3, G4, G5, U27, C29, C30
		−8.48	1	A3, G1, G2, C17, U16
		−8.39	3	G4, G6, C15, G5
	1BNA (DNA)	−10.79	4	G4, A5, G22, A6, C23
		−9.17	5	A6, A17, A18, A5, G16

The results obtained when docking was carried out considering DNA as a macromolecule showed binding energies of −10.79 kcal/mol and the interaction involved hydrogen bonds with Guanine and Adenine (Figure 11d).

3. Discussion

Every year, the global prevalence of cancer rises, as does the resistance to current chemotherapeutic agents like cisplatin; nowadays, one of the main goals of the pharmaceutical industry is the development of more effective drugs for the treatment of cancer [55,56].

This work presents a new structure based on the decavanadate cluster and the organic ligand, 2-aminopyrimidine. In general, pyrimidines are one of the most bioactive classes of compounds, with a wide range of biological activities, including in vitro antiviral, diuretic, antitumor, anti-HIV, and cardiovascular effects. Furthermore, pyrimidine moiety is present in the nucleobases that act as building blocks of nucleic acid, DNA or RNA, thymine, cytosine, and uracil, which might be one possible reason for the vast medicinal applications of pyrimidine-derived compounds [57]. Within the pyrimidines, the aminopyrimidine scaffold is present in the structure of a wide variety of natural and synthetic products, such as Thiamine (vitamin B1), Meridianins (a class of marine alkaloids), or Imatinib, a drug used against leukemia, which displays biological activities such as neuroprotection, antibiotic, antidiabetic, anti-Alzheimer, and anticancer. Among the large family of aminopyrimidines, the 2-aminopyrimidine isomer is the most studied, mainly because of its versatility as a starting material for synthesizing many other bioactive compounds [55].

Vanadium is an essential element, and it is important to keep the average vanadium concentration in the human body at around 0.3 μM by supplementation via food and drinking water. In the organism, vanadium is found primarily in the form of vanadate $H_2VO_4^-$ that, due to the structural and chemical similarity with phosphate, is most likely involved in the regulation of phosphate-dependent processes, such as metabolic pathways involving phosphatases and kinases, as well as phosphate metabolism in general. In addition, organic ligands could aid in the modulation of vanadium's bioavailability, transport, and targeting mechanism, so, nowadays, coordination compounds containing vanadium are gaining in popularity because of their potential in the treatment of diabetes and cancer, leishmaniasis, and HIV [58]. In light of this, among vanadate compounds, some researchers have pointed out decavanadates as alternative antitumor agents with promising findings in tumor growth inhibition. Although the anticancer activity of decavanadate is more recent and not yet fully understood, it is probably due to the inhibition of different enzymes such as alkaline phosphatases, ectonucleotidases, and P-type ATPases [56]. One of the first articles published about decavanadate compounds with potential antitumor activity was $Na_4Co(H_2O)_6V_{10}O_{28} \cdot 18H_2O$ reported by Zhai et al. [15]. This compound in vitro displays higher inhibitory activity to human liver cancer (SMMC-7721) and ovary cancer (SK-OV-3) cell lines than 5-fluorouracil, the antitumoral drug clinically used, while in vivo, it can decrease liver tumor mass in rats. Shortly after, in 2010, Li et al. synthesized two decavanadates compounds with organic ligands that can inhibit proliferation of human lung (A549) and murine leukemia (P388) tumor cell lines in vitro: $(H_2tmen)_3V_{10}O_{28} \cdot 6H_2O$ and $(H_2en)_3V_{10}O_{28} \cdot 2H_2O$ [27]. It is important to highlight that the first compound contains four methyl ($-CH_3$) substituents in the cation moiety. Their presence may enhance the lipophilic effect of this compound, which increases its penetration through the lipid bilayer of the cell membrane, thus showing higher inhibitory activity than the second compound. After that, a few more articles based on decavanadates compounds with potential antitumor activity have been published up to the present [29,53,59–61].

Since hydrogen bonds naturally occur between nucleobases and are of great biological importance for DNA and RNA structures, we chose the 2-aminopyrimidine molecule to obtain a new decavanadate compound with potential anticancer activity. Similar to this organic molecule, some structures with cytosine and decavanadate have been published. In the first compound published by Bošnjaković-Pavlović et al. [36], cytosine forms dimeric cytosine-cytosinium cations that stabilize the charge of the decavanadate anion. A few years later, the second material of decavanadate with cytosine in the literature was published [37]. In this compound, all cytosine (C) molecules are protonated, and the supramolecular structure is dominated by C–C and C–Decavanadate hydrogen bonds, as well as π-stacking interaction among heterocyclic rings with a centroid-to-centroid distance similar to that found in DNA and RNA structures. The fact that in both structures, cytosines interact with one another by hydrogen bonds and π–π stacking similarly as in DNA and RNA polymers lead to two hypotheses: (i) polyoxidometalates could be employed as templates or catalysts for base-base linkages, and (ii) base-base pairing was crucial in the early stages of life [37]. Why nucleobases tend to form this arrangement in DNA and RNA has been studied by Francés-Monerris et al. [62]. The intra- and inter-strand interactions in Watson–Crick base pairs are of great importance for the thermal stability of the double-helix structure. In fact, in their experiments, upon UV light exposure at a B-DNA arrangement, C–C dimers twist towards a face-to-face arrangement to increase the π-stacking interaction and further promote the photostability of the genomic material. On the other hand, structures containing purine-based ligands show the same fact, as in the $(NH_4)_2(C_8H_{10}N_4O_2)_4[H_4V_{10}O_{28}]\cdot2H_2O$ compound, π–π stacking interactions with an interplanar distance of 3.38 Å exist in the purine rings of caffeine [63], and in $[AdH]_6[V_{10}O_{38}]\cdot4H_2O$, Adeninium cations form simultaneous hydrogen bond interactions in a ribbon-like geometry as well as π–π interaction between cationic ribbons [6].

All these facts support the structure of Compound **1** in this paper, since 2-aminopyrimidinium cations are disposed of in the form that they interact with each other by hydrogen bonds and π–π stacking interactions with a centroid-to centroid distance of around 3.6 Å.

Furthermore, to study the potential anticarcinogenic activity of Compound **1**, docking tests were performed to analyze the interaction of the decavanadate anion with some types of microRNAs (miRNAs) and DNA molecules. MiRNAs are small, non-coding RNA molecules with a length of 19–25 nucleotides that regulate various target genes. MiRNAs play a role in the cell cycle, differentiation, proliferation, energy metabolism, and immunological response, among other biological activities, regulating around 30% of human genes, with half of these genes being tumor-related. Recent studies have found that miRNAs play an important role in cancer progression, including tumor growth, differentiation, adhesion, apoptosis, invasion, and metastasis [64,65]. Within the miRNA family, the upregulations of miR-21, as well as lncRNA (long non-coding RNA), are linked to several types of cancer such as malignant B-cell lymphoma or breast cancer, respectively [66,67]. On the other hand, let-7 miRNA (member of the family of let-7 RNA) is known as the keeper of differentiation and has also emerged as a promising therapeutic agent to treat cancer and immune responses [68,69]. Thus, it is interesting to carry out experiments to prove this hypothesis, since our preliminary analyses show great affinity of decavanadate for miRNA fragments.

In addition to the possible antitumoral activity that Compound **1** could exhibit, the interesting structure that it shows, along with all the above examples about structures with decavanadate clusters and organic ligands that arrange remembering the DNA or RNA structures, highlights the idea that these polyoxidovanadates could act as templates or catalysts for base-base pairing. Bernal, in 1949, was the first who proposed the important role of clay minerals in the origin of life [70]. The advantage of these clays could include ordered arrangement, substantial adsorption capacity, UV protection, ability to concentrate organic compounds, and potential to serve as polymerization templates so that clay minerals could have played a key role in chemical evolution and the origin of life [71]. Intercalation of decavanadate into laminar minerals has already been achieved [72], so it is possible to concentrate organic compounds with a high capacity for hydrogen bond formation and catalyze their polymerization. Therefore, it will be worthwhile to explore this idea.

Additionally, it is important to mention that recent work from Aureliano's group about melanoma anticancer activity of a variety of vanadium compounds [73] and potential anti-SARS-CoV-2 activity of vanadium compounds by Scior et al. [74], present a promising future for vanadium compounds as metallodrugs.

4. Materials and Methods

All starting reagents were purchased from Sigma-Aldrich (Merck, Naucalpan de Juarez, Mexico) and were used as received, without additional purification.

4.1. Synthesis of [2-ampymH]$_6$[V$_{10}$O$_{28}$]·5H$_2$O (**1**)

A total of 0.4 mmol (0.047 gr) of ammonium metavanadate (NH$_4$VO$_3$), in situ precursor of decavanadate anion, was dissolved in 10 mL of distilled water at 80 °C and stirred. After cooling to room temperature, the pH was lowered with a dissolution of concentrated HCl (37%) to reach pH = 3. In a separate glass vessel, 0.2 mmol (0.019 gr) of 2-aminopyrimidine ligand was dissolved in 4 mL of distilled water. This solution was gradually added to the orange solution of decavanadate, resulting in a pH close to 4.3. The reaction mixture was then allowed to evaporate slowly at room temperature. After a week, orange crystals of Compound **1** for single-crystal X-ray diffraction were obtained, filtered, and dried in air. Anal. Calcd. for C$_{12}$H$_{23}$N$_9$O$_{16.5}$V$_5$: C, 17.75; H, 2.85; N, 15.52%. Found: C, 16.83; H, 2.98; N, 14.91%.

4.2. Characterization Methods

4.2.1. Physicochemical Characterization

Elemental analyses (C, H, and N) were carried out on a THERMO SCIENTIFIC analyzer model Flash 2000 (Thermo Fisher Scientific S.p.A., Milan, Italy), while thermal analyses were examined with a Shimadzu TGA-50H thermogravimetric analyzer (Shimadzu Corporation, Kyoto, Japan), with a heating rate of 10 °C/min under air atmosphere in the range of 35–950 °C, both of them at the "Centro de Instrumentación Científica" (University of Granada, Granada, Spain). The IR spectra of powdered samples were recorded in the 4000–400 cm^{-1} region with a BRUKER TENSOR 27 FT-IR, using the OPUS program as a data collector (Bruker Corporation, Billerica, MA, USA).

4.2.2. Single-Crystal X-ray Diffraction

X-ray quality orange crystals of Compound **1** were obtained.

The crystals were prepared under inert conditions immersed in perfluoropolyether as protecting oil for manipulation. Then, a suitable crystal was mounted on a MiTeGen Micromount^{TM} (MiTeGen, Ithaca, NY, USA) for data collection at 300(2) K. For Compound **1**, diffraction intensities were collected with a Bruker D8 Venture diffractometer (Bruker Corporation, Billerica, MA, USA), using a photon detector equipped with graphite monochromated *MoKα* radiation (λ = 0.71073 Å). Data reduction was performed with the APEX3 software [75] and corrected for absorption using SADABS [76]. The structure was solved by direct methods using the SHELXS-2013 program [77] and refined by full-matrix least-squares on F^2 with SHELXL-2018 [78]. The positional and anisotropic atomic displacement parameters were refined for all non-hydrogen atoms. Hydrogen atoms were located in difference Fourier maps and included as fixed contributions riding their parent atoms with isotropic thermal factors chosen as 1.2 times or 1.5 times those of their carrier atoms. One water molecule (O2W) is disordered into two dispositions, so it was refined with occupancy of 50% and anisotropic displacement parameters. OLEX2 software [79] was used as a graphical interface. The Addsym routine implemented in the program PLATON [80] supported a $P\bar{1}$ symmetry. Crystallographic data (excluding structure factors) for the structure reported in this paper have been deposited with the Cambridge Crystallographic Data Centre as supplementary publication no. 2099300 for Compound **1**. Copies of the data can be obtained free of charge on application to the Director, CCDC, 12 Union Road, Cambridge, CB2 1EZ, U.K. (Fax: +44-1223-335033; e-mail: deposit@ccdc.cam.ac.uk or http://www.ccdc.cam.ac.uk, accessed on 15 August 2021).

4.2.3. Theoretical Methodology

The structural and electronic structure of Compound **1** were calculated using the density functional theory, DFT [81]. The hybrid functional B3LYP [82] was used, using the Def2SVP basis set [83] for the atoms of C, H, O, and N; and LANL2DZ [84] for vanadium atom with an effective core potential (ECP). The frontier molecular orbitals (highest occupied molecular orbital, HOMO, and lowest unoccupied molecular orbital, LUMO) and the molecular electrostatic potential (MEP) map were analyzed. The calculations were performed with the Gaussian program16 [85], and the visualization of the results was performed with the Gaussian View 6.0.16 program [86]. Additionally, the main non-covalent interactions in Compound **1** were characterized using the atoms in molecules (AIM) approach with AIMAll software [87] and Hirshfeld surface analysis with CrystalExplorer 17.5 software [88].

4.2.4. Docking Analysis

Molecular docking analysis was performed with the semi-flexible methodology, where the RNAs fragments were considered as a rigid entity, while flexibility was allowed for the decavanadate. The preparation of the macromolecule and the ligand was performed through the Autodock Tools 1.5.6 software [89], which includes polar hydrogens and empirical particles of atomic charges (Gasteiger–Marsili method). Different grid box sizes were used for each RNA molecule that encloses the entire fragment: for 6PK9, 60, 106, and 60 Å were used; 70, 126, and 70 Å for 2JXV fragment; and 94, 76, and 66 Å were used for 2MNC. In addition to those fragments, one DNA structure was considered in the docking study: 1BNA, with sizes of 70, 70, and 120 Å. The grid spacing for all the docking calculations was set to the default 0.375 Å value, using the Lamarckian genetic algorithm (LGA) searching methods. The parameters for the vanadium atom were the sum of VDW radii of two similar atoms (3.14 Å), plus the VDW well depth (0.016 kcal/mol), plus the atomic solvation volume (12.0 Å^3), plus the atomic solvation parameter (-0.00110). The H-bond radius of the heteroatom in contact with hydrogen (0.0 Å), the well depth of the H-bond (0.0 kcal/mol) and different integers indicate the type of H-bonding atom and indexes for the generation of the autogrid map (0, -1, -1, 1, respectively).

5. Conclusions

A new decavanadate compound containing 2-aminopyrimidine was synthesized and characterized experimentally through elemental analysis, infrared spectroscopy, thermogravimetric analyses, and single-crystal X-ray diffraction. Using theoretical studies based on DFT calculations and AIM, the non-covalent interactions present in the compound were also studied. Frontier molecular orbital, molecular electrostatic potential, and non-covalent interactions were completely characterized. The hydrogen bond patterns of this molecule, although similar to the already reported Adenine-decavanadate compound, has the interesting feature of resembling the ladder-like structure of DNA and RNA molecules, even with similar distances from the centroid-to-centroid and π–π interactions, it also occurs in triads, which could remind codon an anticodon interactions. A docking study was set to explore more of the resemblance with DNA and RNA to see the possibility of decavanadate interacting with RNA molecules. Thus, a set of test miRNA was considered for docking. The latest development in understanding the roles of non-coding RNAs and microRNAs is worthwhile to try some fitting. Although the negative charge of these molecules could prevent interactions with the anionic decavanadate, the formation of several hydrogen bonds with the nucleobases appears to counteract the charge problem, and relatively good interaction energies were observed. At the moment, this only suggests that experiments should be designed to test whether these interactions are worthwhile to pursue. The catalytic properties to build blocks with the correct orientation and similar types of interactions could also be explored since it could shed some light on the nature of the first replicator. On the other hand, since cancer cells exhibit increased levels of mRNA translation to meet tumor growth requirements [90], it will be interesting to explore the interactions of polyoxidovanadates with mRNA. Therefore, new and fascinating decavanadate chemistry is around the corner.

Supplementary Materials: The following are available online at https://www.mdpi.com/article/10.3390/inorganics9090067/s1, Table S1: Fractional Atomic Coordinates ($\times 10^4$) and Equivalent Isotropic Displacement Parameters ($\text{Å}^2 \times 10^3$) for Compound **1**. U_{eq} is defined as 1/3 of the trace of the orthogonalized U_{IJ} tensor, Table S2: Anisotropic Displacement Parameters ($\text{Å}^2 \times 10^3$) for Compound **1**. The Anisotropic displacement factor exponent takes the form: $-2\pi^2[h^2a^{*2}U_{11} + 2hka^*b^*U_{12} + \ldots]$, Table S3: Bond Lengths for Compound **1**, Table S4: Bond Angles for Compound **1**, Table S5: Torsion Angles for Compound **1**, Table S6: Hydrogen Atom Coordinates ($\text{Å} \times 10^4$) and Isotropic Displacement Parameters ($\text{Å}^2 \times 10^3$) for Compound **1**, Table S7: Atomic Occupancy for Compound **1**, Figure S1: TG spectrum of Compound **1** in the range of 35–950 °C.

Author Contributions: A.G.-G. carried out experimental work (synthesis, crystallization, and experimental characterization). F.J.M.-B., M.E.C. and L.N. carried out the theoretical characterization. A.G.-G., D.C.-L. and A.R.-D. carried out the X-ray diffraction determination. A.G.-G., E.G.-V., M.E.C. and B.L.S.-G., wrote and revised the manuscript. E.G.-V. and A.R.-D. conceived and designed this study. All authors contributed extensively to the work presented in this paper. All authors have read and agreed to the published version of the manuscript.

Funding: Projects that funded this research: 100517029-VIEP, 100233622-VIEP, SEP PRODEP BUAP-PTC_617, and PRODEP Academic Group BUAP-CA-263 (SEP, Mexico). Financial support was also provided by Junta de Andalucía (Spain), project number FQM-394.

Acknowledgments: M.E.C. and F.J.M.-B. wish to thank Laboratorio Nacional de Supercómputo del Sureste de México (LNS-BUAP) and the CONACyT network of national laboratories for the computer resources and provided support. B.L.S.-G. thanks for the grant SEP PRODEP BUAP-PTC_617. We thank the support provided by VIEP-BUAP through Yadira Rosas Bravo for observations and comments to improve this manuscript.

Conflicts of Interest: There are no conflict of interest to declare.

References

1. Proust, A.; Thouvenot, R.; Gouzerh, P. Functionalization of polyoxidometalates: Towards advanced applications in catalysis and materials science. *Chem. Commun.* **2008**, *16*, 1837–1852. [CrossRef]
2. Bijelic, A.; Aureliano, M.; Rompel, A. Polyoxidometalates as Potential Next-Generation Metallodrugs in the Combat against Cancer. *Angew. Chem. Int. Ed.* **2019**, *58*, 2980–2999. [CrossRef]
3. Bijelic, A.; Aureliano, M.; Rompel, A. The antibacterial activity of polyoxidometalates: Structures, antibiotic effects and future perspectives. *Chem. Commun.* **2018**, *54*, 1153–1169. [CrossRef]
4. Evans, H.T. The Molecular Structure of the Isopoly Complex Ion, Decavanadate ($V_{10}O_{28}{}^{6-}$). *Inorg. Chem.* **1966**, *5*, 967–977. [CrossRef]
5. Gumerova, N.I.; Rompel, A. Polyoxometalates in solution: Speciation under spotlight. *Chem. Soc. Rev.* **2020**, *49*, 7568–7601. [CrossRef]
6. Sedghiniya, S.; Soleimannejad, J.; Jahani, Z.; Davoodi, J.; Janczak, J. Crystal engineering of an adenine-decavanadate molecular device towards label-free chemical sensing and biological screening. *Acta Cryst.* **2020**, *B76*, 85–92. [CrossRef] [PubMed]
7. Hartung, S.; Bucher, N.; Chen, H.-Y.; Al-Oweini, R.; Sreejith, S.; Borah, P.; Yanli, Z.; Kortz, U.; Stimming, U.; Hoster, H.E.; et al. Vanadium-based polyoxidometalate as new material for sodium-ion battery anodes. *J. Power Sources* **2015**, *288*, 270–277. [CrossRef]
8. Xie, A.; Ma, C.-A.; Wang, L.; Chu, Y. $Li_6V_{10}O_{28}$, a novel cathode material for Li-ion battery. *Electrochim. Acta* **2007**, *52*, 2945–2949. [CrossRef]
9. Ji, Y.; Liu-Théato, X.; Xiu, Y.; Indris, S.; Njel, C.; Maibach, J.; Ehrenberg, H.; Fichtner, M.; Zhao-Karger, Z. Polyoxidometalate Modified Separator for Performance Enhancement of Magnesium-Sulfur Batteries. *Adv. Funct. Mater.* **2021**, *2100868*, 1–7.
10. Steens, N.; Ramadan, A.M.; Absillis, G.; Parac-Vogt, T.N. Hydrolytic cleavage of DNA-model substrates promoted by polyoxidovanadates. *Dalton Trans.* **2010**, *39*, 585–592. [CrossRef]
11. Wang, C.; Chen, Z.; Yao, X.; Chao, Y.; Xun, S.; Xiong, J.; Fan, L.; Zhu, W.; Li, H. Decavanadates anchored into micropores of graphene-like boron nitride: Efficient heterogeneous catalysts for aerobic oxidative desulfurization. *Fuel* **2018**, *230*, 104–112. [CrossRef]
12. Huang, X.; Gu, X.; Zhang, H.; Shen, G.; Gong, S.; Yang, B.; Wang, Y.; Chen, Y. Decavanadate-based clusters as bifunctional catalysts for efficient treatment of carbon dioxide and simulant sulfur mustard. *J. CO2 Util.* **2021**, *45*, 101419. [CrossRef]
13. Diaz, A.; Muñoz-Arenas, G.; Venegas, B.; Vázquez-Roque, R.; Flores, G.; Guevara, J.; Gonzalez-Vergara, E.; Treviño, S. Metforminium decavanadate (MetfDeca) Treatment Ameliorates Hippocampal Neurodegeneration and Recognition Memory in a Metabolic Syndrome Model. *Neurochem. Res.* **2021**, *46*, 1151–1165. [CrossRef]

14. Pereira, M.J.; Carvalho, E.; Eriksson, J.W.; Crans, D.C.; Aureliano, M. Effects of decavanadate and insulin enhancing vanadium compounds on glucose uptake in isolated rat adipocytes. *J. Inorg. Biochem.* **2009**, *103*, 1687–1692. [CrossRef] [PubMed]

15. Zhai, F.; Wang, X.; Li, D.; Zhang, H.; Li, R.; Song, L. Synthesis and biological evaluation of decavanadate $Na_4Co(H_2O)_6V_{10}O_{28}\cdot18H_2O$. *Biomed. Pharmacother.* **2009**, *63*, 51–55. [CrossRef] [PubMed]

16. Aureliano, M.; Gumerova, N.I.; Sciortino, G.; Garribba, E.; Rompel, A.; Crans, D.C. Polyoxovanadates with emerging biomedical activities. *Coord. Chem. Rev.* **2021**, *447*, 214143. [CrossRef]

17. Cantley, L.C.; Josephson, L.; Warner, R.; Yanagisawa, M.; Lechene, C.; Guidotti, G. Vanadate Is a Potent (Na, K)-ATPase Inhibitor Found in ATP Derived from Muscle. *J. Biol. Chem.* **1977**, *252*, 7421–7423. [CrossRef]

18. Aureliano, M.; Crans, D.C. Decavanadate ($V_{10}O_{28}^{6-}$) and oxovanadates: Oxometalates with many biological activities. *J. Inorg. Biochem.* **2009**, *103*, 536–546. [CrossRef]

19. Aureliano, M. Decavanadate: A journey in a search of a role. *Dalton Trans.* **2009**, *42*, 9093–9100. [CrossRef]

20. García-Vicente, S.; Yraola, F.; Marti, L.; González-Muñoz, E.; García-Barrado, M.J.; Cantó, C.; Abella, A.; Bour, S.; Artuch, R.; Sierra, C.; et al. Oral Insulin-Mimetic Compounds That Act Independently of Insulin. *Diabetes* **2007**, *56*, 486–493. [CrossRef]

21. Treviño, S.; Sánchez-Lara, E.; Sarmiento-Ortega, V.E.; Sánchez-Lombardo, I.; Flores-Hernández, J.Á.; Pérez-Benítez, A.; Brambila-Colombres, E.; González-Vergara, E. Hypoglycemic, lipid-lowering and metabolic regulation activities of metforminium decavanadate $(H_2Metf)_3[V_{10}O_{28}]\cdot8H_2O$ using hypercaloric-induced carbohydrate and lipid deregulation in Wistar rats as biological model. *J. Inorg. Biochem.* **2015**, *147*, 85–92. [CrossRef]

22. Treviño, S.; Díaz, A.; Sánchez-Lara, E.; Sarmiento-Ortega, V.E.; Flores-Hernández, J.Á.; Brambila, E.; Meléndez, F.J.; González-Vergara, E. Pharmacological and Toxicological Threshold of Bisammonium Tetrakis 4-(*N,N*-Dimethylamino)pyridinium Decavanadate in a Rat Model of Metabolic Syndrome and Insulin Resistance. *Bioinorg. Chem. Appl.* **2018**, *2151079*, 1–13. [CrossRef]

23. Turner, T.L.; Nguyen, V.H.; McLauchlan, C.C.; Dymon, Z.; Dorsey, B.M.; Hooker, J.D.; Jones, M.A. Inhibitory effects of decavanadate on several enzymes and *Leishmania tarentolae* In Vitro. *J. Inorg. Biochem.* **2012**, *108*, 96–104. [CrossRef] [PubMed]

24. Missina, J.M.; Gavinho, B.; Postal, K.; Santana, F.S.; Valdameri, G.; de Souza, E.M.; Hughes, D.L.; Ramirez, M.I.; Soares, J.F.; Nunes, G.G. Effects of Decavanadate Salts with Organic and Inorganic Cations on *Escherichia coli, Giardia intestinalis*, and Vero Cells. *Inorg. Chem.* **2018**, *57*, 11930–11941. [CrossRef]

25. Samart, N.; Arhouma, Z.; Kumar, S.; Murakami, H.A.; Crick, D.C.; Crans, D.C. Decavanadate Inhibits Mycobacterial Growth More Potently Than Other Oxovanadates. *Front. Chem.* **2018**, *6*, 519. [CrossRef]

26. Marques-da-Silva, D.; Fraqueza, G.; Lagoa, R.; Vannathan, A.A.; Mal, S.S.; Aureliano, M. Polyoxidovanadate inhibition of *Escherichia coli* growth shows a reverse correlation with Ca^{2+}-ATPase inhibition. *New. J. Chem.* **2019**, *43*, 17577–17587. [CrossRef]

27. Li, Y.-T.; Zhu, C.-Y.; Wu, Z.-Y.; Jiang, M.; Yan, C.-W. Synthesis, crystal structures and anticancer activities of two decavanadate compounds. *Transit. Met. Chem.* **2010**, *35*, 597–603. [CrossRef]

28. Silva-Nolasco, A.M.; Camacho, L.; Saavedra-Díaz, R.O.; Hernández-Abreu, O.; León, I.E.; Sánchez-Lombardo, I. Kinetic Studies of Sodium and Metforminium Decavanadates Decomposition and In Vitro Cytotoxicity and Insulin-Like Activity. *Inorganics* **2020**, *8*, 67. [CrossRef]

29. Louati, M.; Ksiksi, R.; Elbini-Dhouib, I.; Mlayah-Bellalouna, S.; Doghri, R.; Srairi-Abid, N.; Zid, M.-F. Synthesis, structure, and characterization of a novel decavanadate, $Mg(H_2O)_6(C_4N_2H_7)_4V_{10}O_{28}\cdot4H_2O$, with a potential antitumor activity. *J. Mol. Struct.* **2021**, *1242*, 130711. [CrossRef]

30. Sánchez-Lara, E.; Martínez-Valencia, B.; Corona-Motolinia, N.D.; Sanchez-Gaytan, B.L.; Castro, M.E.; Bernès, S.; Méndez-Rojas, M.A.; Meléndez-Bustamante, F.J.; González-Vergara, E. A one-dimensional supramolecular chain based on $[H_2V_{10}O_{28}]^{4-}$ units decorated with 4-dimethylaminopyridinium ions: An experimental and theoretical characterization. *New J. Chem.* **2019**, *43*, 17746–17755. [CrossRef]

31. Bosnjakovic-Pavlovic, N.; Spasojevic-De-Biré, A. Cytosine-Cytosinium Dimer Behavior in a Cocrystal with a Decavanadate Anion as a Function of the Temperature. *J. Phys. Chem. A.* **2010**, *114*, 10664–10675. [CrossRef]

32. Sánchez-Lara, E.; Treviño, S.; Sánchez-Gaytán, B.L.; Sánchez-Mora, E.; Castro, M.E.; Meléndez-Bustamante, F.J.; Méndez-Rojas, M.A.; González-Vergara, E. Decavanadate Salts of Cytosine and Metformin: A Combined Experimental-Theoretical Study of Potential Metallodrugs Against Diabetes and Cancer. *Front. Chem.* **2018**, *6*, 1–18. [CrossRef] [PubMed]

33. Cui, C.; Liu, S.; Zhao, W. The crystal structure of hexakis(2-(pyridin-2-ylamino)pyridin-1-ium) decavanadate(V) dihydrate, $C_{60}H_{64}N_{18}O_{30}V_{10}$. *Z. Kristallogr. NCS* **2021**, *236*, 25–27. [CrossRef]

34. Zarroug, R.; Abdallah, A.H.; Guionneau, P.; Masip-Sánchez, A.; López, X.; Ayed, B. Decavanadate salts of piperidine and triethanolamine: A combined experimental and theoretical study. *J. Mol. Struct.* **2021**, *1241*, 130677. [CrossRef]

35. Sgambellone, M.A.; David, A.; Garner, R.N.; Dunbar, K.R.; Turro, C. Cellular Toxicity Induced by the Photorelease of a Caged Bioactive Molecule: Design of a Potential Dual-Action Ru(II) Complex. *J. Am. Chem. Soc.* **2013**, *135*, 11274–11282. [CrossRef]

36. Amr, A.-G.E.; Mohamed, A.M.; Mohamed, S.F.; Abdel-Hafez, N.A.; Hammam, A.E.-F.G. Anticancer activities of some newly synthesized pyridine, pyrane, and pyrimidine derivatives. *Bioorg. Med. Chem.* **2006**, *14*, 5481–5488. [CrossRef] [PubMed]

37. Eicher, T.; Hauptmann, S.; Speicher, A. *The Chemistry of Heterocycles: Structure, Reactions, Syntheses, and Applications*, 2nd ed.; Wiley-VCH I. Weinheim, Germany, 2003; pp. 269–310, 398–408

38. Sciortino, G.; Aureliano, M.; Garribba, E. Rationalizing the Decavanadate(V) and Oxidovanadium(IV) Binding to G-Actin and the Competition with Decaniobate(V) and ATP. *Inorg. Chem.* **2021**, *60*, 334–344. [CrossRef] [PubMed]

39. Jin, K.P.; Jiang, H.J.; Wang, Y.; Zhang, D.P.; Mei, J.; Cui, S.H. Synthesis and Crystal Structure of Decavanadate-Based Coordination Polymers. *J. Cluster Sci.* **2018**, *29*, 785–792. [CrossRef]
40. Scheinbeim, J.; Schempp, E. 2-Aminopyrimidine. *Acta Cryst.* **1976**, *B32*, 607–609. [CrossRef]
41. Correia, I.; Avecilla, F.; Marcao, S.; Pessoa, J.C. Structural studies of decavanadate compounds with organic molecules and inorganic ions in their crystal packing. *Inorg. Chim. Acta* **2004**, *357*, 4476–4487. [CrossRef]
42. Aissa, T.; Ksiksi, R.; Elbini-Dhouib, I.; Doghri, R.; Srairi-Abid, N.; Zid, M.F. Synthesis of a new vanadium complex (V), hexa[4-methylimidazolium]decavanadate trihydrate $(C_4H_7N_2)_6V_{10}O_{28}\cdot3H_2O$: Physico-chemical and biological characterizations. *J. Mol. Struct.* **2021**, *1236*, 130331. [CrossRef]
43. Mahmoud, G.A.-E.; Ibrahim, A.B.M.; Mayer, P. $(NH_4)_2[Ni(H_2O)_6]_2V_{10}O_{28}\cdot4H_2O$; Structural Analysis and Bactericidal Activity against Pathogenic Gram-Negative Bacteria. *ChemistrySelect* **2021**, *6*, 3782–3787. [CrossRef]
44. Hou, W.; Guo, J.; Wang, Z.; Xu, Y. Synthesis, structural characterization, and properties of two new polyoxidovanadates based on decavanadate $[V_{10}O_{28}]^{6-}$. *J. Coord. Chem.* **2013**, *66*, 2434–2443. [CrossRef]
45. Etter, M.C.; MacDonald, J.C. Graph-Set Analysis of Hydrogen-Bond Patterns in Organic Crystals. *Acta Cryst.* **1990**, *B46*, 256–262. [CrossRef]
46. Janiak, C. A critical account on π–π stacking in metal complexes with aromatic nitrogen-containing ligands. *J. Chem. Soc. Dalton Trans.* **2000**, *21*, 3885–3896. [CrossRef]
47. Chen, J.; Dai, L.; Li, J.; Mohammadnia, M. Pd based on 2-Aminopyrimidine and 1*H*-benzo[*d*]imidazol-2-amine functionalized Fe_3O_4 nanoparticles as novel recyclable magnetic nanocatalysts for Ullmann coupling reaction. *Appl. Organomet. Chem.* **2020**, *34*, e5708. [CrossRef]
48. Gupta, P.K.; Arora, K. Studies on Simulation of Spectra of some Organic Compounds. *Orient. J. Chem.* **2019**, *35*, 1655–1668. [CrossRef]
49. Thangarasu, S.; Athimoolam, S.; Bahadur, S.A.; Manikandan, A. Structural, Spectroscopic Investigation and Computational Study on Nitrate and Hydrogen Oxalate Salts of 2-Aminopyrimidine. *J. Nanosci. Nanotechnol.* **2018**, *18*, 2450–2462. [CrossRef]
50. Ortaboy, S.; Acar, E.T.; Atun, G. The removal of radioactive strontium ions from aqueous solutions by isotopic exchange using strontium decavanadates and corresponding mixed oxides. *Chem. Eng. J.* **2018**, *344*, 194–205. [CrossRef]
51. Omri, I.; Mhiri, T.; Graia, M. Novel decavanadate cluster complex $(HImz)_{12}(V_{10}O_{28})_2\cdot3H_2O$: Synthesis, characterization, crystal structure, optical and thermal properties. *J. Mol. Struct.* **2015**, *1098*, 324–331. [CrossRef]
52. Luo, S.-Y.; Wu, X.-L.; Hu, Q.-P.; Wang, J.-X.; Liu, C.-Z. Structural characterization of a new decavanadate compound with organic molecules and inorganic ions. *J. Struct. Chem.* **2012**, *53*, 915–920. [CrossRef]
53. Kioseoglou, E.; Gabriel, C.; Petanidis, S.; Psycharis, V.; Raptopoulou, C.P.; Terzis, A.; Salifoglou, A. Binary Decavanadate-Betaine Composite Materials of Potential Anticarcinogenic Activity. *Z. Anorg. Allg. Chem.* **2013**, *639*, 1407–1416. [CrossRef]
54. Riou, D.; Roubeau, O.; Férey, G. Evidence for the Solid State Structural Transformation of the Network-Type Decavanadate $(NC_7H_{14})_4[H_2V_{10}O_{28}]$ into a Lamellar Topology $(NC_7H_{14})[V_4O_{10}]$. *Z. Anorg. Allg. Chem.* **1998**, *624*, 1021–1025. [CrossRef]
55. Filho, E.V.; Pinheiro, E.M.C.; Pinheiro, S.; Greco, S.J. Aminopyrimidines: Recent synthetic procedures and anticancer activities. *Tetrahedron* **2021**, *92*, 132256. [CrossRef]
56. Aureliano, M. The Role of Decavanadate in Anti-Tumour Activity. *Glob. J. Cancer Ther.* **2017**, *3*, 12–14. [CrossRef]
57. Sharma, V.; Chitranshi, N.; Agarwal, A.K. Significance and Biological Importance of Pyrimidine in the Microbial World. *Int. J. Med. Chem.* **2014**, *202784*, 1–31. [CrossRef] [PubMed]
58. Rehder, D. The potentiality of vanadium in medicinal applications. *Future Med. Chem.* **2012**, *4*, 1823–1837. [CrossRef]
59. Cheng, M.; Li, N.; Wang, N.; Hu, K.; Xiao, Z.; Wu, P.; Wei, Y. Synthesis, structure and antitumor studies of a novel decavanadate complex with a wavelike two-dimensional network. *Polyhedron* **2018**, *155*, 313–319. [CrossRef]
60. Gu, Y.; Li, Q.; Huang, Y.; Zhu, Y.; Wei, Y.; Ruhlmann, L. Polyoxidovanadate-iodobodipy supramolecular assemblies: New agents for high efficiency cancer photochemotherapy. *Chem. Commun.* **2020**, *56*, 2869–2872. [CrossRef]
61. Ksiksi, R.; Abdelkafi-Koubaa, Z.; Mlayah-Bellalouna, S.; Aissaoui, D.; Marrakchi, N.; Srairi-Abid, N.; Zid, M.F.; Graia, M. Synthesis, structural characterization and antitumoral activity of $(NH_4)_4Li_2V_{10}O_{28}\cdot10H_2O$ compound. *J. Mol. Struct.* **2021**, *1229*, 129492. [CrossRef]
62. Francés-Monerris, A.; Segarra-Martí, J.; Merchán, M.; Roca-Sanjuán, D. Theoretical study on the excited-state π-stacking versus intermolecular hydrogen-transfer processes in the guanine-cytosine/cytosine trimer. *Theor. Chem. Acc.* **2016**, *135*, 31. [CrossRef]
63. Zhai, H.; Liu, S.; Peng, J.; Hu, N.; Jia, H. Synthesis, crystal structure, and thermal property of a novel supramolecular assembly: $(NH_4)_2(C_8H_{10}N_4O_2)_4[H_4V_{10}O_{28}]\cdot2H_2O$, constructed from decavanadate and caffeine. *J. Chem. Crystallogr.* **2004**, *34*, 541–548. [CrossRef]
64. Si, W.; Shen, J.; Zheng, H.; Fan, W. The role and mechanisms of action of microRNAs in cancer drug resistance. *Clin. Epigenet.* **2019**, *11*, 25. [CrossRef]
65. Noda, M.F. MicroRNAs in cancer—From research to the clinical practice. *Rev. Cubana Med.* **2012**, *51*, 325–335.
66. Feng, Y.-H.; Tsao, C.-J. Emerging role of microRNA-21 in cancer (Review). *Biomed. Rep.* **2016**, *5*, 395–402. [CrossRef]
67. Zhao, Z.; Guo, Y.; Liu, Y.; Sun, L.; Chen, B.; Wang, C.; Chen, T.; Wang, Y.; Li, Y.; Dong, Q.; et al. Individualized lncRNA differential expression profile reveals heterogeneity of breast cancer. *Oncogene* **2021**, *40*, 4604–4614. [CrossRef]
68. Gilles, M.-E.; Slack, F.J. *Let-7* microRNA as a potential therapeutic target with implications for immunotherapy. *Expert Opin. Ther. Targets* **2018**, *22*, 929–939. [CrossRef]

69. Chirshev, E.; Oberg, K.C.; Ioffe, Y.J.; Unternaehrer, J.J. *Let-7* as biomarker, prognostic indicator, and therapy for precision medicine in cancer. *Clin. Transl. Med.* **2019**, *8*, 1–14. [CrossRef] [PubMed]

70. Bernal, J.D. The Physical Basis of Life. *Proc. Phys. Soc. A.* **1949**, *62*, 537–558. [CrossRef]

71. Brack, A. Clay Minerals and the Origin of Life. In *Handbook of Clay Science*, 2nd ed.; Bergaya, F., Lagaly, G., Eds.; Elsevier: Amsterdam, The Netherlands, 2013; Volume 5, pp. 507–521.

72. Wu, J.; Peng, D.; He, Y.; Du, X.; Zhang, Z.; Zhang, B.; Li, X.; Huang, Y. In Situ Formation of Decavanadate-Intercalated Layered Double Hydroxide Films on AA2024 and their Anti-Corrosive Properties when Combined with Hybrid Sol Gel Films. *Materials* **2017**, *10*, 426. [CrossRef]

73. Amante, C.; Sousa-Coelho, D.; Luísa, A.; Aureliano, M. Vanadium and Melanoma: A Systematic Review. *Metals* **2021**, *11*, 828. [CrossRef]

74. Scior, T.; Abdallah, H.H.; Mustafa, S.F.Z.; Guevara-García, J.A.; Rehder, D. Are vanadium complexes druggable against the main protease Mpro of SARS-CoV-2?—A computational approach. *Inorg. Chim. Acta* **2021**, *519*, 120287. [CrossRef]

75. Bruker. *Bruker AXS Inc. V2019.1*; Bruker: Madison, WI, USA, 2019.

76. Sheldrick, G.M. *SADABS, Program for Empirical Absorption Correction of Area Detector Data*; Institute for Inorganic Chemistry, University of Göttingen: Göttingen, Germany, 1996.

77. Sheldrick, G.M. A short history of *SHELX*. *Acta Cryst.* **2008**, *A64*, 112–122. [CrossRef] [PubMed]

78. Sheldrick, G.M. Crystal Structure Refinement with SHELXL. *Acta Cryst.* **2015**, *C71*, 3–8.

79. Dolomanov, O.; Bourhis, L.J.; Gildea, R.; Howard, J.A.; Puschmann, H. OLEX2: A complete structure solution, refinement, and analysis program. *J. Appl. Crystallogr.* **2009**, *42*, 339–341. [CrossRef]

80. Spek, A.L. Single-crystal structure validation with the program PLATON. *J. Appl. Crystallogr.* **2003**, *36*, 7–11. [CrossRef]

81. Parr, R.G.; Yang, W. *Density-Functional Theory of Atoms and Molecules. International Series of Monograph on Chemistry-16*; Oxford University Press: New York, NY, USA, 1989.

82. Becke, A.D. Density-functional thermochemistry. III. The role of exact exchange. *J. Chem. Phys.* **1993**, *98*, 5648–5652. [CrossRef]

83. Weigend, F.; Ahlrichs, R. Balanced basis sets of split valence and quadruple zeta valence quality for H to Rn: Design and assessment of accuracy. *Phys. Chem. Chem. Phys.* **2005**, *7*, 3297–3305. [CrossRef]

84. Hay, P.J.; Wadt, W.R. Ab initio effective core potentials for molecular calculations. Potentials for the transition metal atoms Sc to Hg. *J. Chem. Phys.* **1985**, *82*, 270–283. [CrossRef]

85. Frisch, M.J.; Trucks, G.W.; Schlegel, H.B.; Scuseria, G.E.; Robb, M.A.; Cheeseman, J.R.; Scalmani, G.; Barone, V.; Petersson, G.A.; Nakatsuji, H.; et al. *Gaussian 16, Revision, B.*; Gaussian Inc.: Wallingford, CT, USA, 2016.

86. Dennington, R.; Keith, T.; Millam, J. *Gauss View, Version 6.0.16*; Semichem Inc.: Shawnee Mission, KS, USA, 2016.

87. Keith, T.A. *TK Gristmill Software, Version 19.02.13*; AIMAll: Overland Park, KS, USA, 2019.

88. Turner, M.J.; MacKiinnon, J.J.; Wolff, S.K.; Grimwood, D.J.; Spackman, P.R.; Jayatilaka, D.; Spackman, M.A. CrystalExplorer17. Available online: https://crystalexplorer.scb.uwa.edu.au/ (accessed on 24 July 2021).

89. Morris, G.M.; Huey, R.; Lindstrom, W.; Sanner, M.F.; Belew, R.K.; Goodsell, D.S.; Olson, A.J. AutoDock4 and AutoDockTools4: Automated docking with selective receptor flexibility. *J. Comput. Chem.* **2009**, *30*, 2785–2791. [CrossRef]

90. Ortega, E.; Vigueras, G.; Ballester, F.J.; Ruiz, J. Targeting translation: A promising strategy for anticancer metallodrugs. *Coord. Chem. Rev.* **2021**, *446*, 214129. [CrossRef]

Article

Study of DNA Interaction and Cytotoxicity Activity of Oxidovanadium(V) Complexes with ONO Donor Schiff Base Ligands

Gurunath Sahu [1], Edward R. T. Tiekink [2] and Rupam Dinda [1,*]

[1] Department of Chemistry, National Institute of Technology, Rourkela, Odisha 769008, India; 518cy1003@nitrkl.ac.in
[2] Research Centre for Crystalline Materials, School of Medical and Life Sciences, Sunway University, Bandar Sunway 47500, Selangor Darul Ehsan, Malaysia; edwardt@sunway.edu.my
* Correspondence: rupamdinda@nitrkl.ac.in

Abstract: Two new oxidovanadium(V) complexes, $(HNEt_3)[V^VO_2L]$ (**1**) and $[(V^VOL)_2\mu\text{-}O]$ (**2**), have been synthesized using a tridentate Schiff base ligand H_2L [where H_2L = 4-((E)-(2-hydroxy-5-nitrophenylimino)methyl)benzene-1,3-diol] and $VO(acac)_2$ as starting metal precursor. The ligand and corresponding metal complexes are characterized by physicochemical (elemental analysis), spectroscopic (FT-IR, UV–Vis, and NMR), and spectrometric (ESI–MS) methods. X-ray crystallographic analysis indicates the anion in salt **1** features a distorted square-pyramidal geometry for the vanadium(V) center defined by imine-N, two phenoxide-O, and two oxido-O atoms. The interaction of the compounds with CT–DNA was studied through UV–Vis absorption titration and circular dichroism methods. The results indicated that complexes showed enhanced binding affinity towards DNA compared to the ligand molecule. Finally, the in vitro cytotoxicity studies of H_2L, **1**, and **2** were evaluated against colon cancer (HT-29) and mouse embryonic fibroblast (NIH-3T3) cell lines by MTT assay. The results demonstrated that the compounds manifested a cytotoxic potential comparable with clinically referred drugs and caused cell death by apoptosis.

Keywords: Oxidovanadium(V); Schiff base; X-ray crystallography; DNA interaction; cytotoxicity

Citation: Sahu, G.; Tiekink, E.R.T.; Dinda, R. Study of DNA Interaction and Cytotoxicity Activity of Oxidovanadium(V) Complexes with ONO Donor Schiff Base Ligands. *Inorganics* **2021**, *9*, 66. https://doi.org/10.3390/inorganics9090066

Academic Editor: Dinorah Gambino

Received: 26 July 2021
Accepted: 24 August 2021
Published: 27 August 2021

Publisher's Note: MDPI stays neutral with regard to jurisdictional claims in published maps and institutional affiliations.

1. Introduction

In the family of the vanadium complexes, the oxoidovanadium Schiff base complexes are the most rapidly growing class owing to their rich underlying features and vital role during the process of interaction with various biomolecules [1]. Although vanadium exists in different oxidation states from −III to +V, for the higher oxidation states (+IV and +V), vanadium is highly stable and can form oxophilic complexes [2–4]. Oxidovanadium complexes have various roles in biochemical processes, such as nitrogen fixation, haloperoxidation, and glycogen metabolism [5,6]. In recent years, the investigation of the antifungal, antibacterial, and anticancer activities of these complexes has become the main subject of many studies. Recently, there is a growing interest in the in vitro and in vivo studies of vanadium complexes towards the treatments of diabetes and cancer [7,8]. After the discovery of many oxidovanadium drugs, bis(maltolato)oxovanadium(IV), BMOV, as glucose and lipid-lowering insulin mimetics, the focus on these types of compounds was stimulated [9]. Additionally, the anticancer activity of vanadium complexes has been widely examined on trial carcinogenesis and tumor-bearing animals [10,11]. The anticancer activity of several oxidovanadium complexes has recently received attention due to physiochemical changes in the solution medium leading to reduced systemic toxicity with beneficial effects [12,13]. There are also reports that vanadium accumulates in cancerous cells and tissues more than in normal cells [11,14]. For this reason, vanadium complexes have displayed promising cytotoxicity against various human cancer cell lines and these complexes were found to show better selectivity and higher cytotoxicity with reduced side effects. Therefore, attempts are being made to develop anticancer drugs using these oxidovanadium(V) complexes as suitable alternatives to platinum-based drugs [15].

In addition, Schiff bases play an immense role in coordination chemistry due to their ability to stabilize metal ions in various oxidation states, and their participation in numerous catalytic applications and biological activities [16–18]. The formation of stable metal complexes is due to the nitrogen lone pair of electrons present in the azomethine (−N=CH) backbone of the ligand molecule [18,19]. Various types of Schiff base ligands have been explored for their fascinating and significant properties, for example, complexing ability towards a wide range of transition metals, and applications in biological activity [20,21]. Certain oxidovanadium(V) Schiff base complexes have been reported earlier as model compounds, displaying biomolecular interactions with proteins and bio-ligands such as DNA [8,22,23]. Additionally, it is demonstrated that with increase in substitution and planarity of ligands, DNA interactions are enhanced [24]. Furthermore, metal complexes which can effectively interact with DNA under physiological conditions are considered to be possible contenders for use as therapeutic agents in medicinal applications and for genomic research [24,25]. Therefore, attempts are being made to develop anticancer drugs using these oxidovanadium(V) complexes as suitable alternatives to platinum-based drugs. Reportedly, polyphenolic/polyhydroxy compounds can prevent oxidative damage as they can scavenge reactive oxygen species such as hydroxyl radicals and superoxide anions. The prooxidant properties of polyphenolic compounds may contribute to tumor cell apoptosis [26–28]. In consideration of the inherent property of phenols and other polyhydroxy compounds particularly for medicinal and pharmacological applications [29,30], their corresponding complexes might be crucial for investigation for anticancer activity.

In continuation of our previous work on the synthesis, characterization, and biological studies of vanadium(V/IV) complexes [8,13,31–45], here we report a new mononuclear dioxidovanadium(V) (**1**) as well as an oxido-bridged dinuclear oxidovanadium(V) (**2**) complex, each with a tridentate ONO donor Schiff base ligand derived from 2,4-dihydroxybenzaldehyde and 2-amino-4-nitrophenol. Considering the therapeutic potential of the synthesized polyphenolic ligand molecule [27,30,46], corresponding oxidovanadium(V) complexes were synthesized to further investigate their pharmacological activities such as DNA interaction and anticancer activities. The primary objective of this current work was to investigate the significant characteristics of these ligand(H_2L) and oxidovanadium(V) complexes in terms of their applications as anticancer agents. The synthesized ligand and respective complexes were characterized by various spectroscopic (FT-IR, UV–Vis, and NMR), spectrometric (ESI–MS) techniques and the purity of the compounds were confirmed by CHN analysis. Furthermore, the single-crystal X-ray crystal structure of **1** was determined. The binding of the complexes toward CT–DNA was studied by UV–Vis absorption titration and circular dichroism. Finally, the cytotoxicity of the synthesized compounds was determined against HT-29 cell lines by MTT assay and for comparison a normal cell line, mouse embryonic fibroblast (NIH-3T3), was used.

2. Results and Discussion

2.1. Synthesis

New mononuclear dioxidovanadium(V) (**1**) and oxido-bridged dinuclear oxidovanadium(V) (**2**) complexes were synthesized by the reaction of the metal precursor $[V^{IV}O(acac)_2]$ with a tridentate ONO donor Schiff base ligand (H_2L) derived from condensation of 2,4-dihydroxybenzaldehyde and 2-amino-4-nitrophenol under reflux conditions. Scheme 1 depicts the synthetic methods of preparation of the complexes. The compounds were characterized by several spectroscopic (FT-IR, UV–Vis, and NMR) and spectrometric (ESI–MS) methods, and their purity was further confirmed by CHN elemental analysis. The structure of **1** was determined by single-crystal X-ray crystallography.

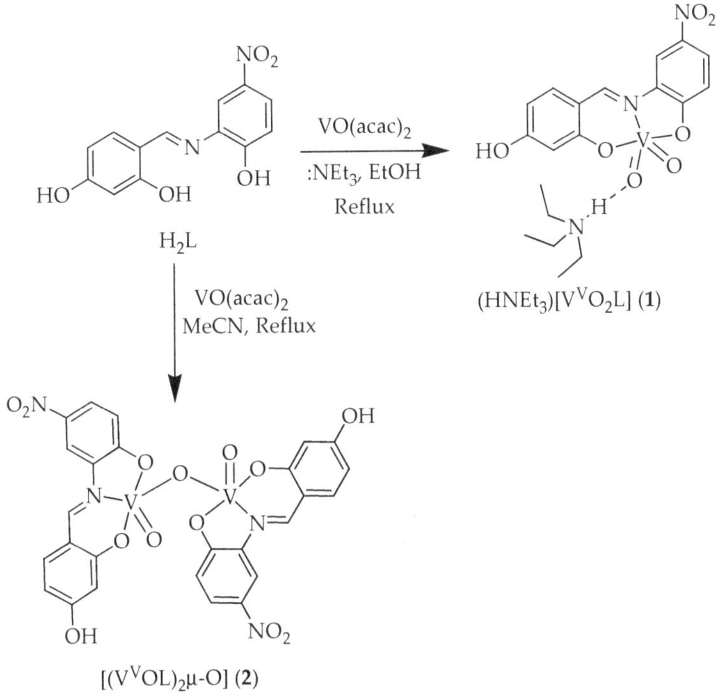

Scheme 1. Outline of the pathways for the synthesis of $(HNEt_3)[V^VO_2L]$ (**1**) and $[(V^VOL)_2\mu\text{-}O]$ (**2**).

2.2. Spectral Characteristics

2.2.1. IR Spectroscopy

Selected spectroscopic data of ligand (H_2L) and respective complexes (**1 and 2**) have been compiled in the Experimental Section. The IR spectrum of the free ligand (H_2L) exhibits one sharp band in the region 3205 cm^{-1} due to ν(O–H) stretching vibrations, which is absent in the corresponding metal complexes due to deprotonation of phenolic hydrogen [40]. Furthermore, the stretching band found in the region 1632–1607 cm^{-1} clearly indicates the presence of ν(C=N) in the ligand as well in the complexes [38]. In addition, two additional new stretching bands appeared in the region 888 and 947 cm^{-1} assigned to the two ν(V=O) stretching of cis-ν(V=O) groups in **1** whereas for **2** it is observed in the region 891 and 975 cm^{-1}. These stretching vibrations are in agreement with the terminal V=O groups present in related oxidovanadium(V) complexes [43]. Additionally, a new stretching band observed at 819 cm^{-1} assigned to the ν(V–O–V) residue of complex **2** which further indicates the existence of a dinuclear species [45]. The representative IR spectra of the ligand (H_2L) and its corresponding complex **1** are depicted in Figure S1.

2.2.2. Electronic Spectra

The UV-visible spectra of the ligand (H_2L) and its complexes (**1 and 2**) were recorded in DMSO with a complex concentration of 1.6×10^{-4} M (Figure 1). The spectrum of the free ligand shows two strong absorptions in the region 317 and 278 nm whereas their respective complexes show three strong absorption bands in the region 446–264 nm. The low energy absorption bands observed for the complexes in the region 422 and 446 nm could be attributed to ligand to metal charge transfer (LMCT) transition whereas the high energy bands appeared in the UV region (361–264 nm) are likely to be due to ligand center transitions [34].

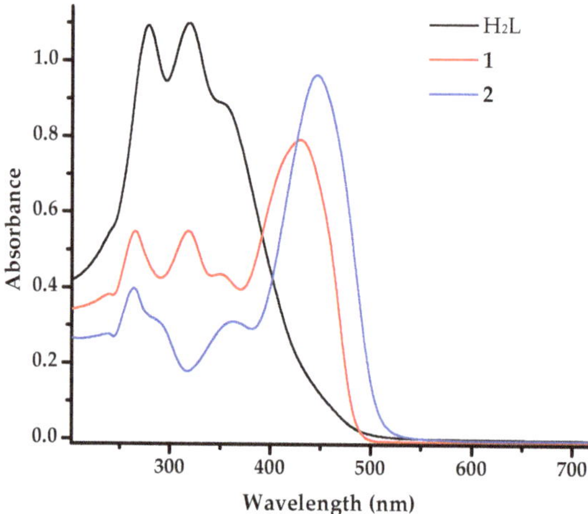

Figure 1. UV–Vis spectra of H_2L, **1**, and **2** (1.6×10^{-4} M) in DMSO.

2.2.3. NMR Spectra

The ^{1}H and ^{13}C{^{1}H} NMR data of ligand was recorded in DMSO-d_6. The spectrum of H_2L exhibits two compounds connected via intermolecular hydrogen bonding as shown in Figure S2 due to which two equivalent sets of protons obtained in the NMR. The ^{1}H NMR spectra of H_2L show singlet resonances in the downfield region in the range δ = 10.92–8.95, and 8.21 ppm due to –OH, and –CH (azomethine) protons, respectively [42]. The aromatic protons were observed in the expected range between δ = 8.20-6.27 ppm [37]. However, in **1**, the spectra suggests a mononuclear vanadium(V) complex and it exhibits a singlet for each –OH and –HC=N in the region 10.34 and 9.39, ppm respectively [42]. The aromatic protons were observed in the expected range between δ = 8.59-6.77 ppm and additionally two sets of resonances that is, a quartet at δ = 3.08 and a triplet at 1.15 ppm were observed for N–CH$_2$- and –CH$_3$, respectively in the aliphatic region, which are attributed to the presence of a triethylammonium counterion [42]. In the case of **2** two equivalent sets of protons are observed which are attributed to dimerization of the complex through μ_2-oxido-bridging. The spectra exhibit singlets in the regions δ = 10.44-9.39 and 8.66 ppm for –OH and –HC=N, respectively. The aromatic protons were observed in the expected range between δ = 8.66-6.18 ppm [45]. The representative spectra of **1** (^{1}H, ^{13}C, and ^{51}V NMR) and **2** (^{1}H, and ^{51}V NMR) are depicted in the ESI section (Figures S3–S7).

2.2.4. ESI Mass Spectra

The mass spectral data for **1** and **2** were recorded in acetonitrile solution (Figures S8 and S9). The ESI mass spectra display characteristic molecular ion peaks at m/z 480.10 and 694.97 for **1** and **2**, respectively. In addition to the molecular ion peak, the complex **1** shows a peak at m/z 467.19 corresponding to the [M + H$^+$ + 0.5 H$_2$O]$^+$ aggregate.

2.3. Single-Crystal X-ray Crystallography of 1

Crystals of salt **1** were obtained enabling a structure determination by X-ray crystallography. Salt **1** crystallizes in the triclinic space group P$^-$1 with two independent triethylammonium cations and two complex anions comprising the crystallographic asymmetric unit. The molecular structure of the first independent anion is shown in Figure 2a while those of the other constituents of the asymmetric unit are shown in Figure S10. Selected geometric parameters for the independent anions are listed in Table 1. The vanadium atom is penta-coordinated within a NO$_4$ donor set provided by an imine-N1, two phenoxide-O1, O2, and two oxido-O3, O4 atoms. The five-coordinate geometry is distorted from the ideal square-pyramidal and trigonal-bipyramidal geometries as quantified in the values of τ [47]. For the ideal geometries, τ = 0.0 and 1.0, respectively, whereas in the experimental structures τ computes to 0.26 (anion "a") and 0.24 (anion "b"). In this description, the V1 atom lies 0.4691(6) Å above the least-squares plane through the O1, O2, O4 and N1 atoms [r.m.s. deviation = 0.1663 Å] in the direction of the oxido-O3a atom; the comparable parameters for the V2-anion are 0.4935(6) and 0.0939 Å, respectively. The bond valency for the vanadium atoms, as calculated in PLATON [48], amount to 5.07 and 5.11, respectively, consistent with the assignment of vanadium(V) centers.

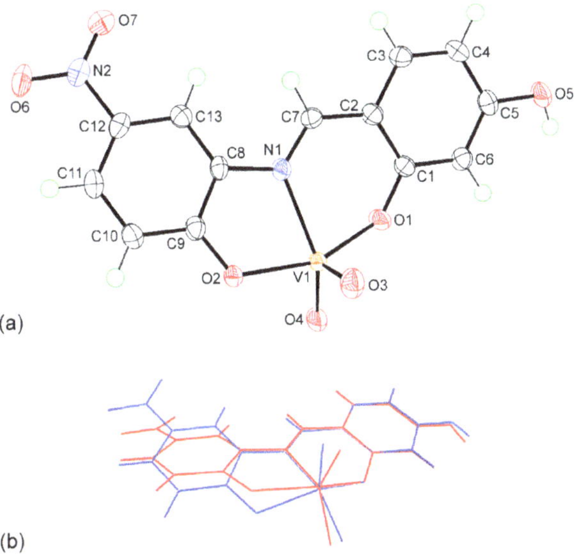

Figure 2. (**a**) Molecular structure of the first independent complex anion of salt **1** showing atom labelling scheme and displacement ellipsoids at the 70% probability level and (**b**) overlay diagram of the independent complex anions of **1**: red image, the molecule shown in (**a**). The molecule anions have been overlapped so the O1, C1, and C2 atoms are coincident.

Table 1. Selected geometric parameters (Å, °) for the independent anions in salt **1**.

Parameter	Anion "a"	Anion "b"
V–O1	1.9152(9)	1.8942(9)
V–O2	1.9735(9)	1.9191(9)
V–O3	1.6219(10)	1.6356(10)
V–O4	1.6463(9)	1.6473(10)
V–N1	2.1658(11)	2.1968(11)
C7–N1	1.3033(17)	1.2942(17)
O1–V–O2	155.41(4)	140.66(4)
O3–V–O4	109.05(5)	108.51(5)
N1–V–O4	140.02(5)	155.17(5)

The tridentate mode of coordination of the Schiff base dianion leads to the formation of six- and five-membered chelate rings. The best description of the six-membered ring is based on an envelope with the V1 atom being the flap atom. Here, the V1 atom lies 0.5624(15) Å out of the plane of the remaining atoms [r.m.s. deviation = 0.0348 Å]; the equivalent parameters for the second independent anion are 0.5256(15) and 0.0429 Å, respectively. By contrast, the five-membered ring for anion "a" is essentially planar exhibiting a r.m.s. deviation of 0.0239 Å with the maximum deviation of 0.0311(6) Å being for the N1a atom. However, an envelope conformation is the best description for the five-membered chelate ring of anion "b" whereby the V2 atom lies 0.5978(19) Å out of the plane of the remaining four atoms [r.m.s. deviation of 0.0066 Å]. The conformational differences between the molecules are highlighted in the overlay diagram of Figure 2b. Some significant differences in geometric parameters are apparent, especially, the elongation of the V1–O2(phenoxide) bond length compared with the other comparable bonds, and the elongation of the V2–N1(imine) bond as well as differences of up to 15° in the O1–V–O2 and N1–V–O4 bond angles. Although these may relate to conformational disparities, the influence of hydrogen bonding interactions cannot be discounted.

The presence of hydroxyl-O–H··· O(oxido) hydrogen bonds link the two independent anions into a two-molecule aggregate as shown in Figure 3; the geometric parameters characterizing the identified hydrogen bonding interactions in **1** are listed in Table 2. These hydrogen bonding interactions are consistent with the lengthening of the V–O4 bond lengths compared with the V–O3 bonds. Appended to the two-molecule aggregate are the triethylammonium cations which form charge-assisted N–H··· O3 hydrogen bonds. The N3a-cation also forms a hydrogen bond to the O2 atom indicating the H1n atom is bifurcated; this interaction accounts, at least partially, for the lengthening of the V1–O2 bond (see above). As illustrated in Supplementary Figure S11, the four-molecule aggregates are assembled into a three-dimensional architecture featuring hydroxyphenyl-C–H··· O(phenoxide), nitrophenyl-C–H··· O(oxide, hydroxyl), methylene-C–H··· O(phenoxide, oxide and nitro) and methyl-C–H··· O(oxide) interactions, as detailed in Table S1.

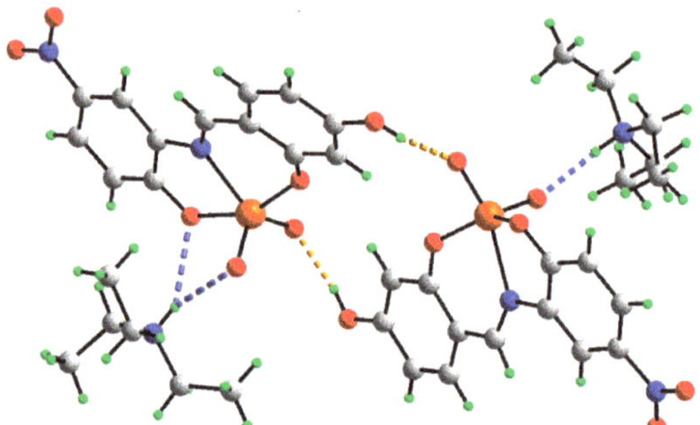

Figure 3. The four-molecule aggregate in **1** features hydroxyl-O–H··· O(oxido) and charge-assisted N–H··· O3 hydrogen bonds shown as orange and blue dashed lines, respectively.

Table 2. Geometric parameters (Å, °) characterizing the identified hydrogen bonding contacts between the constituents of the asymmetric unit of salt **1** leading to a four-molecule aggregate.

A	H	B	H···B	A···B	A–H···B
O5a	H1o	O4b	1.825(12)	2.6548(14)	171.3(18)
O5b	H2o	O4a	1.830(15)	2.6640(14)	171.7(17)
N3a	H1n	O2a	2.466(13)	3.2244(15)	145.4(13)
N3a	H1n	O3a	2.224(12)	2.9473(15)	139.8(14)
N3b	H2n	O3b	1.918(12)	2.7904(16)	170.7(14)

There are relatively few structural precedents for **1** in the crystallographic literature. Arguably the most closely related structure is that of [Et₃NH][VO₂L] where L is the 1-(((5-chloro-2-oxidophenyl)imino)methyl)naphthalen-2-olate dianion [11]. Here, a very similar square-pyramidal coordination geometry is noted with each chelate ring having an envelope conformation as seen for anion "b" in **1**.

2.4. DNA-Binding Studies

2.4.1. UV–Vis Absorption Studies

The interactions of metal complexes with DNA provide the binding information of metal complexes with the DNA helix [49]. Therefore, the absorption study was performed by maintaining the concentration of the complexes constant (25 μM) with varying DNA concentrations from 0 to 100 μM. Upon increasing the CT–DNA concentration, hypochromic shifts are observed in both the complexes for the maximal peaks (Figure 4). Generally, hypochromism or hyperchromism shifts often are observed in the absorption spectrum of a metal complex when the complex interacts with DNA [50]. Hypochromism in absorption spectra is generally associated with the binding of complexes to DNA through the intercalation mode [51]. To compare the DNA-binding affinity of these compounds quantitatively, their intrinsic binding constants were calculated with the aid of the following equation: [52]

$$\frac{[DNA]}{\varepsilon_a - \varepsilon_f} = \frac{[DNA]}{\varepsilon_b - \varepsilon_f} + \frac{1}{K_b\left(\varepsilon_b - \varepsilon_f\right)} \tag{1}$$

where $[DNA]$ is the concentration of DNA base pairs, K_b is binding constant and ε_a, ε_f, and ε_b are the apparent extinction coefficients for the complex i.e., Abs/[complex] in the presence of DNA, in absence of DNA and fully bound of DNA, respectively. A plot of $[DNA]/(\varepsilon_a - \varepsilon_f)$ vs. $[DNA]$ gave a slope and an intercept equal to $1/(\varepsilon_b - \varepsilon_f)$ and $1/K_b(\varepsilon_b - \varepsilon_f)$, respectively, while the binding constant K_b was calculated from the ratio of the slope to the intercept. The intrinsic binding constants K_b were found to be 2.81×10^4 and 2.35×10^4 M^{-1} for **1** and **2**, respectively (Table 3). From the binding constant values, it is clear that complex **1** interacts more strongly with CT–DNA. However, the free ligand H$_2$L itself shows good binding activity (1.59×10^4 M^{-1}) with DNA molecules [53,54].

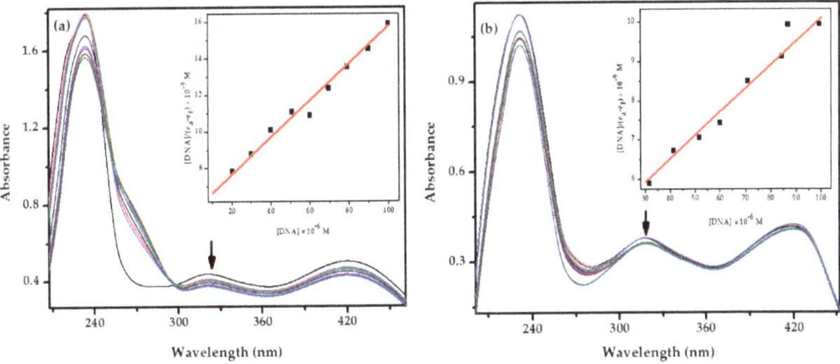

Figure 4. Absorption spectroscopic study of complex **1** (a) and **2** (b) (25 μM) with increasing concentrations of CT–DNA (0–100 μM). The inset shows the plots of $[DNA]/(\varepsilon_a - \varepsilon_f)$ versus $[DNA]$ for the titration of the prepared compounds with CT–DNA.

Table 3. DNA-binding parameters for **1**, **2**, and H$_2$L.

Complex	Binding Constants (K_b) (M^{-1})
1	2.81×10^4
2	2.35×10^4
H$_2$L	1.59×10^4

2.4.2. Circular Dichroism Studies

Circular dichroism (CD) studies were performed to investigate the conformational changes in CT–DNA upon interaction with the new compounds. The spectra show two significant CD bands in the UV region, a positive band at 275 nm due to base stacking whereas a negative band at 245 nm is due to right handed helicity [38,51]. In intercalation mode of small molecules, there occurs perturbation in the spectra whereas for groove binding and electrostatic interaction there will be minimal or no perturbation [38,51]. However, from Figure 5, it is observed that there are significant changes in the CD spectra of CT–DNA which further suggest that the tested compounds bind to CT–DNA in an intercalating mode.

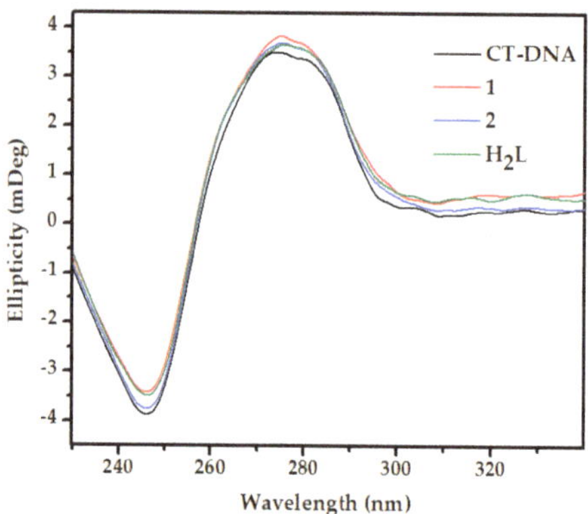

Figure 5. Circular dichroism spectra of CT–DNA (150 µM) in the presence and absence H$_2$L, and complexes (**1** and **2**) in 50 mM Tris–HCl buffer (pH 7.4). The path length of the cuvette was 5 mm.

2.5. Cytotoxicity

According to GLOBOCAN 2020, disease occurrence and mortality due to colorectal cancer has increased to 1.9 million. Moreover, it has been placed as the third most and second most incidences in terms of mortality in both male and female, respectively. Furthermore, the incidence of colorectal cancer has been reported to be the most in eastern Asian countries such as India. Therefore, in this study we have chosen a most aggressive colorectal cancer cell line HT-29 [55]. Hence, the cytotoxicity of the oxidovandium(V) complexes (**1** and **2**) were measured against HT-29 cancer and NIH-3T3 normal cells by MTT assay and were compared to the ligand (H$_2$L) alone. The IC$_{50}$ values are listed in Table 4, and the cell viability percentage diagrams are depicted in Figure 6. Specifically, the cytotoxicity of **1** and **2** exhibited IC$_{50}$ values 8.56 ± 0.62, and 9.09 ± 0.03 µM, respectively, while that of the ligand alone was determined to be 7.75 ± 0.53 µM against HT-29 cancer cell line. These findings suggest that with respect to the ligand, the coordination to vanadium did not improve its activity; in fact, H$_2$L is marginally more active than both of the complexes. As mentioned earlier, the polyphenol groups in ligand molecule (H$_2$L) induce apoptosis in cancer cells which is the primary reason towards the enhanced toxicity [28,53].

Furthermore, the cytotoxicity of the tested complexes was studied against the normal cell line NIH-3T3. The NIH-3T3 cell line is one of the most frequently used cell lines as the results provided by these tests can easily be compared with data published in the literature [56,57]. The results indicated there was a decrease in the cell viability upon the same exposure of the compounds. This result indicates that tested complexes were less damaging towards NIH-3T3 as compared to cancer cell lines. Additionally, it was observed that the ligand precursor H$_2$L is more selective against HT-29 cells than both complex molecules with selectivity index (SI) = 8.94, whereas SI values for **1** and **2** are 7.92 and 8.77, respectively. On the other hand, under the same experimental conditions the tested compounds exhibited comparable cytotoxicity against HT-29 compared with commonly used chemotherapeutic drugs such as cisplatin [58]. Furthermore, the results obtained on the present study may be also compared with previously reported oxidovanadium(V) complexes of Schiff base ligands such as [VO(sal-L-tryp)(Me-ATSC)], [VO(sal-Ltryp)(N-ethhymethohcarbthio)]·H$_2$O) and, [VO(sal-L-tryp)(acetylethTSC)]·C$_2$H$_5$OH with IC$_{50}$ > 47 µM against HT-29 cell lines [59]. Additionally, we can also compare our results with recent work where two oxido-bridged vanadium(V) complexes of Schiff base ligands [{V^{V}O(R-salval)(H$_2$O)}(µ$_2$-O){V^{V}O(R-salval)}] and [{V^{V}O(R-vanval)(CH$_3$OH)}$_2$(µ$_2$-O)] (val = valine, sal = salicylaldehyde, and van = o-vanillin) were investigated for cytotoxic studies against human hepatoma cell line with IC_{50} values > 200 µM [7].

Table 4. IC_{50} values of H_2L and Complexes (**1** and **2**) taking 10, 50, and 100 µM concentrations.

Compound	IC_{50} (µM)	
	HT-29	NIH-3T3
1	8.56 ± 0.62	67.85 ± 5.48
2	9.09 ± 0.03	79.77 ± 4.00
H_2L	7.75 ± 0.53	69.32 ± 4.42

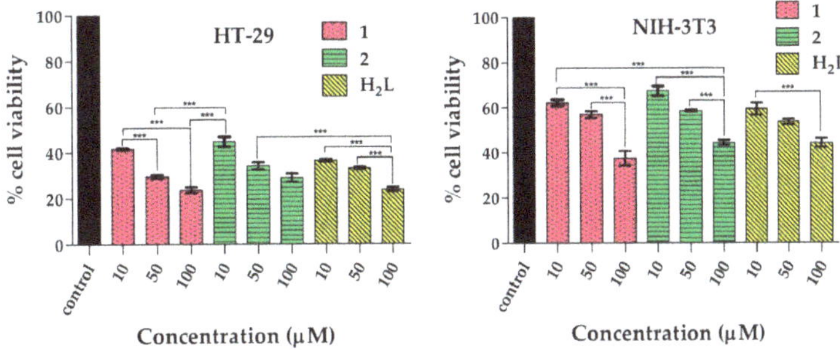

Figure 6. The effect of **1**, **2**, and H_2L on the cell viability of HT-29 and NIH-3T3 cells after 48 h of exposure, taking 10, 50, and 100 µM concentrations of the compounds. The cell viability was calculated by MTT assay. Data were reported as the mean $\pm$ SD for n = 4. *** $p < 0.0001$ was considered statistically significant.

2.6. Nuclear DAPI Staining Assay

To examine the apoptotic potential of test compounds in HT-29 cells, DAPI staining assay was conducted. Chromatin condensation, cell shrinkage, and nuclear fragmentation during the process of apoptosis (type I programmed cell death) is a distinguishing marker of nuclear change [51]. For this assay HT-29 cells were treated with 20 µM of H_2L, **1** and **2**, respectively and then the cells were incubated for 24 h before DAPI nuclear staining. Later, the image of cells was captured under fluorescent microscope fitted with a DAPI filter. The nuclear blebbings and brightly condensed chromatin bodies were marked by arrows in Figure 7.

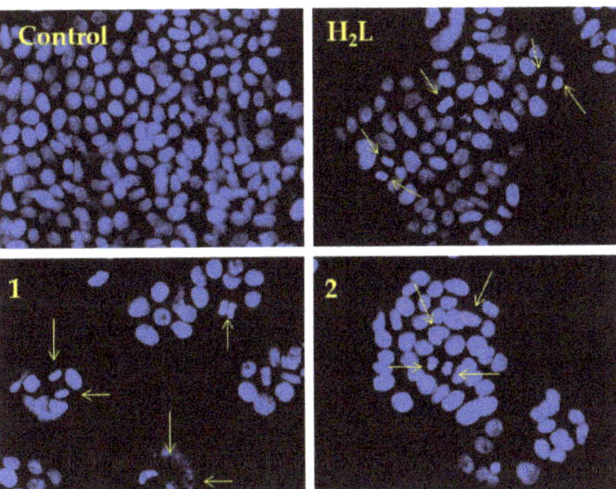

Figure 7. Study of apoptosis by morphological changes in nuclei of HT-29 cells. Arrows show the morphological changes in the nuclei of HT-29 cells observed on applying H_2L, **1**, and **2** in comparisons to control.

3. Materials and Methods

3.1. Materials

All the starting materials such as 2-amino-4-nitrophenol, and 2,4-dihydroxybenzaldehyde were purchased from Sigma Aldrich and used without further purification. Reagent grade solvents were dried and distilled prior to use. [VO(acac)$_2$] was prepared by the reported method [60]. MTT (3-[4, 5-dimethylthiazol-2-yl]-2, 5-diphenyltetrazolium), DAPI (4′, 6-diamidino-2-phenylindoledihydrochloride), and CT–DNA were purchased from Sigma Aldrich (St. Louis, MO, USA). HT-29 and NIH-3T3 cell lines were procured from National Centre for Cell Science (NCCS), Pune, India. CHN analyses were carried out on a Vario ELcube CHNS Elemental analyzer. ESI–MS data of the complexes were recorded on a Waters XEVO G2-XS QTOF MS instrument operating in positive ion mode. IR spectra were recorded on a Perkin-Elmer Spectrum RXI spectrophotometer. ^{1}H and ^{13}C NMR spectra were recorded on a Bruker Ultrashield 400 MHz spectrometer in the presence of SiMe$_4$ as the internal standard. Electronic spectra were recorded on a Shimadzu spectrophotometer (UV-2450).

3.2. Synthesis of Ligands

The Schiff base ligand H$_2$L was synthesized by refluxing 2,4-dihydroxy-benzaldehyde and 2-amino-4-nitrophenol in equimolar ratio in ethanol following a standard procedure [13]. The resulting compound was isolated through filtration, washed with ethanol, and dried over fused CaCl$_2$ under desiccator conditions. The molecular structure of the compound was confirmed by elemental and spectroscopic (FT-IR, UV–Vis, and NMR (^{1}H, ^{13}C{^{1}H})) analysis.

H$_2$L: Yield: 67%. Anal. calcd. for C$_{13}$H$_{10}$N$_2$O$_5$ (274.23): C, 56.94; H, 3.68; N, 10.22. Found: C, 56.81; H, 3.61; N, 10.18. IR (KBr pellet, cm^{-1}): 3205 ν(O–H); 1632 ν(C=N). UV–Vis (DMSO) [λ_{max}, nm (ε, M^{-1} cm^{-1})]: 317 (6817), 278 (6812). ^{1}H NMR (400 MHz, DMSO-d$_6$): δ (ppm) = 10.92 (s, 1H, –OH), 10.35 (s, 2H, –OH), 9.25 (s, 1H, –OH), 8.95 (s, 2H, –OH), 8.21 (s, 2H, HC=N–), 8.20–6.27 (m, 12H, aromatic). ^{13}C{^{1}H} NMR (100 MHz, DMSO-d$_6$): δ (ppm) = 191.51, 167.47, 165.67, 165.39, 163.71, 162.04, 156.52, 153.69, 140.32, 140.19, 137.87, 133.30, 131.57, 130.21, 124.65, 118.13, 116.72, 115.64, 115.53, 114.73, 112.83, 110.52, 110.16, 109.11, 103.01, 102.67.

3.3. Synthesis of Oxidovanadium(V) Complexes

(HNEt$_3$)[V^{V}O$_2$L] (**1**). This was synthesized by refluxing the H$_2$L (0.27 g, 1 mmol) and VO(acac)$_2$ (0.265 g, 1 mmol) in hot absolute ethanol (20 mL) using triethylamine as a base for 4 h. Dark brown crystals were obtained from the filtrate after 2–4 days. The crystals were filtered and washed with ethanol for X-ray structure determination. (HNEt$_3$)[V^{V}O$_2$L] (**1**): Yield: 0.32 g (70%). Anal. calcd. for C$_{19}$H$_{24}$N$_3$O$_7$V (457.35): C, 49.90; H, 5.29; N, 9.19; found C, 49.87; H, 5.22; N, 9.12. IR (KBr pellet, cm^{-1}): 2986 ν(O–H), 1607 ν(C=N), 947, 890 ν(V=O). UV–Vis (DMSO) [λ_{max}, nm (ε, M^{-1} cm^{-1})]: 428 (6242), 317 (4331), 265 (4356). ^{1}H NMR (400 MHz, DMSO-d$_6$): δ (ppm) = 10.34 (s, 1H,–OH), 9.39 (s, 1H, HC=N–), 8.59–6.17 (m, 6H, aromatic), 3.09 (m, 6H, N-CH$_2$), 1.17 (m, 9H, –CH$_3$). ^{13}C{^{1}H} NMR (100 MHz, DMSO-d$_6$): δ (ppm) = 169.41, 165.60, 159.26, 136.92, 136.81, 136.33, 133.50, 125.03, 124.63, 116.14, 115.22, 114.86, 111.05, 46.26, 9.31. ^{51}V NMR (DMSO-d$_6$): δ (ppm) = −527.86. ESI–MS: *m/z* 480.1007 [M + Na]$^+$.

[(V^{V}OL)$_2$μ-O] (**2**). This complex was synthesized by refluxing the H$_2$L (0.27 g, 1 mmol) and VO(acac)$_2$ (0.265 g, 1 mmol) in hot MeCN (20 mL) for 4 h. Dark brown crystalline materials were obtained from the filtrate after 2–4 days. The crystalline materials were filtered and washed with ethanol. [(V^{V}OL)$_2$μ-O] (**2**): Yield: 0.45 g (66%). Anal. calcd. for C$_{26}$H$_{16}$N$_4$O$_{13}$V$_2$ (694.30): C, 44.98; H, 2.32; N, 8.07; found C, 44.92; H, 2.22; N, 8.13. IR (KBr pellet, cm−1): 2986 ν(O–H), 1607 ν(C=N), 975, 891 ν(V=O). UV–Vis (DMSO) [λ_{max}, nm (ε, M^{-1} cm^{-1})]: 446 (6037), 361 (1956), 264 (2450). ^{1}H NMR (400 MHz, DMSO-d$_6$): δ (ppm) = 10.44 (s, 1H, –OH), 8.63 (s, 2H, –OH), 8.01–6.18 (m, 12H, aromatic). ^{13}C{^{1}H} NMR (100 MHz, DMSO-d$_6$): δ (ppm) = 168.18, 163.71, 159.21, 156.82, 143.28, 141.33, 131.72, 121.78,

118.32, 117.56, 111.07, 108.71, 102.88. ^{51}V NMR (DMSO-d$_6$): δ (ppm) = -575.27, -530.96. ESI–MS: *m/z* 694.9722 [M + H]$^+$.

3.4. Single-Crystal X-ray Crystallography

X-ray Intensity data for a brown crystal of **1** (0.06 × 0.19 × 0.24 mm) were measured at 100 K on Rigaku/Oxford Diffraction XtaLAB Synergy diffractometer (Dualflex, AtlasS2) fitted with CuKα radiation (α = 1.54178 Å) so that θ$_{max}$ = 67.1° (= 100% completeness). Data reduction, including Gaussian absorption correction, was accomplished with CrysAlisPro [61]. The structure was solved by direct-methods [62] and refined (anisotropic displacement parameters and H atoms in the riding model approximation) on F^2 [63]. The O- and N-bound H atoms were located from Fourier difference maps and refined with distance constraints of O–H = 0.84 ± 01 Å and N–H = 0.88 ± 01 Å, respectively. A weighting scheme of the form w = 1/[σ^2($F_o{}^2$) + (0.043P)^2 + 0.792P], where P = ($F_o{}^2$ + 2$F_c{}^2$)/3, was introduced towards the end of the refinement. The molecular structure diagrams were generated with ORTEP for Windows [64] with 70% displacement ellipsoids, and the packing diagrams were drawn with DIAMOND [65]. Crystal data and refinement details are given in Table 5.

Table 5. Crystallographic data and refinement details for salt **1**.

Formula	[C$_6$H$_{16}$N][C$_{13}$H$_8$N$_2$O$_7$V]
Molecular weight	457.35
Crystal system	triclinic
Space group	$P\bar{1}$
a/Å	10.8226(1)
b/Å	10.9399(2)
c/Å	17.3389(3)
α/°	79.412(1)
β/°	78.905(1)
γ/°	86.861(1)
V/Å^3	1979.82(5)
Z	4
D_c/g cm^{-3}	1.534
μ/mm^{-1}	4.621
Measured data	46,958
Unique data	7061
Observed data ($I \geq 2.0\sigma(I)$)	6958
No. parameters	559
R, obs. data; all data	0.025; 0.026
R_w, obs. data; all data	0.072; 0.072
Range of residual electron density peaks/eÅ^{-3}	$-0.58–0.23$

3.5. DNA-Binding Experiments

3.5.1. UV–Vis Absorption Studies

The interaction of the ligand and its respective oxidovanadium(V) complexes (**1** and **2**) with CT–DNA was investigated by the absorption titration method with a Shimadzu spectrophotometer (UV–2450) [39,40,42,51]. The absorption titration of DNA was conducted by using a fixed concentration of metal complex (25 μM) in 50 mM Tris–HCl buffer (pH = 7.4), with gradual increases in concentration of the CT–DNA from 0 to 100 μM. Each of the above experiments was performed in triplicate at room temperature and the incubation time was 5 min after the subsequent addition of CT–DNA for each time to equilibrate DNA and the complexes properly.

3.5.2. Circular Dichroism Studies

Circular dichroism study was performed in a JASCO J-1500 CD Spectrophotometer at 25 °C using a quartz cell with 5 mm path length [51]. CD spectra of CT–DNA (150 μM) were collected both in the presence and absence of complexes (25 μM) at a wavelength range of 230–350 nm in 50 mM Tris–HCl buffer (pH 7.4) (HiMedia), after averaging three accumulations and a scan speed of 200 nm/min.

3.6. Cytotoxicity Analysis through MTT Assay

The cytotoxicity of the ligand molecule and its respective vanadium(V) complexes was evaluated against colon cancer (HT-29) and mouse embryonic fibroblast (NIH-3T3) cells using the MTT (3-[4, 5-dimethylthiazol-2-yl]-2, 5-diphenyltetrazolium) assay [42]. All cell lines were cultured in DMEM (Dulbecco's phosphate-buffered saline) medium supplemented with 10% FBS (Fetal bovine serum) and maintained at 37 °C in a CO_2 incubator (5% CO_2) and humidified atmosphere (95% humidity). During the MTT assays, cells were seeded in 96-well plates at a density of 6×10^3 cells per well after cell counting in a hemocytometer and allowed for 70–80% confluence. Then complexes and the ligand were dissolved in DMSO at a concentration of 100 mM and suitably diluted in DMEM media to achieve final working concentrations of 10, 50, and 100 μM. After 12 h of initial seeding, the HT-29 and NIH-3T3 cells were treated with the above prepared concentrations of each complex and further subjected to incubation for 48 h. MTT was dissolved in the DPBS (Dulbecco's Modified Eagle Medium) solution and was added to the culture medium. After additional 3 h incubation at 37 °C, the media were carefully removed and 200 μL of DMSO was added to each well and the absorbance values were determined by spectrophotometry at 595 nm with a microplate reader spectrophotometer (Perkin-Elmer 2030). The results were expressed as percentages of the control.

$$\% \text{ cell viability} = [\text{mean OD of the treated cell}/\text{mean OD of the control}] \times 100$$

IC_{50} value of the compounds was calculated from the absorbance concentration plot following standard procedure [51,66].

3.7. Nuclear DAPI Staining Assay

The morphology of nucleus during the cell death of cells in response to treatment with the complexes was investigated using fluorescence microscopy (Olympus IX 71). DAPI (4′, 6-diamidino-2-phenylindoledihydrochloride) stain was used for this nuclear staining assay and was performed according to a standard procedure previously reported [67]. Accordingly, HT-29 cells were treated with treated (20 μM of compound for 24 h) and un-treated cells were fixed with 4% paraformaldehyde for 15 min. Then the cells were stained with DAPI and incubated for 5 min at 37 °C after washing two times with DPBS. Finally, again after washing with DPBS, the cells were examined by fluorescence microscopy.

4. Conclusions

In this work, two new oxidovanadium(V) complexes $(HNEt_3)[V^VO_2L]$ (**1**), and $[(V^VOL)_2\mu\text{-O}]$ (**2**) have been synthesized using a Schiff base ligand (H_2L) derived from 2,4-dihydroxybenzaldehyde and 2-amino-4-nitrophenol. The ligand and complexes were characterized by FT-IR, UV–Vis, and NMR (^{1}H, ^{13}C, and ^{51}V) spectroscopy, ESI–MS, and the purity was confirmed by CHN analysis. The molecular structure of **1** was determined by X-ray crystallography indicating a distorted square-pyramidal geometry for the vanadium(V) center defined by imine-N, two phenoxide-O, and two oxido-O atoms. DNA-binding experiments were conducted using a UV–Vis absorption titration method and circular dichroism studies and the results suggested that the ligand as well as the complexes have considerable DNA-binding propensity. From the results, complex **1**, displayed maximum DNA-binding activity with $K_b = 2.81 \times 10^4$ M^{-1}. From circular dichroism studies it was further confirmed that the synthesized molecules interacted with DNA through the intercalation mode. Finally, from the results of cytotoxicity studies it is confirmed that all the tested compounds including the ligand molecule induce cell death against HT-29 cells mainly through the apoptotic mode. However, the ligand molecule with $IC_{50} = 7.75 + 0.53$ μM was found more cytotoxic than its corresponding complexes.

In addition, the cytotoxicity of ligand and complexes was also studied against NIH-3T3 normal cells, and it was found to be relatively less damaging towards them. In summary, the present group of compounds should stimulate further in vitro and in vivo studies of related compounds as part of the quest to develop new drugs for the treatment of cancer.

Supplementary Materials: The following are available online at https://www.mdpi.com/article/10.3390/inorganics9090066/s1, Figure S1: IR spectra of H_2L and **1**, Figure S2: ^1H NMR spectra of H_2L, Figure S3-S5: ^1H, ^{13}C, and ^{51}V of **1**, Figure S6-S7: ^1H and ^{51}V NMR spectra of **2**, Figure S8-S9: ESI–MS of **1** and **2**, Figure S10: independent molecular structure of cation and anion of **1**, Figure S11. A view of the unit-cell contents for salt **1**, Table S1: Geometric parameters (Å, °) of **1**.

Author Contributions: R.D.: Conceptualization, Visualization, Supervision, Funding acquisition. G.S.: Investigation, Validation, Writing-original draft, review, and editing. E.R.T.T.: Investigation, Writing-original draft, review, and editing. All authors have read and agreed to the published version of the manuscript.

Funding: This research was funded by CSIR, Govt. of India [Grant No. 01(2963)/18/EMR-II] and Sunway University Sdn Bhd [Grant No. GRTIN-IRG-01-2021].

Institutional Review Board Statement: Not applicable.

Informed Consent Statement: Not applicable.

Data Availability Statement: Not applicable.

Conflicts of Interest: The authors declare no conflict of interest.

References

1. Pessoa, J.C.; Garribba, E.; Santos, M.F.A.; Santos-Silva, T. Vanadium and proteins: Uptake, transport, structure, activity and function. *Coord. Chem. Rev.* **2015**, *301–302*, 49–86. [CrossRef]
2. Rehder, D. The future of/for vanadium. *Dalton Trans.* **2013**, *42*, 11749–11761. [CrossRef]
3. Thompson, K. Coordination chemistry of vanadium in metallopharmaceutical candidate compounds. *Coord. Chem. Rev.* **2001**, *219–221*, 1033–1053. [CrossRef]
4. Baran, E.J. Oxovanadium(IV) and oxovanadium(V) complexes relevant to biological systems. *J. Inorg. Biochem.* **2000**, *80*, 1–10. [CrossRef]
5. Domingues, N.; Pelletier, J.; Ostenson, C.-G.; Castro, M.M.C.A. Therapeutic properties of VO(dmpp)$_2$ as assessed by in vitro and in vivo studies in type 2 diabetic GK rats. *J. Inorg. Biochem.* **2014**, *131*, 115–122. [CrossRef]
6. Pillai, S.I.; Subramanian, S.P.; Kandaswamy, M. A novel insulin mimetic vanadium-flavonol complex: Synthesis, characterization and in vivo evaluation in STZ-induced rats. *Eur. J. Med. Chem.* **2013**, *63*, 109–117. [CrossRef] [PubMed]
7. Turtoi, M.; Anghelache, M.; Patrascu, A.A.; Maxim, C.; Manduteanu, I.; Calin, M.; Popescu, D.-L. Synthesis, Characterization, and In Vitro Insulin-Mimetic Activity Evaluation of Valine Schiff Base Coordination Compounds of Oxidovanadium(V). *Biomedicines* **2021**, *9*, 562. [CrossRef] [PubMed]
8. Dash, S.P.; Pasayat, S.; Bhakat, S.; Roy, S.; Dinda, R.; Tiekink, E.R.T.; Mukhopadhyay, S.; Bhutia, S.K.; Hardikar, M.R.; Joshi, B.N.; et al. Highly stable hexacoordinated nonoxidovanadium(IV) complexes of sterically constrained ligands: Syntheses, structure, and study of antiproliferative and insulin mimetic activity. *Inorg. Chem.* **2013**, *52*, 14096–14107. [CrossRef] [PubMed]
9. Orvig, C.; Caravan, P.; Gelmini, L.; Glover, N.; Herring, F.G.; Li, H.; McNeill, J.H.; Rettig, S.J.; Setyawati, I.A. Reaction chemistry of BMOV, bis(maltolato)oxovanadium(IV), a potent insulin mimetic agent. *J. Am. Chem. Soc.* **1995**, *117*, 12759–12770. [CrossRef]
10. Evangelou, A.M. Vanadium in cancer treatment. *Crit. Rev. Oncol.* **2002**, *42*, 249–265. [CrossRef]
11. Ebrahimipour, S.Y.; Sheikhshoaie, I.; Kautz, A.C.; Ameri, M.; Pasban-Aliabadi, H.; Amiri Rudbari, H.; Bruno, G.; Janiak, C. Mono- and dioxido-vanadium(V) complexes of a tridentate ONO Schiff base ligand: Synthesis, spectral characterization, X-ray crystal structure, and anticancer activity. *Polyhedron* **2015**, *93*, 99–105. [CrossRef]
12. Levina, A.; Pires Vieira, A.; Wijetunga, A.; Kaur, R.; Koehn, J.T.; Crans, D.C.; Lay, P.A. A Short-Lived but Highly Cytotoxic Vanadium(V) Complex as a Potential Drug Lead for Brain Cancer Treatment by Intratumoral Injections. *Angew. Chem. Int. Ed. Engl.* **2020**, *59*, 15834–15838. [CrossRef]
13. Banerjee, A.; Dash, S.P.; Mohanty, M.; Sahu, G.; Sciortino, G.; Garribba, E.; Carvalho, M.F.N.N.; Marques, F.; Costa Pessoa, J.; Kaminsky, W.; et al. New V^{IV}, $V^{IV}O$, V^VO, and V^VO_2 Systems: Exploring their Interconversion in Solution, Protein Interactions, and Cytotoxicity. *Inorg. Chem.* **2020**, *59*, 14042–14057. [CrossRef]
14. Bishayee, A.; Waghray, A.; Patel, M.A.; Chatterjee, M. Vanadium in the detection, prevention and treatment of cancer: The in vivo evidence. *Cancer Lett.* **2010**, *294*, 1–12. [CrossRef]

15. Ni, L.; Zhao, H.; Tao, L.; Li, X.; Zhou, Z.; Sun, Y.; Chen, C.; Wei, D.; Liu, Y.; Diao, G. Synthesis, in vitro cytotoxicity, and structure-activity relationships (SAR) of multidentate oxidovanadium(IV) complexes as anticancer agents. *Dalton Trans.* **2018**, *47*, 10035–10045. [CrossRef] [PubMed]
16. Sutradhar, M.; Pombeiro, A.J.L. Coordination chemistry of non-oxido, oxido and dioxidovanadium(IV/V) complexes with azine fragment ligands. *Coord. Chem. Rev.* **2014**, *265*, 89–124. [CrossRef]
17. Rezaeivala, M.; Keypour, H. Schiff base and non-Schiff base macrocyclic ligands and complexes incorporating the pyridine moiety—The first 50 years. *Coord. Chem. Rev.* **2014**, *280*, 203–253. [CrossRef]
18. Gupta, K.C.; Sutar, A.K. Catalytic activities of Schiff base transition metal complexes. *Coord. Chem. Rev.* **2008**, *252*, 1420–1450. [CrossRef]
19. Liu, X.; Hamon, J.-R. Recent developments in penta-, hexa- and heptadentate Schiff base ligands and their metal complexes. *Coord. Chem. Rev.* **2019**, *389*, 94–118. [CrossRef]
20. Khan, M.I.; Khan, A.; Hussain, I.; Khan, M.A.; Gul, S.; Iqbal, M.; Inayat-Ur-Rahman; Khuda, F. Spectral, XRD, SEM and biological properties of new mononuclear Schiff base transition metal complexes. *Inorg. Chem. Commun.* **2013**, *35*, 104–109. [CrossRef]
21. Zhang, N.; Fan, Y.-h.; Zhang, Z.; Zuo, J.; Zhang, P.-f.; Wang, Q.; Liu, S.-B.; Bi, C.-F. Syntheses, crystal structures and anticancer activities of three novel transition metal complexes with Schiff base derived from 2-acetylpyridine and l-tryptophan. *Inorg. Chem. Commun.* **2012**, *22*, 68–72. [CrossRef]
22. Tsuchida, E. Oxovanadium(III–V) mononuclear complexes and their linear assemblies bearing tetradentate Schiff base ligands: Structure and reactivity as multielectron redox catalysts. *Coord. Chem. Rev.* **2003**, *237*, 213–228. [CrossRef]
23. Patra, S.; Chatterjee, S.; Si, T.K.; Mukherjea, K.K. Synthesis, structural characterization, VHPO mimicking peroxidative bromination and DNA nuclease activity of oxovanadium(V) complexes. *Dalton Trans.* **2013**, *42*, 13425–13435. [CrossRef]
24. Sathyadevi, P.; Krishnamoorthy, P.; Butorac, R.R.; Cowley, A.H.; Bhuvanesh, N.S.P.; Dharmaraj, N. Effect of substitution and planarity of the ligand on DNA/BSA interaction, free radical scavenging and cytotoxicity of diamagnetic Ni(II) complexes: A systematic investigation. *Dalton Trans.* **2011**, *40*, 9690–9702. [CrossRef]
25. Raj Kumar, R.; Mohamed Subarkhan, M.K.; Ramesh, R. Synthesis and structure of nickel(II) thiocarboxamide complexes: Effect of ligand substitutions on DNA/protein binding, antioxidant and cytotoxicity. *RSC Adv.* **2015**, *5*, 46760–46773. [CrossRef]
26. Galati, G.; Chan, T.; Wu, B.; O'Brien, P.J. Glutathione-dependent generation of reactive oxygen species by the peroxidase-catalyzed redox cycling of flavonoids. *Chem. Res. Toxicol.* **1999**, *12*, 521–525. [CrossRef]
27. Galati, G. Prooxidant activity and cellular effects of the phenoxyl radicals of dietary flavonoids and other polyphenolics. *Toxicology* **2002**, *177*, 91–104. [CrossRef]
28. Rice-Evans, C.A.; Miller, N.J.; Paganga, G. Structure-antioxidant activity relationships of flavonoids and phenolic acids. *Free Radic. Biol. Med.* **1996**, *20*, 933–956. [CrossRef]
29. Prior, R.L.; Wu, X.; Gu, L. Flavonoid metabolism and challenges to understanding mechanisms of health effects. *J. Sci. Food Agric.* **2006**, *86*, 2487–2491. [CrossRef]
30. Kessler, M.; Ubeaud, G.; Jung, L. Anti- and pro-oxidant activity of rutin and quercetin derivatives. *J. Pharm. Pharmacol.* **2003**, *55*, 131–142. [CrossRef]
31. Pasayat, S.; Böhme, M.; Dhaka, S.; Dash, S.P.; Majumder, S.; Maurya, M.R.; Plass, W.; Kaminsky, W.; Dinda, R. Synthesis, Theoretical Study and Catalytic Application of Oxidometal (Mo or V) Complexes: Unexpected Coordination Due to Ligand Rearrangement through Metal-Mediated C-C Bond Formation. *Eur. J. Inorg. Chem.* **2016**, *2016*, 1604–1618. [CrossRef]
32. Dinda, R.; Majhi, P.K.; Sengupta, P.; Pasayat, S.; Ghosh, S.; Falvello, L.R.; Mak, T.C.W. Alkali metal (Na^+ and K^+)-mediated supramolecular assembly of oxovanadium(V) complexes: Synthesis and structural characterization. *Polyhedron* **2010**, *29*, 248–253. [CrossRef]
33. Dash, S.P.; Pasayat, S.; Saswati; Dash, H.R.; Das, S.; Butcher, R.J.; Dinda, R. Oxovanadium(V) complexes incorporating tridentate aroylhydrazoneoximes: Synthesis, characterizations and antibacterial activity. *Polyhedron* **2012**, *31*, 524–529. [CrossRef]
34. Dash, S.P.; Panda, A.K.; Dhaka, S.; Pasayat, S.; Biswas, A.; Maurya, M.R.; Majhi, P.K.; Crochet, A.; Dinda, R. A study of DNA/BSA interaction and catalytic potential of oxidovanadium(V) complexes with ONO donor ligands. *Dalton Trans.* **2016**, *45*, 18292–18307. [CrossRef]
35. Saswati; Adão, P.; Majumder, S.; Dash, S.P.; Roy, S.; Kuznetsov, M.L.; Pessoa, J.C.; Gomes, C.S.B.; Hardikar, M.R.; Tiekink, E.R.T.; et al. Synthesis, structure, solution behavior, reactivity and biological evaluation of oxidovanadium(IV/V) thiosemicarbazone complexes. *Dalton Trans.* **2018**, *47*, 11358–11374. [CrossRef]
36. Banerjee, A.; Dash, S.P.; Mohanty, M.; Sanna, D.; Sciortino, G.; Ugone, V.; Garribba, E.; Reuter, H.; Kaminsky, W.; Dinda, R. Chemistry of mixed-ligand oxidovanadium(IV) complexes of aroylhydrazones incorporating quinoline derivatives: Study of solution behavior, theoretical evaluation and protein/DNA interaction. *J. Inorg. Biochem.* **2019**, *199*, 110786. [CrossRef]
37. Roy, S.; Böhme, M.; Dash, S.P.; Mohanty, M.; Buchholz, A.; Plass, W.; Majumder, S.; Kulanthaivel, S.; Banerjee, I.; Reuter, H.; et al. Anionic Dinuclear Oxidovanadium(IV) Complexes with Azo Functionalized Tridentate Ligands and μ-Ethoxido Bridge Leading to an Unsymmetric Twisted Arrangement: Synthesis, X-ray Structure, Magnetic Properties, and Cytotoxicity. *Inorg. Chem.* **2018**, *57*, 5767–5781. [CrossRef]
38. Dash, S.P.; Panda, A.K.; Pasayat, S.; Dinda, R.; Biswas, A.; Tiekink, E.R.T.; Mukhopadhyay, S.; Bhutia, S.K.; Kaminsky, W.; Sinn, E. Oxidovanadium(V) complexes of aroylhydrazones incorporating heterocycles: Synthesis, characterization and study of DNA binding, photo-induced DNA cleavage and cytotoxic activities. *RSC Adv.* **2015**, *5*, 51852–51867. [CrossRef]

39. Lima, S.; Banerjee, A.; Mohanty, M.; Sahu, G.; Kausar, C.; Patra, S.K.; Garribba, E.; Kaminsky, W.; Dinda, R. Synthesis, structure and biological evaluation of mixed ligand oxidovanadium(IV) complexes incorporating 2-(arylazo)phenolates. *New J. Chem.* **2019**, *43*, 17711–17725. [CrossRef]

40. Banerjee, A.; Mohanty, M.; Lima, S.; Samanta, R.; Garribba, E.; Sasamori, T.; Dinda, R. Synthesis, structure and characterization of new dithiocarbazate-based mixed ligand oxidovanadium(IV) complexes: DNA/HSA interaction, cytotoxic activity and DFT studies. *New J. Chem.* **2020**, *44*, 10946–10963. [CrossRef]

41. Dash, S.P.; Panda, A.K.; Pasayat, S.; Majumder, S.; Biswas, A.; Kaminsky, W.; Mukhopadhyay, S.; Bhutia, S.K.; Dinda, R. Evaluation of the cell cytotoxicity and DNA/BSA binding and cleavage activity of some dioxidovanadium(V) complexes containing aroylhydrazones. *J. Inorg. Biochem.* **2015**, *144*, 1–12. [CrossRef]

42. Mohanty, M.; Maurya, S.K.; Banerjee, A.; Patra, S.A.; Maurya, M.R.; Crochet, A.; Brzezinski, K.; Dinda, R. In vitro cytotoxicity and catalytic evaluation of dioxidovanadium(V) complexes in an azohydrazone ligand environment. *New J. Chem.* **2019**, *43*, 17680–17695. [CrossRef]

43. Dash, S.P.; Panda, A.K.; Pasayat, S.; Dinda, R.; Biswas, A.; Tiekink, E.R.T.; Patil, Y.P.; Nethaji, M.; Kaminsky, W.; Mukhopadhyay, S.; et al. Syntheses and structural investigation of some alkali metal ion-mediated LV^{V}O^{2-} (L^{2-} = tridentate ONO ligands) species: DNA binding, photo-induced DNA cleavage and cytotoxic activities. *Dalton Trans.* **2014**, *43*, 10139–10156. [CrossRef] [PubMed]

44. Dash, S.P.; Roy, S.; Mohanty, M.; Carvalho, M.F.N.N.; Kuznetsov, M.L.; Pessoa, J.C.; Kumar, A.; Patil, Y.P.; Crochet, A.; Dinda, R. Versatile Reactivity and Theoretical Evaluation of Mono- and Dinuclear Oxidovanadium(V) Compounds of Aroylazines: Electro-generation of Mixed-Valence Divanadium(IV,V) Complexes. *Inorg. Chem.* **2016**, *55*, 8407–8421. [CrossRef]

45. Dash, S.P.; Majumder, S.; Banerjee, A.; Carvalho, M.F.N.N.; Adão, P.; Pessoa, J.C.; Brzezinski, K.; Garribba, E.; Reuter, H.; Dinda, R. Chemistry of Monomeric and Dinuclear Non-Oxido Vanadium(IV) and Oxidovanadium(V) Aroylazine Complexes: Exploring Solution Behavior. *Inorg. Chem.* **2016**, *55*, 1165–1182. [CrossRef] [PubMed]

46. Nemeikaite-Ceniene, A.; Imbrasaite, A.; Sergediene, E.; Cenas, N. Quantitative structure-activity relationships in prooxidant cytotoxicity of polyphenols: Role of potential of phenoxyl radical/phenol redox couple. *Arch. Biochem. Biophys.* **2005**, *441*, 182–190. [CrossRef]

47. Addison, A.W.; Rao, T.N.; Reedijk, J.; van Rijn, J.; Verschoor, G.C. Synthesis, structure, and spectroscopic properties of copper(II) compounds containing nitrogen–sulphur donor ligands; the crystal and molecular structure of aqua[1,7-bis(*N*-methylbenzimidazol-2′-yl)-2,6-dithiaheptane]copper(II) perchlorate. *J. Chem. Soc. Dalton Trans.* **1984**, 1349–1356. [CrossRef]

48. Spek, A.L. checkCIF validation ALERTS: What they mean and how to respond. *Acta Crystallogr. E Crystallogr. Commun.* **2020**, *76*, 1–11. [CrossRef]

49. Krishnamoorthy, P.; Sathyadevi, P.; Butorac, R.R.; Cowley, A.H.; Bhuvanesh, N.S.P.; Dharmaraj, N. Copper(I) and nickel(II) complexes with 1:1 vs. 1:2 coordination of ferrocenyl hydrazone ligands: Do the geometry and composition of complexes affect DNA binding/cleavage, protein binding, antioxidant and cytotoxic activities? *Dalton Trans.* **2012**, *41*, 4423–4436. [CrossRef] [PubMed]

50. Sirajuddin, M.; Ali, S.; Badshah, A. Drug-DNA interactions and their study by UV-Visible, fluorescence spectroscopies and cyclic voltametry. *J. Photochem. Photobiol. B* **2013**, *124*, 1–19. [CrossRef] [PubMed]

51. Saswati; Mohanty, M.; Banerjee, A.; Biswal, S.; Horn, A.; Schenk, G.; Brzezinski, K.; Sinn, E.; Reuter, H.; Dinda, R. Polynuclear zinc(II) complexes of thiosemicarbazone: Synthesis, X-ray structure and biological evaluation. *J. Inorg. Biochem.* **2020**, *203*, 110908. [CrossRef] [PubMed]

52. Ganeshpandian, M.; Loganathan, R.; Suresh, E.; Riyasdeen, A.; Akbarsha, M.A.; Palaniandavar, M. New ruthenium(II) arene complexes of anthracenyl-appended diazacycloalkanes: Effect of ligand intercalation and hydrophobicity on DNA and protein binding and cleavage and cytotoxicity. *Dalton Trans.* **2014**, *43*, 1203–1219. [CrossRef]

53. Tan, J.; Wang, B.; Zhu, L. DNA binding and oxidative DNA damage induced by a quercetin copper(II) complex: Potential mechanism of its antitumor properties. *J. Biol. Inorg. Chem.* **2009**, *14*, 727–739. [CrossRef]

54. Perron, N.R.; Hodges, J.N.; Jenkins, M.; Brumaghim, J.L. Predicting how polyphenol antioxidants prevent DNA damage by binding to iron. *Inorg. Chem.* **2008**, *47*, 6153–6161. [CrossRef]

55. Sung, H.; Ferlay, J.; Siegel, R.L.; Laversanne, M.; Soerjomataram, I.; Jemal, A.; Bray, F. Global Cancer Statistics 2020: GLOBOCAN Estimates of Incidence and Mortality Worldwide for 36 Cancers in 185 Countries. *CA Cancer J. Clin.* **2021**, *71*, 209–249. [CrossRef]

56. Krishnamoorthy, P.; Sathyadevi, P.; Cowley, A.H.; Butorac, R.R.; Dharmaraj, N. Evaluation of DNA binding, DNA cleavage, protein binding and in vitro cytotoxic activities of bivalent transition metal hydrazone complexes. *Eur. J. Med. Chem.* **2011**, *46*, 3376–3387. [CrossRef]

57. Raja, D.S.; Paramaguru, G.; Bhuvanesh, N.S.P.; Reibenspies, J.H.; Renganathan, R.; Natarajan, K. Effect of terminal *N*-substitution in 2-oxo-1,2-dihydroquinoline-3-carbaldehyde thiosemicarbazones on the mode of coordination, structure, interaction with protein, radical scavenging and cytotoxic activity of copper(II) complexes. *Dalton Trans.* **2011**, *40*, 4548–4559. [CrossRef]

58. Chen, Y.; Qin, M.-Y.; Wu, J.-H.; Wang, L.; Chao, H.; Ji, L.-N.; Xu, A.-L. Synthesis, characterization, and anticancer activity of ruthenium(II)-β-carboline complex. *Eur. J. Med. Chem.* **2013**, *70*, 120–129. [CrossRef]

59. Mohamadi, M.; Yousef Ebrahimipour, S.; Torkzadeh-Mahani, M.; Foro, S.; Akbari, A. A mononuclear diketone-based oxido-vanadium(V) complex: Structure, DNA and BSA binding, molecular docking and anticancer activities against MCF-7, HPG-2, and HT-29 cell lines. *RSC Adv.* **2015**, *5*, 101063–101075. [CrossRef]

60. Moeller, T. (Ed.) *Inorganic Syntheses*; Mc-Graw Hill Book Company: New York, NY, USA, 1957; Volume 5, pp. 113–116.

61. Rigaku Oxford Diffraction. *CrysAlisPRO*; Rigaku Oxford Diffraction Ltd.: Oxfordshire, UK, 2017.
62. Sheldrick, G.M. A short history of SHELX. *Acta Crystallogr. A* **2008**, *64*, 112–122. [CrossRef]
63. Sheldrick, G.M. Crystal structure refinement with SHELXL. *Acta Crystallogr. C Struct. Chem.* **2015**, *71*, 3–8. [CrossRef]
64. Farrugia, L.J. WinGX and ORTEP for Windows: An update. *J. Appl. Crystallogr.* **2012**, *45*, 849–854. [CrossRef]
65. Brandenburg, K.; Berndt, M. *DIAMOND, Version 3.2k*; GbR: Bonn, Germany, 2006.
66. Roy, S.; Mohanty, M.; Miller, R.G.; Patra, S.A.; Lima, S.; Banerjee, A.; Metzler-Nolte, N.; Sinn, E.; Kaminsky, W.; Dinda, R. Probing CO Generation through Metal-Assisted Alcohol Dehydrogenation in Metal-2-(arylazo)phenol Complexes Using Isotopic Labeling (Metal = Ru, Ir): Synthesis, Characterization, and Cytotoxicity Studies. *Inorg. Chem.* **2020**, *59*, 15526–15540. [CrossRef]
67. Bhutia, S.K.; Mallick, S.K.; Maiti, S.; Maiti, T.K. Antitumor and proapoptotic effect of Abrus agglutinin derived peptide in Dalton's lymphoma tumor model. *Chem. Biol. Interact.* **2008**, *174*, 11–18. [CrossRef]

Article

Acute Toxicity Evaluation of Non-Innocent Oxidovanadium(V) Schiff Base Complex

Lidiane M. A. Lima [1], Heide Murakami [2], D. Jackson Gaebler [2], Wagner E. Silva [1], Mônica F. Belian [1], Eduardo C. Lira [3,*] and Debbie C. Crans [2,4]

1 Departamento de Química, Universidade Federal Rural de Pernambuco, Recife-PE 52171-900, Brazil; lidianelimaa@gmail.com (L.M.A.L.); wesdqr@gmail.com (W.E.S.); mfbelian@gmail.com (M.F.B.)
2 Department of Chemistry, Colorado State University, Fort Collins, CO 80523, USA; heideam@gmail.com (H.M.); jackson.Gaebler@rams.colostate.edu (D.J.G.); debbie.crans@coloState.edu (D.C.C.)
3 Centro de Biociências, Departamento de Fisiologia e Farmacologia, Universidade Federal de Pernambuco, Recife-PE 50670-901, Brazil
4 Cell and Molecular Biology Program, Colorado State University, Fort Collins, CO 80513, USA
* Correspondence: eduardo.clira2@ufpe.br; Tel./Fax: +55-(81)-996368873

Abstract: The vanadium(V) complexes have been investigated as potential anticancer agents which makes it essential to evaluate their toxicity for safe use in the clinic. The large-scale synthesis and the acute oral toxicity in mice of the oxidovanadium(V) Schiff base catecholate complex, abbreviated as [VO(HSHED)dtb] containing a redox-active ligand with tridentate Schiff base (HSHED = N-(salicylide neaminato)-N'-(2-hydroxyethyl)-1,2-ethylenediamine) and dtb = 3,5-di-(t-butyl)catechol ligands were carried out. The body weight, food consumption, water intake as well biomarkers of liver and kidney toxicity of the [VO(HSHED)dtb] were compared to the precursors, sodium orthovanadate, and free ligand. The 10-fold scale-up synthesis of the oxidovanadium(V) complex resulting in the preparation of material in improved yield leading to 2–3 g (79%) material suitable for investigating the toxicity of vanadium complex. No evidence of toxicity was observed in animals when acutely exposed to a single dose of 300 mg/kg for 14 days. The toxicological results obtained with biochemical and hematological analyses did not show significant changes in kidney and liver parameters when compared with reference values. The low oral acute toxicity of the [VO(HSHED)dtb] is attributed to redox chemistry taking place under biological conditions combined with the hydrolytic stability of the oxidovanadium(V) complex. These results document the design of oxidovanadium(V) complexes that have low toxicity but still are antioxidant and anticancer agents.

Keywords: oxidovanadium(V); vanadium Schiff base coordination complex; low acute toxicity

Citation: Lima, L.M.A.; Murakami, H.; Gaebler, D.J.; Silva, W.E.; Belian, M.F.; Lira, E.C.; Crans, D.C. Acute Toxicity Evaluation of Non-Innocent Oxidovanadium(V) Schiff Base Complex. *Inorganics* **2021**, 9, 42. https://doi.org/10.3390/inorganics9060042

Academic Editor: Isabel Correia

Received: 20 February 2021
Accepted: 13 May 2021
Published: 24 May 2021

Publisher's Note: MDPI stays neutral with regard to jurisdictional claims in published maps and institutional affiliations.

1. Introduction

A wide range of vanadium(IV) and (V) coordination complexes and salts display desirable biological effects such as antidiabetic and anticancer agents [1–19]. Vanadium coordination complexes such as bis(maltolato)oxidovanadium(IV) (BMOV) [5,9] and bis(allixinato)oxidovanadium(IV) ([VO(alx)$_2$]) (Figure 1) [20] have been found to result in glucose-lowering levels in streptozotocin (STZ)-induced rats [21–26]. Furthermore, vanadium Schiff base complexes such as dioxidovanadium(V)dipicolinate ([VO$_2$dipic]$^-$) [18,27,28] and V(V)-catecholate substituted complexes as [VO(HSHED)dtb] [19] and [VO(naph-L-Pheol-im)(8HQ)] [1] have demonstrated anticancer properties against human ovarian, prostate and brain cells as well as enhancing the effects of oncolytic viruses (Figure 1).

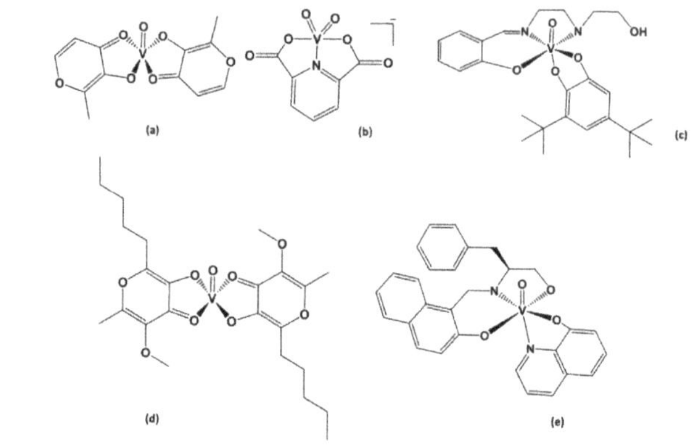

Figure 1. Structures of vanadium complexes with antidiabetic and/or anticancer properties, (**a**) bis(maltolato)oxidovanadium(IV) [BMOV], (**b**) dioxidovanadium(V)dipicolinate $[VO_2(dipic)]^-$, (**c**) V(V)-catecholate substituted [VO(HSHED)dtb] (Hshed = N-(salicylideneaminato)-N′-(2-hydroxyethyl)-1,2-ethylenediamine and dtb = 3,5-di(t-butyl)catechol), (**d**) bis(allixinato) oxidovanadium(IV) $[VO(alx)_2]$, and (**e**) V(V)-Schiff base substituted [VO(naph-L-Pheol-im)(8HQ)] (L-pheol-im = L-phenylalaninol, 8HQ = hydroxyquinoline).

Due to the prospective application of various compounds as therapeutic agents, significant effort has been directed toward demonstrating that such compounds have no toxic effects in vivo and in vitro [29–31]. In the case of vanadium compounds, few studies have been carried out to determine the toxicity of the coordination complexes [32], although literature reports exist for vanadium salts and simple vanadium oxides [32–34]. It has been known that an excess of vanadate induces toxic effects in cells by oxidative stress increasing [35–37]. In vivo and in vitro studies show that high levels of reactive oxygen species are often implicated in vanadium deleterious effects [38–41]. However, the ability of V-complexes to inhibit protein phosphatases enhances their potential application as a therapeutic agent and has been an area of extensive research [42].

The presence of catechol-moieties is an established feature of leading anticancer agents used in the clinic including doxorubicin, daunorubicin, and mitomycin C (quinone-based compounds) [43–45]. Furthermore, the Schiff base vanadium complexes containing catecholates display a particularly interesting redox-chemistry in cell environments and have been investigated due to antidiabetic [46,47] and anticancer properties [19,48–53], however, there is no study with a focus on their toxicity. Thus, the evaluation of the toxicity of vanadium Schiff base catecholate complexes in vivo and in vitro is fundamentally important for potential medicinal applications [54].

In this manuscript, the focus is to evaluate the toxicity of the vanadium Schiff base di-t-butyl substituted catecholate complex—[VO(HSHED)dtb]—compared with the vanadate and the free catecholate ligand. The toxicological analysis was performed as recommended by the OECD (Organization for Economic Cooperation and Development) 423 guidelines characterizing the acute toxicity in mice.

2. Results

2.1. Synthesis of [VO(HSHED)dtb] Complex

Considering that the amount of material to carry out in vivo animal studies was 100 times greater than the scale of the vanadium catecholate previously prepared, it was necessary to scale up the reactions to prepare the target compounds for the biological studies. Furthermore, considering that V(V)-Schiff base vanadium complexes are non-innocent coordination complexes and thus redox-active, attempts to change the reaction scale was non-trivial. Impure side products were avoided maintaining the reaction under argon atmosphere and the solvent to reactant ratio was kept as that described in the originally reported reaction [19,55,56]. Increasing the literature reported milligrams starting material reactions by a factor of about 7–10 but keeping the reactant–solvent ratios (1:70) constant, gram–scale of the target compound was obtained in each run. The major change to the scaled-up reactions was increasing the reaction time from 3 h to 48 h. This increased the amounts of products, from 200 mg to 2 g for each reaction. Interestingly, despite the scale-up of the reactions, the yields increased to 79% from the first reported 40% [55]. The improved yield is due to the longer reaction times in addition to minimizing oxidation by keeping the reaction under argon. ^{51}V NMR of the scaled-up reaction product was identical to the reported previously [55,56].

2.2. Stability of [VO(HSHED)dtb] Complex

Studies were previously carried out with [VO(HSHED)dtb] and were found to remain stable for few hours in cell culture media so the compound could be taken up by cells [19,56]. Even though the compound has low water solubility, it was found to be readily absorbed into cells from cell culture studies [56].

Furthermore, a low concentration was observed in membrane model studies. Although upon extended periods of time (5–24 h) contact with water will result in complex hydrolysis. The compound hydrolyzes to form starting materials, vanadate, catechol, Schiff base ligand which then degrades into salicylaldehyde and amine as reported previously [19,56]. However, the [VO(HSHED)dtb] has a longer lifetime than other Schiff base catecholate vanadium complexes and remains intact for a few hours before hydrolysis in various aqueous environments.

The experimental design called for administration of [VO(HSHED)dtb] by oral gavage, the concerns with compound solubility are less. We point to the fact the hydrophobic nature of this compound would increase the absorption of the complex once it was administered. Furthermore, if the solutions are freshly prepared and used immediately the amount of hydrolysis is minimized. That is, this mode of administration would allow delivery of compounds with low water solubility and lower stability because compounds would be delivered directly to the stomach of the animal.

To determine how much of the [VO(HSHED)dtb] was intact during the treatment, the UV-Vis (Ultraviolet-visible spectroscopy) under the conditions the compound was administered for the toxicological studies were investigated. In Figure 2, is shown the UV-Vis spectra of [VO(HSHED)dtb] dissolved in 5%, 10%, and 20% Tween 80 compared to a spectrum where the compounds were dissolved in DMSO and added to an aqueous solution. We carried out these studies as a function of time ranging from t = 0 (black curves) to t = 24 h (turquoise curves) at three different Tween 80 concentrations. As observed in Figure 2 all samples were sufficiently stable for the short periods of time needed for the administration, and importantly the addition of Tween 80 did not reduce the stability of the compounds significantly. However, significant amounts of compound decomposed after 24 h, particularly in the presence of the Tween 80. We conclude that Tween 80 destabilizes [VO(HSHED)dtb] but that over 24 h of time the presence of Tween 80 facilitates the complete hydrolysis and more rapidly than the control aqueous solution.

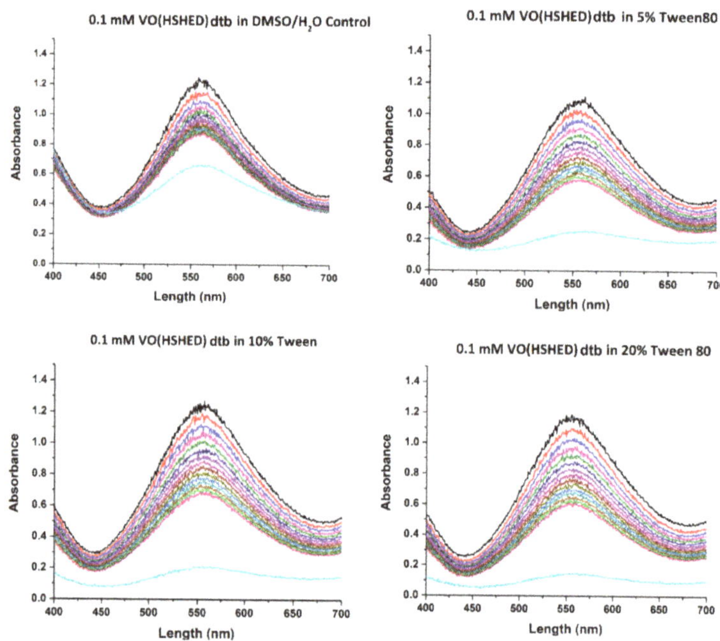

Figure 2. The UV-Vis spectra are shown of 0.1 mM [VO(HSHED)dtb] in DMSO, H$_2$O, and in 5%, 10%, and 20% Tween 80 at various time increments, from t = 0 h to t = 24 h.

Administration of compounds in animals is always subjected to the pharmacokinetic and ADME (Absorption, Distribution, Metabolism, and Excretion) as well as the pharmacodynamic, and as such it is of interest what forms of the compounds are present. However, since the compounds are rapidly distributed making concentrations low, and it is non-trivial to measure since this compound undergoes both hydrolytic and redox chemistry [19,56], methods are being developed to measure such different forms [57,58]. However, at present it seems pertinent and more profitable for the determination of whether the compound is toxic, to measure the impact on toxicity markers. Such an approach is particularly beneficial if these parameters are compared to marker formation in animals treated with the potentially toxic hydrolysis products, vanadate, and catecholate.

2.3. Acute Toxicity in Mice

Neither animal mortality, severe toxicity nor detrimental side effects were observed during the entire 14 days experimental period in all groups, except for the very high concentration (2000 mg/kg) of [VO(HSHED)dtb]. At the high dose administration of vanadium(V) Schiff base complex resulted in the death of all animals during the first 24 h, and as a result, it was not possible to collect the data on the 14th day. In contrast, no signs of toxicity were observed after the administration of [VO(HSHED)dtb] in a single dose of 300 mg/kg, i.e., no-observed adverse-effect level (NOAEL). Neither behavior alterations like lethargy, sleep, tremors, salivation, convulsion, nor common side effects for V-compounds, such as diarrhea, were observed in all groups for 14 days [59].

The acute toxicity was performed using OECD 423 guideline. The OECD is an international organization that works to build better policies for better lives, and they have developed protocols for animal studies [60]. This guideline allows the characterization of substances according to the Globally Harmonized System (GHS) for the classification of chemicals that cause acute toxicity. The classification of the test substance is based on the determination of the mortality dose(s), when there are no effects observed this concentration will be the highest dose tested by gavage in a single dose. Thus, the compounds could be rated as categories 1 to 5, from the most to least toxic, according to the LD_{50} (median lethal dose) values in mg/kg b.w from each category [60]. [VO(HSHED)dtb] and 3,5-di(t-butyl)catechol was found to be category 4, with median lethal dose LD_{50} estimated to be between 300 mg/kg and 2000 mg/kg b.w. and orthovanadate is more toxic than both (category 3) with LD_{50} estimated to be between 50 mg/kg and 300 mg/kg b.w. Furthermore, the use of female *Swiss* mice was chosen in this study because literature surveys of conventional LD_{50} tests show that usually there is little difference in sensitivity between the sexes, but in these cases where differences are observed, females are generally slightly more sensitive [61].

2.4. Clinical Observations

The observations were made on the animals treated with the vanadium coordination complex, orthovanadate salt, the catechol ligand, and control group. Figure 3 shows that there is no statistical difference in body weight, food consumption, and water intake between the control treated with water and treated groups ($p > 0.05$) throughout 14 days and at the end of treatment (Table 1).

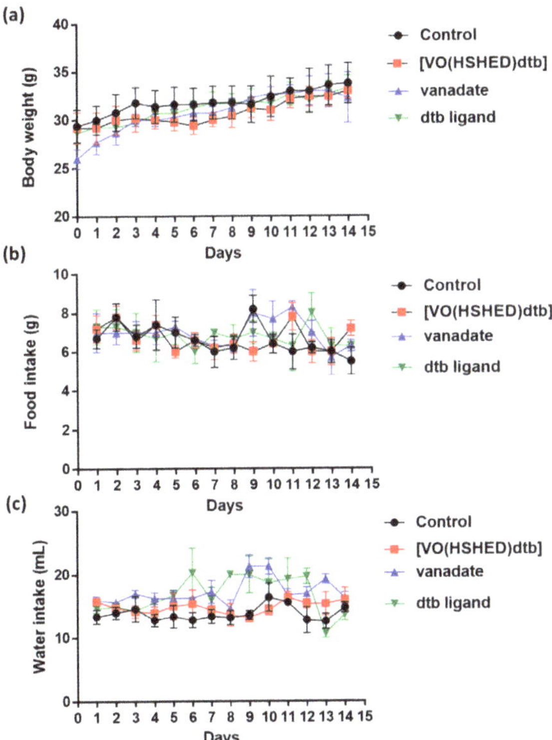

Figure 3. Effects on (**a**) body weight, (**b**) food intake, and (**c**) fluid intake throughout 14 days in mice administered acute levels of 300 mg/kg [VO(HSHED)dtb], 50 mg/kg orthovanadate and 300 mg/kg dtb ligand (3,5-di(t-butyl)catechol) groups by oral gavage. The data are presented as the mean ± SD ($n = 5$).

Table 1. Effects by acute oral administration of 300 mg/kg [VO(HSHED)dtb], 50 mg/kg orthovanadate, and 300 mg/kg dtb ligand, in different mice groups with regards to body weight, weight gain, food intake, and water intake at day 14.

Parameter	Study Group			
	Control	[VO(HSHED)dtb]	Orthovanadate	3,5-di(t-butyl)catechol
Body weight (g)	31.8 ± 1.8	30.7 ± 1.0	30.8 ± 0.9	31.2 ± 0.7
Weight gain (g)	0.3 ± 0.0	0.3 ± 0.1	0.20 ±.1	0.2 ± 0.1
Food intake (g/24 h)	6.8 ± 0.3	6.7 ± 0.3	7.0 ± 0.1	6.8 ± 0.3
Water intake (mL/24 h)	13.8 ± 0.8	15.0 ± 0.6	17.3 ± 0.5 *	16.7 ± 1.5

The data are presented as the mean ± SD ($n = 5$). Values are statistically significant at * $p < 0.05$ compared to the control group.

Once the [VO(HSHED)dtb] was found to cause no changes in clinical parameters monitored at the 300 mg/kg level, the biochemical and hematological parameters were monitored to investigate further toxic effects of the compounds as shown in Tables 2 and 3. Furthermore, the macroscopic analysis did not show any changes in tissue weight such as liver, kidney, heart, spleen, and lung in animals treated with 300 mg/kg [VO(HSHED)dtb] for the 14 days as shown by the data summarized in Table 4.

Table 2. Effects by acute oral administration of 300 mg/kg [VO(HSHED)dtb], 50 mg/kg orthovanadate, and 300 mg/kg dtb ligand in different mice groups with regard to AST/ALT ratio, albumin, total proteins, globulin, and A/G ratio parameters at day 14.

Biochemical Parameters	Study Group			
	Control	[VO(HSHED)dtb]	Orthovanadate	3,5-di(t-butyl)catechol
AST/ALT ratio	1.1 ± 0.01	0.9 ± 0.02	1.0 ± 0.01	1.1 ± 0.01
Albumin (g/dL)	2.2 ± 0.2	2.1 ± 0.1	2.1 ± 0.1	2.1 ± 0.1
Total proteins (g/dL)	5.7 ± 0.4	5.5 ± 0.2	6.5 ± 0.6	5.8 ± 0.3
Globulin (g/dL)	3.5 ± 0.3	3.4 ± 0.2	4.3 ± 0.5 *	3.8 ± 0.3
A/G ratio	0.6 ± 0.01	0.6 ± 0.01	0.5 ± 0.02	0.6 ± 0.02

The data are presented as the mean ± SD ($n = 5$). Values are statistically significant at * $p < 0.05$ compared to the control group.

Table 3. Effects by acute oral administration of 300 mg/kg [VO(HSHED)dtb], 50 mg/kg orthovanadate, and 300 mg/kg dtb ligand in different mice groups with regard to hematological parameters at day 14.

Hematological Parameters	Study Group			
	Control	[VO(HSHED)dtb]	Orthovanadate	3,5-di(t-butyl)catechol
Red blood cell (10^{12}/L)	9.5 ± 0.2	9.3 ± 0.2	8.9 ± 0.2	9.7 ± 0.3
Mean corpuscular volume (Fl)	57.6 ± 0.2	56.2 ± 0.8	57.0 ± 0.6	56.3 ± 0.3
Hemoglobin (g/dL)	12.4 ± 0.5	12.5 ± 0.1	12.3 ± 0.1	13.4 ± 0.2
Mean corpuscular hemoglobin (ρg)	13.5 ± 0.5	15.4 ± 2.0	13.7 ± 0.3	13.8 ± 0.1
mean corpuscular hemoglobin concentration (g/dL)	23.5 ± 0.6	23.9 ± 0.2	23.9 ± 0.2	24.4 ± 0.1
Hematocrit (%)	52.6 ± 0.9	52.5 ± 0.5	51.6 ± 0.8	55.0 ± 1.6
White blood cell (10^9/L)	12.6 ± 1.3	8.6 ± 1.3 *	12.7 ± 1.9	10.9 ± 1.1
Lymphocytes (%)	82.6 ± 0.8	81.8 ± 0.9	84.3 ± 1.5	83.2 ± 1.3

The data are presented as the mean ± SD ($n = 5$). Values are statistically significant at * $p < 0.05$ compared to the control group.

Table 4. Effects by acute oral administration of 300 mg/kg [VO(HSHED)dtb], 50 mg/kg orthovanadate, and 300 mg/kg dtb ligand in different mice groups with regard to the heart, kidney, liver, spleen, and lung weight at day 14.

Organs (g/100 g b.w)	Study Group			
	Control	[VO(HSHED)dtb]	Orthovanadate	3,5-di(t-butyl)catechol
Heart	0.54 ± 0.03	0.53 ± 0.02	0.59 ± 0.04	0.60 ± 0.07
Kidney	5.75 ± 0.29	6.23 ± 0.38	6.11 ± 0.11	5.50 ± 0.54
Liver	1.36 ± 0.10	1.52 ± 0.08	1.70 ± 0.20 *	1.61 ± 0.18
Spleen	0.48 ± 0.05	0.47 ± 0.03	0.66 ± 0.06 *	0.59 ± 0.13
Lung	0.88 ± 0.11	0.79 ± 0.03	0.86 ± 0.14	0.75 ± 0.15

The weight of body organs was measured in mice after 14 days of treatment. The data are presented as the mean $\pm$ SD ($n = 5$). Values are statistically significant at * $p < 0.05$ compared to the control group.

2.5. Hematology and Biochemical Analysis

Treatment with neither the dtb (3,5-di(t-butyl)catechol) ligand nor orthovanadate administration caused any alteration in biochemical and hematology parameters, in contrast to treatment with [VO(HSHED)dtb] which slightly decreased the WBC (white blood cell) level ($p < 0.05$) as shown in Table 3.

The alanine and aspartate aminotransferases (abbreviated ALT and AST, respectively) are commonly used as a marker in the diagnosis of liver injury and disease [62]. High levels of serum ALT activity reflects damage to hepatocyte and is considered to be a highly sensitive and fairly specific preclinical and clinical biomarker of hepatotoxicity [63]. In the present study, there were no significant changes in hepatic enzymes as shown in Figure 4, suggesting that the [VO(HSHED)dtb] did not cause any damage to the hepatic functions. In addition, hepatic-cellular damage leads to a reduction in albumin accompanied by a relative increase in globulins, which decreases the A/G ratio [64]. As shown in Table 2, no experimental significant difference between the control group and 300 mg/kg [VO(HSHED)dtb] ($p > 0.05$) at protein levels suggested the vanadium-catecholate complex did not cause hepatotoxicity in vivo.

As shown in Figure 4, no significant changes were observed in kidney biomarkers such as BUN (blood urea nitrogen) and creatinine which suggested that [VO(HSHED)dtb] did not cause nephrotoxicity [65]. However, the treatment with the free catechol ligand did increase significantly creatine levels ($p < 0.05$) documenting some toxicity is observed of the free ligand at 300 mg/kg single dose. These results provide evidence that the vanadium Schiff base catecholate complex does not exert toxicity in vivo up to therapeutic doses.

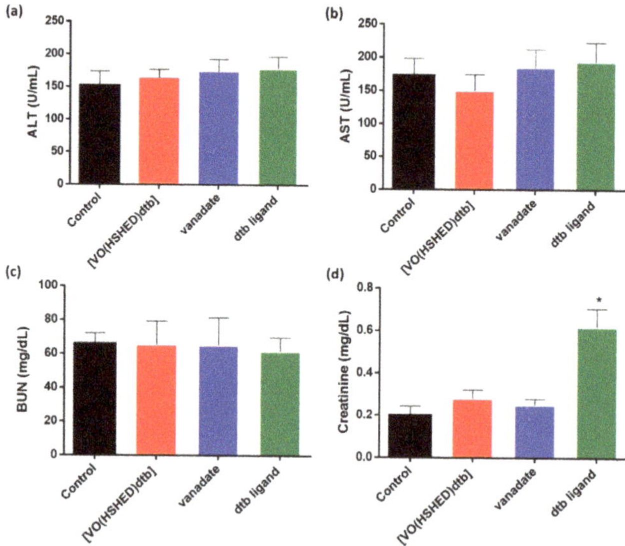

Figure 4. Effects by acute oral administration of 300 mg/kg [VO(HSHED)dtb], 50 mg/kg ortho-vanadate, and 300 mg/kg dtb ligand on mice groups with regard to biomarkers (**a**) ALT (Alanine aminotransferase), (**b**) AST (aspartate aminotransferase), (**c**) BUN (blood urea nitrogen), and (**d**) plasmatic creatinine levels on day 14. The data are presented as the mean $\pm$ SD (n = 5). Values are statistically significant at * $p < 0.05$ compared to the control group.

3. Discussion

The evaluation of the toxic effects of an oxidovanadium(V) Schiff base complex was carried out as a result of the recently discovered anticancer effects of this complex. This complex has anticancer activity against prostate, brain, and breast cancer cells [19,56], however, no previous toxicity studies of [VO(HSHED)dtb] in vivo have been reported. Previous toxicity studies were mainly carried out with simple salts and oxides, including metavanadate and orthovanadate, vanadyl sulfate, and vanadium pentoxide [32,66,67]. A few coordination complexes have also been investigated, and those include vanadium dipicolinate complexes [28,68]. This [VO(HSHED)dtb] complex was chosen because it is a non-innocent vanadium complex and thus can undergo redox chemistry both at the ligand and the metal site. This chemical reactivity is thus fundamentally different from the vanadium coordination compounds and salts that have previously been subjected to toxicity studies. Furthermore, the [VO(HSHED)dtb] compound deviates from these previous compounds in that it is more hydrolytically stable, and that results in it being significantly more readily taken up by cells. Finally, it was recently reported with anticancer properties that exceeded cisplatin [19].

In order to generate enough materials for toxicological studies, the preparation of the redox-active target compound was scaled up to a 2 g scale. The preparation of the compound was done without loss in yields. This was accomplished by keeping the reaction under argon, keeping the same substrate to solvent ratio (1:70) but increasing the reaction time. This approach saved much time in preparation of the 20-plus gram pure vanadium complex needed for the animal studies and suggests that although these compounds are not trivial to prepare there is potential for scale-up.

The toxicity of vanadium(IV) and (V) compounds with antidiabetic effects have been evaluated after acute parenteral administration but the effects are less compared to oral administration [69]. This is presumably because the compound is poorly absorbed by the gastrointestinal tract [9,69]. Nevertheless, attempts have been made to develop vanadium derivatives with biological effects and low toxicity like the coordination complex [bis(maltolato)oxidovanadium(IV)] (BMOV) [5,70]. This complex has high lipophilicity and was selected for clinical trials where it demonstrated glucose-lowering effects in animals and humans [5]. Similarly, as reported previously, the hydrophobic nature of this [VO(HSHED)dtb] would enhance the absorption of the complex and exert its anticarcinogenic effects [56].

Acute toxicity evaluation of [VO(HSHED)dtb] using a dose of 300 mg/kg did not induce mortality or symptoms of severe toxicity throughout the period of the experiment. Administration of oxidovanadium(V) Schiff base complex did not affect hematological indices, hepatic, and renal biomarkers levels. Furthermore, the most common side-effects of vanadium treatment, such as weight loss and diarrhea [59], were not observed after acute exposure to a dosage of 300 mg/kg [VO(HSHED)dtb] for 14 days.

Vanadium compounds are known to often convert to other species upon administration in biological environments [71]. The toxicity of the [VO(HSHED)dtb] should be determined and compared to the effects of its components, i.e., free ligand catechol, Schiff base, and vanadate before applications as a therapeutic compound [31,72–74]. The hydrolysis of the complex and the Schiff base generates the salicylaldehyde and the amine [19,56]. Since previous works have reported that simple aldehyde and amine-based compounds are safe and used for designing new drugs [75,76], accordingly, to investigate the most toxic component of the ligands we focused on catechols and designed an experiment investigating the toxicity of the catechol.

Although the parent catechol is known to show some toxicity, the 3,5-di(t-butyl)catechol is much less toxic with an LD_{50} value in mice of 1.040 g/kg b.w. by OG (oral gavage) [77]. However, our results showed a slight increase of creatinine levels in animals treated with 3,5-di(t-butyl) catechol suggesting nephrotoxicity. Thus, both [VO(HSHED)dtb] and 3,5-di(t-butyl)catechol were rated at the same toxicity level according to the Globally Harmonized System (GHS) with LD_{50} values between 300 and 2000 mg/kg.

The sodium orthovanadate salt has been reported to cause insulin-mimetic activity [9,10], reduction in glucose levels in STZ-diabetic animals [9,10] and be potent phosphatase inhibitors [2,3], and previous toxic studies showed LD_{50} values in rats of 36.3 mg/kg and 330 mg/kg [74]. Similar experiments with $NaVO_3$ showed LD_{50} values in mice of 74.6 mg/kg and 35.9 mg/kg for oral and intraperitoneal (i.p) administration, respectively [32,34,78]. The toxicity of the [VO(HSHED)dtb] was found to be lower than vanadate. This result confirmed some previous reports that have shown the vanadium will be less toxic when coordinated to a ligand [5–8,79]. However, other complexes show similar effects between the vanadium coordination complex and salt which is explained the vanadium compound readily hydrolyze under physiological conditions [71].

Attributing the biological effects to coordination complexes is less trivial than simple organic compounds, because depending on the metal complex and the conditions such complexes may undergo ligand exchange, hydrolysis, or formation of oxidation products [80]. Therefore, it is important that the stability of the compounds under investigation be determined under the conditions of the biological system [71,80–83]. In this study, the stability of the [VO(HSHED)dtb] in the solutions used for oral gavage treatments was characterized. Since [VO(HSHED)dtb] has limited solubility aqueous solution and to be able to solubilize enough for oral gavage administration, 10% Tween 80 was used. Accordingly, we measured out the stability of [VO(HSHED)dtb] in 5%, 10%, and 20% Tween, and we found that the Tween barely affected the stability of [VO(HSHED)dtb]. The toxicity results we obtained showed that the [VO(HSHED)dtb] was less toxic, and this is consistent with the interpretation that the compound was intact for some time after administration.

Vanadium compounds undergo redox chemistry under physiological conditions, can act as a strong pro-oxidant, and interact synergistically with other oxidants enhancing oxidative stress [84,85]. These findings are consistent with redox chemistry taking place under biological conditions and may impact the low toxicity properties of the [VO(HSHED)dtb]. These results are very encouraging because they demonstrate that low toxicity of a vanadium complex is possible and importantly lower toxicity than the salt ($H_2VO_4^-$) and hence lend support for future potential applications on cancer therapy. These findings suggest that it will be possible to design vanadium complexes that have even lower toxicity even when containing redox-active components.

4. Material and Methods

4.1. Reagents and Chemical Analysis

Catechol, 3,5-di(t-butyl)catechol, salicylaldehyde, N-(2-hydroxyethyl) ethylenediamine, vanadyl sulfate were purchased from Sigma Aldrich. Chemicals were used without purification. HPLC (high performance liquid chromatography) grade solvents were used for the synthesis and characterization of the oxidovanadium(V) complex. The compounds were characterized in the Colorado State Central Instrumentation Facility. Nuclear Magnetic Resonance (NMR) measurements were performed using a Bruker spectrometer at 78.9 MHz for ^{51}V and 400.13 MHz for ^{1}H, using 4096 scans and a window from -53 to -1043 ppm as reported previously [86–88]. V chemical shifts were measured in parts per million from $VOCl_3$ as external standard at 0.00 ppm with upfield shifts considered negative.

4.2. Synthesis of [VO₂(HSHED)] Complex

The precursor [VO$_2$(HSHED)] complex was synthesized using a condensation reaction between salicylaldehyde and N-(2-hydroxyethyl)ethylenediamine followed by a coupling reaction to vanadyl sulfate using previously reported methods [89,90]. The NMR spectra have shown the complex to have ^{51}V NMR chemical shifts consistent with the literature (-529 ppm vs. $VOCl_3$) [89].

4.3. Synthesis of [VO(HSHED)dtb] Complex

To synthesize [VO(HSHED)dtb] complex, the 3,5-di(t-butyl)catechol (1.11 g, 5.00 mmol) was added to a solution of [VO$_2$(HSHED)] (1.45 g, 5.00 mmol) and stirred in 350 mL of acetone for 48 h under an argon atmosphere (Scheme 1) [89,90]. A dark purple solution formed after 10 min. After 48 h, the solution was concentrated to dryness, and then the solution was vacuum filtered. A minimal amount of acetone (27 mL) was used to dissolve the crude product, followed by the addition of 300 mL of hexane. The solution was left to precipitate overnight at $-20\ ^\circ$C. The precipitate was filtered, washed with cold hexane (75 mL), and dried on a pump for 4 days. Yield 1.90 g (79%). The solid has similar characteristics as reported [VO(HSHED)dtb] [19,89,91].

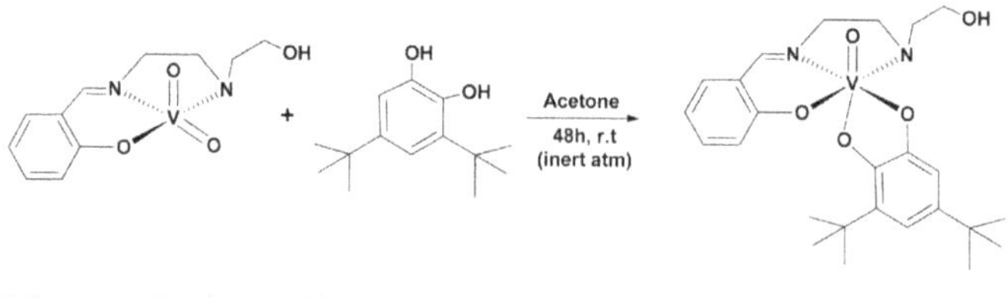

Scheme 1. Synthesis of [VO(HSHED)dtb] using precursor [VO$_2$(HSHED)] complex and 3,5-di(t-butyl) catechol.

4.4. Stability of [VO(HSHED)dtb] Complex

The stability of the [VO(HSHED)dtb] was measured using UV-Vis spectroscopy. [VO(HSHED)dtb] was dissolved in DMSO and added to an aqueous solution at the final concentration of 0.1 mM. The UV-Vis spectra were recorded at t = 0 h, and every 15 s following, and then again at 24 h. Corresponding experiments were done with [VO(HSHED)dtb] dissolved in 5%, 10% and 20% Tween 80 to a final concentration of 0.1 mM.

4.5. Animals

The protocols for these experiments were approved by the Animal Ethics Committee of the Universidade Federal de Pernambuco (process number #004-2019) and conducted in accordance with the Ethical Principles in Animal Research. Twenty-five female Swiss mice (30 ± 5 g) were obtained from the Laboratory of the immunopathology of Keizo Asami (LIKA) at Universidade Federal de Pernambuco (UFPE) and the animal experiments were performed at the Laboratory of Neuroendocrinology and Metabolism at UFPE. The mice were housed in individual cages with a 1–12 light-dark cycle, 22 ± 3 °C, and were given free access to water and conventional lab chow diet ad libitum (Purina, Labina®, Ribeirão Preto, Brazil) during the study period. All experiments were performed between 8:00 and 10:00 am.

4.6. Acute Toxicity in Mice

Acute oral toxicity effect of [VO(HSHED)dtb] was performed in accordance with the Organization for Economic Cooperation and Development (OECD) [60]. After 5 days of acclimatization in individual cages, non-pregnant and nulliparous female mice fasted overnight before the administration of compounds. The [VO(HSHED)dtb] was administered by gavage in mice at a single dose using animal feeding needles (100 µL/100 g b.w). The [VO(HSHED)dtb] was dissolved in warm water (50 °C) with 10 µL of 10% the surfactant Tween 80 (v/v) because of its poor solubility in water. Animals were randomly divided into 5 groups with 5 animals each: (a) negative control group which received only vehicle (distilled water), (b) group treated with low dose 300 mg/kg of [VO(HSHED)dtb], (c) group treated with high dose 2000 mg/kg of [VO(HSHED)dtb], (d) positive control group treated with 50 mg/kg of sodium orthovanadate, and (e) free ligand group treated with 300 mg/kg of 3,5-di(t-butyl)catechol.

Since there is no toxicological information about [VO(HSHED)dtb], in accordance with OECD 423 [60], [VO(HSHED)dtb] was evaluated at 2 doses and vanadate and free ligand groups were used in single dose because acute toxicity tests have been previously described in the literature [74,77]. After the administration period, animals were observed for mortality and clinical symptoms of toxicity daily for 14 days. The symptoms of toxicity analyzed were alterations on skin and eyes, mucous membrane toxic effects and behavior patterns, lethargy, sleep, diarrhea, tremors, salivation, convulsion, coma, motor activity, hypo-activity, abdominal rigidity, breathing difficulty, cyanosis, and death following the Hippocratic screening protocol. The consumption of water and feed, as well as the body weight of each animal, were recorded daily. In addition, animals were daily observed for general health conditions and clinical evidence of toxicity [92]. On the 14th day of experiment, fasting mice were anesthetized using ketamine (90% b.w) and xilazin (10% b.w), and the blood was drawn from the retro-orbital route with or without heparin for hematological and serum biochemical analysis, respectively. The following organs were removed, cleaned, weighed on an analytical scale, and macroscopically analyzed: liver, kidney, spleen, heart, and lung [93].

4.7. Biochemical Analysis: Determination of Serum Biomarkers for Liver and Kidney Functions

The blood samples on the 14th day were collected in dry tubes, were centrifuged (3000 rpm, 15 min) and the obtained serums were analyzed for liver and kidney functions by colorimetric assay using commercial kits (Lab Test Diagnostic SA, Santa Lagoa, Brazil). Optical densities were measured by spectrophotometry (Varioskan TM Lux multimode microplate reader, Thermo Scientific®, Waltham, MA, USA) at wavelengths specific for each biochemical parameter described on datasheets [93]. Baseline measurements were obtained by comparing the optical densities of the samples with the respective standards, available in the kits. Biochemical analysis was performed for determining the following biomarkers parameters: alanine (ALT) and aspartate aminotransferase (AST), total protein (TP), albumin (ALB), AST/ALT and albumin/globulin ratio (A/G), blood urea nitrogen (BUN), and creatinine (CRE). Globulin was obtained from the difference between total protein and albumin [94]. Data were expressed by U/mL (AST and ALT) and mg/dL for the others.

4.8. Hematology Analysis

Hematological parameters included red blood cells (RBC), mean corpuscular volume (MCV), hemoglobin (Hb), mean corpuscular hemoglobin (MCH), mean corpuscular hemoglobin concentration (MCHC), hematocrit (HCT), white blood cells (WBC), lymphocytes (LYM) were performed using a multiparameter hematology analyzer (The Sysmex® XE-2100D, Sysmex®, Curitiba, Brazil) designed for hematology testing samples with ethylenediaminetetraacetic acid (EDTA).

4.9. Statistical Analysis

In vivo data was expressed as mean $\pm$ standard deviation (SD). The one-way analysis of variance (ANOVA) was employed to analyze the data between treated groups and their respective control groups followed. Tukey's Multiple Comparison Test was used to analyze the statistical comparisons. The p values less than 0.05 were considered statistically significant among the groups. Graph Pad Prism® (GraphPad Software, San Diego, CA, USA), version 5.0 software was used for all statistical analysis.

5. Conclusions

In summary, the administration of [VO(HSHED)dtb] complex in mice did not show any signs of toxicity up to a dose of 300 mg/kg. The complex was found to be less toxic than orthovanadate salt consistent with the compound being at least partially intact during the administration. The hematology, liver, and kidney biomarkers parameters demonstrate that the vanadium Schiff base complex exerts neither hepatotoxicity nor nephrotoxicity in mice. Although this compound contains both vanadium and a catechol which individually are known to be redox-active, and exert some toxicity, administration of this complex was tolerated at a low level (0.6 mol L^{-1}, 300 mg/kg). It was found to be toxic at a high level (4.0 mol L^{-1}, 2000 mg/kg) giving it an estimated LD_{50} between 300 and 2000 mg/kg, the latter being significantly higher than the usual therapeutic doses.

The low toxicity is attributed to the redox properties obtained when combining the redox-active ligand 3,5-di(tert-butyl)catechol with the hydrolytic stability of the [VO(HSHED)dtb], which prevent the formation of the vanadate and catechol ligand. The stability test in aqueous media with Tween 80 demonstrated that although the complex may not be stable in an aqueous solution for more than a few hours, this stability is not changed significantly in the presence of Tween 80 used for the administration of the compound. Hence, the compound stability is sufficient for the complex to exerts its action. Therefore, these studies are very encouraging and demonstrate that a vanadium compound such as the [VO(HSHED)dtb] complex does not exert toxicity to the same degree as vanadate even in a complex with a ligand that has potential for toxicity.

Author Contributions: H.M., D.J.G., and D.C.C. carried out the chemical methodology. L.M.A.L., M.F.B., W.E.S., and E.C.L. performed the biological studies. M.F.B., W.E.S., E.C.L., H.M., and D.C.C. reviewed and edited the manuscript; L.M.A.L. wrote drafting preparation and carried out the biological studies. E.C.L. and D.C.C. supervised the project. All authors have read and agreed to the published version of the manuscript.

Funding: This research was funded by CAPES, CNPq, and FACEPE. D.C.C. thanks the Arthur P. Cope Foundation for partial funding.

Institutional Review Board Statement: The protocols for these experiments were approved by the Animal Ethics Committee of the Universidade Federal de Pernambuco (process number #004-2019) and conducted in accordance with the Ethical Principles in Animal Research.

Informed Consent Statement: Not applicable.

Acknowledgments: L.M.A.L would like to thank FACEPE for the scholarship (AMD 0060-1.06/2019) at Colorado State University—Colorado/EUA with D.C.C.

Conflicts of Interest: The authors declare no conflict of interest.

Abbreviations

ADME	Absorption, Distribution, Metabolism and Excretion
ALB	albumin
ALT	alanine aminotransferase
Alx	allixinate
ANOVA	one-way analysis of variance
AST	aspartate aminotransferase
BMOV	bis(maltolato)oxidovanadium(IV)
BUN	blood urea nitrogen
CRE	creatinine
Dipic	dipicolinate
Dtb	3,5-di(t-butyl)catechol
EDTA	ethylenediamine tetraacetic acid
GHS	Globally Harmonized System
GLB	globulin
Hb	hemoglobin
HCT	hematocrit
HPLC	high performance liquid chromatography
i.p	intraperitoneal administration
LD_{50}	median lethal dose
LIKA	Laboratory of the immunopathology of Keizo Asami
L-Pheol-im	L-phenylalaninol
LYM	lymphocytes
MCH	mean corpuscular hemoglobin
MCHC	mean corpuscular hemoglobin concentration
NMR	nuclear magnetic ressonance
NOAEL	no-observed adverse-effect level
O.G	oral gavage
OECD	Organization for Economic Cooperation and Development
RBC	red blood cells
Rpm	rotation per minute
SD	standard deviation
STZ	streptozotocin
TP	total protein
UV-Vis	Ultraviolet-visible spectroscopy
WBC	white blood cells
8HQ	hydroxyquinoline

References

1. Correia, I.; Adao, P.; Roy, S.; Wahba, M.; Matos, C.; Maurya, M.R.; Marques, F.; Pavan, F.R.; Leite, C.Q.F.; Avecilla, F.; et al. Hydroxyquinoline Derived Vanadium (IV and V) and Copper (II) Complexes as Potential Anti-Tuberculosis and Anti-Tumor Agents. *J. Inorg. Biochem.* **2014**, *141*, 83–93.
2. Crans, D.C. Antidiabetic, Chemical, and Physical Properties of Organic Vanadates as Presumed Transition-State Inhibitors for Phosphatases. *J. Org. Chem.* **2015**, *80*, 11899–11915.
3. McLauchlan, C.C.; Peters, B.J.; Willsky, G.R.; Crans, D.C. Vanadium–Phosphatase Complexes: Phosphatase Inhibitors Favor the Trigonal Bipyramidal Transition State Geometries. *Coord. Chem. Rev.* **2015**, *301*, 163–199.
4. Pessoa, J.C.; Etcheverry, S.; Gambino, D. Vanadium Compounds in Medicine. *Coord. Chem. Rev.* **2015**, *301*, 24–48.
5. Thompson, K.H.; Liboiron, B.D.; Sun, Y.; Bellman, K.D.; Setyawati, I.A.; Patrick, B.O.; Karunaratne, V.; Rawji, G.; Wheeler, J.; Sutton, K.; et al. Preparation and Characterization of Vanadyl Complexes with Bidentate Maltol-Type Ligands; in vivo Comparisons of Anti-Diabetic Therapeutic Potential. *J. Biol. Inorg. Chem.* **2003**, *8*, 66–74.
6. Wei, Y.B.; Yang, X.D. Synthesis, Characterization and Anti-Diabetic Therapeutic Potential of a New Benzyl Acid-Derivatized Kojic Acid Vanadyl Complex. *BioMetals* **2012**, *25*, 1261–1268.

7. Kioseoglou, E.; Petanidis, S.; Gabriel, C.; Salifoglou, A. The Chemistry and Biology of Vanadium Compounds in Cancer Therapeutics. *Coord. Chem. Rev.* **2015**, *301*, 87–105.
8. Heyliger, C.E.; Tahiliani, A.G.; McNeill, J.H. Effect of Vanadate on Elevated Glucose and Depressed Cardiac Performance of Diabetic Rats. *Science* **1985**, *227*, 1474–1477.
9. Thompson, K.H.; Orvig, C. Vanadium in Diabetes: 100 Years from Phase 0 to Phase I. *J. Inorg. Biochem.* **2006**, *100*, 1925–1935.
10. Crans, D.C.; Henry, L.R.; Cardiff, G.; Posner, B.I. Developing Vanadium as an Antidiabetic or Anticancer Drug: A Clinical and Historical Perspective. *Met. Ions Life Sci.* **2019**, *19*, 203–230.
11. Willsky, G.R.; Goldfine, A.B.; Kostyniak, P.J.; McNeill, J.H.; Yang, L.Q.; Khan, H.R.; Crans, D.C. Effect of Vanadium (IV) Compounds in the Treatment of Diabetes: In vivo and in vitro Studies with Vanadyl Sulfate and Bis(Maltolato)Oxovandium(IV). *J. Inorg. Biochem.* **2001**, *85*, 33–42.
12. Mjos, K.D.; Orvig, C. Metallodrugs in Medicinal Inorganic Chemistry. *Chem. Rev.* **2014**, *114*, 4540–4563.
13. Tolman, E.L.; Barris, E.; Burns, M.; Pansini, A.; Partridge, R. Effects of Vanadium on Glucose Metabolism in vitro. *Life Sci.* **1979**, *25*, 1159–1164.
14. Crans, D.C.; Gambino, D.; Etcheverry, S.B. Vanadium Science: Chemistry, Catalysis, Materials, Biological and Medicinal Studies. *New J. Chem.* **2019**, *43*, 17535–17537.
15. Ramanadham, S.; Mongold, J.J.; Brownsey, R.W.; Cros, G.H.; McNeill, J.H. Oral Vanadyl Sulfate in Treatment of Diabetes Mellitus in Rats. *Am. J. Physiol. Circ. Physiol.* **1989**, *257*, 904–911.
16. Cam, M.C.; Pederson, R.A.; Brownsey, R.W.; McNeill, J.H. Long-Term Effectiveness of Oral Vanadyl Sulphate in Streptozotocin-Diabetic Rats. *Diabetologia* **1993**, *36*, 218–224.
17. León, I.E.; Cadavid-Vargas, J.F.; Tiscornia, I.; Porro, V.; Castelli, S.; Katkar, P.; Desideri, A.; Bollati-Fogolin, M.; Etcheverry, S.B. Oxidovanadium(IV) Complexes with Chrysin and Silibinin: Anticancer Activity and Mechanisms of Action in a Human Colon Adenocarcinoma Model. *J. Biol. Inorg. Chem.* **2015**, *20*, 1175–1191.
18. Selman, M.; Rousso, C.; Bergeron, A.; Son, H.H.; Krishnan, R.; El-sayes, N.A.; Varette, O.; Chen, A.; Le Boeuf, F.; Tzelepis, F.; et al. Multi-Modal Potentiation of Oncolytic Virotherapy by Vanadium Compounds. *Mol. Ther.* **2018**, *26*, 56–69.
19. Levina, A.; Pires Vieira, A.; Wijetunga, A.; Kaur, R.; Koehn, J.T.; Crans, D.C.; Lay, P.A. A Short-Lived but Highly Cytotoxic Vanadium(V) Complex as a Potential Drug Lead for Brain Cancer Treatment by Intratumoral Injections. *Angew. Chemie Int. Ed.* **2020**, *59*, 15834–15838.
20. Hiromura, M.; Adachi, Y.; Machida, M.; Hattori, M.; Sakurai, H. Glucose Lowering Activity by Oral Administration of Bis (Allixinato)Oxidovanadium (Iv) Complex in Streptozotocin-Induced Diabetic Mice and Gene Expression Profiling in Their Skeletal Muscles. *Metallomics* **2009**, *1*, 92–100.
21. Thompson, K.H.; Lichter, J.; LeBel, C.; Scaife, M.C.; McNeill, J.H.; Orvig, C. Vanadium Treatment of Type 2 Diabetes: A View to the Future. *J. Inorg. Biochem.* **2009**, *103*, 554–558.
22. Crans, D.C. Chemistry and Insulin-like Properties of Vanadium (IV) and Vanadium (V) Compounds. *J. Inorg. Biochem.* **2000**, *80*, 123–131.
23. Boden, G.; Chen, X.; Ruiz, J.; Van Rossum, G.D.V.; Turco, S. Effects of Vanadyl Sulfate on Carbohydrate and Lipid Metabolism in Patients with Non-Insulin-Dependent Diabetes Mellitus. *Metab. Clin. Exp.* **1996**, *45*, 1130–1135.
24. Willsky, G.R.; Chi, L.H.; Godzala, M.; Kostyniak, P.J.; Smee, J.J.; Trujillo, A.M.; Alfano, J.A.; Ding, W.; Hu, Z.; Crans, D.C. Anti-Diabetic Effects of a Series of Vanadium Dipicolinate Complexes in Rats with Streptozotocin-Induced Diabetes. *Coord. Chem. Rev.* **2011**, *255*, 2258–2269.
25. Koyuturk, M.; Tunali, S.; Bolkent, S.; Yanardag, R. Effects of Vanadyl Sulfate on Liver of Streptozotocin-Induced Diabetic Rats. *Biol. Trace Elem. Res.* **2005**, *104*, 233–247.
26. Sakurai, H.; Tsuchiya, K.; Nukatsuka, M.; Kawada, J.; Ishikawa, S.; Yoshida, H.; Komatsu, M. Insulin-Mimetic Action of Vanadyl Complexes. *J. Clin. Biochem. Nutr.* **1990**, *8*, 193–200.
27. Crans, D.; Trujillo, A.; Pharazyn, P.; Cohen, M. How Environment Affects Drug Activity: Localization, Compartmentalization and Reactions of a Vanadium Insulin-Enhancing Compound, Dipicolinatooxovanadium(V). *Coord. Chem. Rev.* **2011**, *255*, 2178–2192.
28. Bergeron, A.; Kostenkova, K.; Selman, M.; Murakami, H.; Owens, E.; Haribabu, N.; Arulanandam, R.; Diallo, J.S.; Crans, D. Enhancement of Oncolytic Virotherapy by Vanadium(V) Dipicolinates. *BioMetals* **2019**, *32*, 545–561.
29. Chinedu, E.; David, A.; Fidelis, S. A New Method for Determining Acute Toxicity in Animal Models. *Toxicol. Int.* **2013**, *20*, 224–226.
30. Erhirhie, E.O.; Ihekwereme, C.P.; Ilodigwe, E.E. Advances in Acute Toxicity Testing: Strengths, Weaknesses and Regulatory Acceptance. *Interdiscip. Toxicol.* **2018**, *11*, 5–12.
31. Alhaji Saganuwan, S. Toxicity Studies of Drugs and Chemicals in Animals: An Overview. *Bulg. J. Vet. Med.* **2017**, *20*, 291–318.
32. Llobet, J.M.; Domingo, J.L. Acute Toxicity of Vanadium Compounds in Rats and Mice. *Toxicol. Lett.* **1984**, *23*, 227–231.
33. Mongold, J.J.; Cros, G.H.; Vian, L.; Tep, A.; Ramanadham, S.; Siou, G.; Diaz, J.; McNeill, J.H.; Serrano, J.J. Toxicological Aspects of Vanadyl Sulphate on Diabetic Rats: Effects on Vanadium Levels and Pancreatic B-Cell Morphology. *Pharmacol. Toxicol.* **1990**, *67*, 192–198.
34. Domingo, J.L.; Gomez, M.; Sanchez, D.J.; Llobet, J.M.; Keen, C.L. Toxicology of Vanadium Compounds in Diabetic Rats: The Action of Chelating Agents on Vanadium Accumulation. *Mol. Cell. Biochem.* **1995**, *153*, 233–240.

35. Doucette, K.A.; Hassell, K.N.; Crans, D.C. Selective Speciation Improves Efficacy and Lowers Toxicity of Platinum Anticancer and Vanadium Antidiabetic Drugs. *J. Inorg. Biochem.* **2016**, *165*, 56–70.
36. Naso, L.; Ferrer, E.G.; Lezama, L.; Rojo, T.; Etcheverry, S.B.; Williams, P. Role of Oxidative Stress in the Antitumoral Action of a New Vanadyl(IV) Complex with the Flavonoid Chrysin in Two Osteoblast Cell Lines: Relationship with the Radical Scavenger Activity. *J. Biol. Inorg. Chem.* **2010**, *15*, 889–902.
37. Treviño, S.; Díaz, A.; Sánchez-Lara, E.; Sanchez-Gaytan, B.L.; Perez-Aguilar, J.M.; González-Vergara, E. Vanadium in Biological Action: Chemical, Pharmacological Aspects, and Metabolic Implications in Diabetes Mellitus. *Biol. Trace Elem. Res.* **2019**, *188*, 68–98.
38. Yilmaz-Ozden, T.; Kurt-Sirin, O.; Tunali, S.; Akev, N.; Can, A.; Yanardag, R. Ameliorative Effect of Vanadium on Oxidative Stress in Stomach Tissue of Diabetic Rats. *Bosn. J. Basic Med. Sci.* **2014**, *14*, 105–109.
39. Cortizo, A.M.; Bruzzone, L.; Molinuevo, S.; Etcheverry, S.B. A Possible Role of Oxidative Stress in the Vanadium-Induced Cytotoxicity in the MC3T3E1 Osteoblast and UMR106 Osteosarcoma Cell Lines. *Toxicology* **2000**, *147*, 89–99.
40. Kowalski, S.; Wyrzykowski, D.; Inkielewicz-Stępniak, I. Molecular and Cellular Mechanisms of Cytotoxic Activity of Vanadium Compounds against Cancer Cells. *Molecules* **2020**, *25*, 1757.
41. Rojas-Lemus, M.; Patricia, B.N.; Nelly, L.V.; Gonzalez-Villalva, A.; Gabriela, G.P.; Eugenia, C.V.; Otto, T.C.; Norma, R.F.; Brenda, C.T.; Martha, U.C.; et al. Oxidative Stress and Vanadium. In *Antimutagens–Mechanisms of DNA Protection*; Intech Open: London, UK, 2020.
42. Irving, E.; Stoker, A.W. Vanadium Compounds as PTP Inhibitors. *Molecules* **2017**, *22*, 2269.
43. Saibu, M.; Sagar, S.; Green, I.; Ameer, F.; Meyer, M. Evaluating the Cytotoxic Effects of Novel Quinone Compounds. *Anticancer Res.* **2014**, *34*, 4077–4086.
44. Sanna, D.; Ugone, V.; Fadda, A.; Micera, G.; Garribba, E. Behavior of the Potential Antitumor VIVO Complexes Formed by Flavonoid Ligands. 3. Antioxidant Properties and Radical Production Capability. *J. Inorg. Biochem.* **2016**, *161*, 18–26.
45. Dankhoff, K.; Ahmad, A.; Weber, B.; Biersack, B.; Schobert, R. Anticancer Properties of a New Non-Oxido Vanadium (IV) Complex with a Catechol-Modified 3,3′-Diindolylmethane Ligand. *J. Inorg. Biochem.* **2019**, *194*, 1–6.
46. Li, M.; Wei, D.; Ding, W.; Baruah, B.; Crans, D.C. Anti-Diabetic Effects of Cesium Aqua (N,N′-Ethylene(Salicylideneiminato)-5-Sulfonato) Oxovanadium (IV) Dihydrate in Streptozotocin-Induced Diabetic Rats. *Biol. Trace Elem. Res.* **2008**, *121*, 226–232.
47. Crans, D.C.; Mahroof-Tahir, M.; Johnson, M.D.; Wilkins, P.C.; Yang, L.; Robbins, K.; Johnson, A.; Alfano, J.A.; Godzala, M.E.; Austin, L.T.; et al. Vanadium (IV) and Vanadium (V) Complexes of Dipicolinic Acid and Derivatives. Synthesis, X-Ray Structure, Solution State Properties and Effects in Rats with STZ-Induced Diabetes. *Inorg. Chim. Acta* **2003**, *356*, 365–378.
48. Tadele, K.T.; Tsega, T.W. Schiff Bases and Their Metal Complexes as Potential Anticancer Candidates: A Review of Recent Works. *Anti Cancer Agents Med. Chem.* **2019**, *19*, 1786–1795.
49. Kowalski, S.; Wyrzykowski, D.; Hac, S.; Rychlowski, M.; Radomski, M.W.; Inkielewicz-Stepniak, I. New Oxidovanadium(IV) Coordination Complex Containing 2-Methylnitrilotriacetate Ligands Induces Cell Cycle Arrest and Autophagy in Human Pancreatic Ductal Adenocarcinoma Cell Lines. *Int. J. Mol. Sci.* **2019**, *20*, 261.
50. Lewis, N.A.; Liu, F.; Seymour, L.; Magnusen, A.; Erves, T.R.; Arca, J.F.; Beckford, F.A.; Venkatraman, R.; González-Sarrías, A.; Fronczek, F.R.; et al. Synthesis, Characterization, and Preliminary in vitro Studies of Vanadium(IV) Complexes with a Schiff Base and Thiosemicarbazones as Mixed-Ligands. *Eur. J. Inorg. Chem.* **2012**, *2012*, 664–677.
51. Leon, I.E.; Cadavid-Vargas, J.F.; Di Virgilio, A.L.; Etcheverry, S.B. Vanadium, Ruthenium and Copper Compounds: A New Class of Nonplatinum Metallodrugs with Anticancer Activity. *Curr. Med. Chem.* **2017**, *24*, 112–148.
52. Leon, I.E.; Díez, P.; Etcheverry, S.; Fuentes, M. Deciphering the Effect of an Oxovanadium(IV) Complex with the Flavonoid Chrysin (VOChrys) in Intracellular Cell Signalling Pathways in an Osteosarcoma Cell Line. *Metallomics* **2016**, *8*, 739–749.
53. León, I.E.; Butenko, N.; Di Virgilio, A.L.; Muglia, C.I.; Baran, E.J.; Cavaco, I.; Etcheverry, S.B. Vanadium and Cancer Treatment: Antitumoral Mechanisms of Three Oxidovanadium(IV) Complexes on a Human Osteosarcoma Cell Line. *J. Inorg. Biochem.* **2014**, *134*, 106–117.
54. Nica, S.; Rudolph, M.; Lippold, I.; Buchholz, A.; Görls, H.; Plass, W. Vanadium(V) Complex with Schiff-Base Ligand Containing a Flexible Amino Side Chain: Synthesis, Structure and Reactivity. *J. Inorg. Biochem.* **2015**, *147*, 193–203.
55. Cornman, C.R.; Colpas, G.J.; Hoeschele, J.D.; Kampf, J.; Pecoraro, V.L. Implications for the Spectroscopic Assignment of Vanadium Biomolecules: Structural and Spectroscopic Characterization of Monooxovanadium(V) Complexes Containing Catecholate and Hydroximate Based Noninnocent Ligands. *J. Am. Chem. Soc.* **1992**, *114*, 9925–9933.
56. Crans, D.C.; Koehn, J.T.; Petry, S.M.; Glover, C.M.; Wijetunga, A.; Kaur, R.; Levina, A.; Lay, P.A. Hydrophobicity May Enhance Membrane Affinity and Anti-Cancer Effects of Schiff Base Vanadium(v) Catecholate Complexes. *Dalt. Trans.* **2019**, *48*, 6383–6395.
57. Boukhobza, I.; Crans, D.C. Application of HPLC to Measure Vanadium in Environmental, Biological and Clinical Matrices. *Arabian J. Chem.* **2020**, *13*, 1198–1228.
58. Ugone, V.; Sanna, D.; Sciortino, G.; Crans, D.C.; Garribba, E. ESI-MS Study of the Interaction of Potential VIV Drugs. *Inorg. Chem.* **2020**, *59*, 9739–9755.
59. Srivastava, A. Anti-Diabetic and Toxic Effects of Vanadium Compounds. *Mol. Cell. Biochem.* **2000**, *206*, 177–182.
60. OECD. *Test No. 423: Acute Oral Toxicity—Acute Toxic Class Method*; OECD Guidelines for the Testing of Chemicals, Section 4; OECD Publishing: Paris, France, 2002.

61. Lipnick, R.L.; Cotruvo, J.A.; Hill, R.N.; Bruce, R.D.; Stitzel, K.A.; Walker, A.P.; Chu, I.; Goddard, M.; Segal, L.; Springer, J.A.; et al. Comparison of the Up-and-down, Conventional LD50, and Fixed-Dose Acute Toxity Procedures. *Food Chem. Toxicol.* **1995**, *33*, 223–231.

62. Ozer, J.; Ratner, M.; Shaw, M.; Bailey, W.; Schomaker, S. The Current State of Serum Biomarkers of Hepatotoxicity. *Toxicology* **2008**, *245*, 194–205.

63. McGill, M.R. The Past and Present of Serum Aminotransferases and the Future of Liver Injury Biomarkers. *EXCLI J.* **2016**, *15*, 817–828.

64. Pan, X.; Chang, F.; Liu, Y.; Li, D.; Xu, A.; Shen, Y.; Huang, Z. Mouse Toxicity of Anabaena Flos-Aquae from Lake Dianchi, China. *Environ. Toxicol.* **2009**, *24*, 10–18.

65. Edelstein, C.L. Biomarkers of Acute Kidney Injury. *Adv. Chronic Kidney Dis.* **2008**, *15*, 222–234.

66. Cooper, R.G. Vanadium Pentoxide Inhalation. *Indian J. Occup. Environ. Med.* **2007**, *11*, 97–102.

67. Crans, D.C.; Postal, K.; MacGregor, J.A. Vanadium–speciation Chemistry is Important when Assessing Health Effects on Living Systems. In *Metal Toxicology Handbook*; CRC Press: Boca Raton, FL, USA, 2020; Chapter 6.

68. Gajens, J.; Meier, B.; Adachi, Y.; Sakurai, H.; Rehder, D. Characterization and Insulin-Mimetic Potential of Oxidovanadium (IV) Complexes Derived from Monoesters and -carboxylates of 2,5-Dipicolinic Acid. *Eur. J. Inorg. Chem.* **2006**, *18*, 3575–3585.

69. Brichard, S.M.; Henquin, J.C. The Role of Vanadium in the Management of Diabetes. *Trends Pharmacol. Sci.* **1995**, *16*, 265–270.

70. Hanson, G.R.; Sun, Y.; Orvig, C. Characterization of the Potent Insulin Mimetic Agent Bis (Maltolato) Oxovanadium (IV) (BMOV) in Solution by EPR Spectroscopy. *Inorg. Chem.* **1996**, *35*, 6507–6512.

71. Pessoa, J.C.; Correa, I. Misinterpretations in Evaluating Interactions of Vanadium Complexes with Proteins and Other Biological Targets. *Inorganics* **2021**, *9*, 17.

72. Wang, B.; Tanaka, K.; Morita, A.; Ninomiya, Y.; Maruyama, K.; Fujita, K.; Hosoi, Y.; Nenoi, M. Sodium Orthovanadate (Vanadate), a Potent Mitigator of Radiation-Induced Damage to the Hematopoietic System in Mice. *J. Radiat. Res.* **2013**, *54*, 620–629.

73. Roy, S.; Majumdar, S.; Singh, A.K.; Ghosh, B.; Ghosh, N.; Manna, S.; Chakraborty, T.; Mallick, S. Synthesis, Characterization, Antioxidant Status, and Toxicity Study of Vanadium-Rutin Complex in Balb/c Mice. *Biol. Trace Elem. Res.* **2015**, *166*, 183–200.

74. Sanchez, D.; Ortega, A.; Domingo, J.L.; Corbella, J. Developmental Toxicity Evaluation of Orthovanadate in the Mouse. *Biol. Trace Elem. Res.* **1991**, *30*, 219–226.

75. Montaser, A.S.; Wassel, A.R.; Al-Shaye'a, O.N. Synthesis, characterization and antimicrobial activity of Schiff bases from chitosan and salicylaldehyde/TiO$_2$ nanocomposite membrane. *Int. J. Biol. Macromol.* **2019**, *124*, 802–809.

76. Kenkyūjo, K.G.S. *Toxic and Hazardous Industrial Chemicals Safety Manual or Handling and Disposal with Toxicity and Hazard Data*; The International Technical Information Institute: Tokyo, Japan, 1988; p. 591.

77. Miadzvedski, I.; Nikolayuk, O.; Dubovik, B. Acute Toxicity of Spatially Hindered Derivatives of Aminophenol and Catechol. In Proceedings of the Actual Problems of Medicine, Grodno, Belarus, January 2013.

78. Boehm, O.; Zur, B.; Koch, A.; Tran, N.; Freyenhagen, R.; Hartmann, M.; Zacharowski, K. Clinical Chemistry Reference Database for Wistar Rats and C57/BL6 Mice. *Biol Chem.* **2007**, *388*, 547–554.

79. Reul, B.A.; Amin, S.S.; Buchet, J.; Ongemba, L.N.; Crans, D.C.; Brichard, S.M. Effects of Vanadium Complexes with Organic Ligands on Glucose Metabolism: A Comparison Study in Diabetic Rats. *Br. J. Pharmacol.* **1999**, *126*, 467–477.

80. Levina, A.; Crans, D.C.; Lay, P.A. Speciation of Metal Drugs, Supplements and Toxins in Media and Bodily Fluids Controls in vitro Activities. *Coor. Chem. Rev.* **2017**, *352*, 473–498.

81. Elvingson, K.; Gonzalez Baro, A.; Pettersson, L. Speciation in Vanadium Bioinorganic Systems. 2. An NMR, ESR, and Potentiometric Study of the Aqueous H+-Vanadate-Maltol System. *Inorg. Chem.* **1996**, *35*, 3388–3393.

82. Samart, N.; Arhouma, Z.; Kumar, S.; Murakami, H.A. Decavanadate Inhibits Mycobacterial Growth More Potently Than Other Oxovanadates. *Front. Chem.* **2018**, *6*, 1–16.

83. Althumairy, D.; Postal, K.; Barisas, G.B.; Nunes, G.G.; Roess, D.A.; Crans, D.C. Polyoxometalates Function as Indirect Activators of a G Protein-Coupled Receptor. *Metallomics* **2020**, *12*, 1044–1061.

84. Yoshikawa, Y.; Sakurai, H.; Crans, D.C.; Micera, G.; Garribba, E. Structural and Redox Requirements for the Action of Anti-Diabetic Vanadium Compounds. *Dalt. Trans.* **2014**, *43*, 6965–6972.

85. Ścibior, A.; Pietrzyk, Ł.; Plewa, Z.; Skiba, A. Vanadium: Risks and Possible Benefits in the Light of a Comprehensive Overview of Its Pharmacotoxicological Mechanisms and Multi-Applications with a Summary of Further Research Trends. *J. Trace Elem. Med. Biol.* **2020**, *61*, 126508.

86. Lima, L.M.A.; Belian, M.F.; Silva, W.E.; Postal, K.; Kostenkova, K.; Crans, D.C.; Rossiter, A.K.F.F.; da Silva Júnior, V.A. Vanadium (IV)-Diamine Complex with Hypoglycemic Activity and a Reduction in Testicular Atrophy. *J. Inorg. Biochem.* **2020**, *216*, 111312.

87. Crans, D.C.; Shin, P.K.; Armstrong, K.B. Characterization of vanadium(V) complexes in aqueous solutions: Ethanolamine and glycine derived complexes. *J. Am. Chem. Soc.* **1994**, *116*, 1305–1315.

88. Rehder, D.; Polenova, T.; Bühl, M. Vanadium-51 NMR. *Annu. Rep. NMR Spectrosc.* **2007**, *62*, 49–114.

89. Li, X.; Lah, M.S.; Pecoraro, V.L. Vanadium Complexes of the Tridentate Schiff Base Ligand N-Salicylidene-N'-(2-Hydroxyethyl) Ethylenediamine: Acid-Base and Redox Conversion between Vanadium (IV) and Vanadium (V) Imino Phenolates. *Inorg. Chem.* **1988**, *27*, 4657–4664.

90. Rajendiran, V.; Karthik, R.; Palaniandavar, M.; Stoeckli-Evans, H.; Periasamy, V.S.; Akbarsha, M.A.; Srinag, B.S.; Krishnamurthy, H. Mixed-ligand Copper (II)-phenolate Complexes: Effect of Coligand on Enhanced DNA and Protein Binding, DNA Cleavage, and Anticancer Activity K. *Inorg. Chem.* **2007**, *46*, 8208–8221.

91. Chatterjee, P.B.; Goncharov-Zapata, O.; Quinn, L.L.; Hou, G.; Hamaed, H.; Schurko, R.W.; Polenova, T.; Crans, D.C. Characterization of Non-innocent Metal Complexes Using Solid–state NMR Spectroscopy: O-dioxolene Vanadium Complexes. *Inorg. Chem.* **2011**, *50*, 9794–9803.

92. Hasegawa, R.; Nakaji, Y.; Kurokawa, Y.; Tobe, M. Acute Toxicity Tests on 113 Environmental Chemicals. *Sci. Rep. Res. Inst. Tohoku Univ. Med.* **1989**, *36*, 10–16.

93. Barbosa, H.M.; Do Nascimento, J.N.; Araújo, T.A.S.; Duarte, F.S.; Albuquerque, U.P.; Vieira, J.R.C.; De Santana, E.R.B.; Yara, R.; Lima, C.S.A.; Gomes, D.A.; et al. Acute Toxicity and Cytotoxicity Effect of Ethanolic Extract of Spondias Tuberose Arruda Bark: Hematological, Biochemical and Histopathological Evaluation. *An. Acad. Bras. Cienc.* **2016**, *88*, 1993–2004.

94. Howard, W.; Robinson, J.W.P.; Hogden, C.G. The Estimation of Albumin and Globulin in Blood. *J. Bio. Chem.* **1937**, *120*, 481–498.

 *inorganics*

Review

Misinterpretations in Evaluating Interactions of Vanadium Complexes with Proteins and Other Biological Targets

João Costa Pessoa * and Isabel Correia

Centro de Química Estrutural and Departamento de Engenharia Química, Instituto Superior Técnico, Universidade de Lisboa, Av. Rovisco Pais, 1049-001 Lisboa, Portugal; icorreia@tecnico.ulisboa.pt
* Correspondence: joao.pessoa@tecnico.ulisboa.pt

Abstract: In aqueous media, V^{IV}- and V^V-ions and compounds undergo chemical changes such as hydrolysis, ligand exchange and redox reactions that depend on pH and concentration of the vanadium species, and on the nature of the several components present. In particular, the behaviour of vanadium compounds in biological fluids depends on their environment and on concentration of the many potential ligands present. However, when reporting the biological action of a particular complex, often the possibility of chemical changes occurring has been neglected, and the modifications of the complex added are not taken into account. In this work, we highlight that as soon as most vanadium(IV) and vanadium(V) compounds are dissolved in a biological media, they undergo several types of chemical transformations, and these changes are particularly extensive at the low concentrations normally used in biological experiments. We also emphasize that in case of a biochemical interaction or effect, to determine binding constants or the active species and/or propose mechanisms of action, it is essential to evaluate its speciation in the media where it is acting. This is because the vanadium complex no longer exists in its initial form.

Keywords: vanadium; proteins; DNA; fluorescence; binding constants; mechanism of action

Citation: Pessoa, J.C.; Correia, I. Misinterpretations in Evaluating Interactions of Vanadium Complexes with Proteins and Other Biological Targets. *Inorganics* 2021, 9, 17. https://doi.org/10.3390/inorganics9020017

Academic Editor: Peter Faller
Received: 19 January 2021
Accepted: 4 February 2021
Published: 9 February 2021

Publisher's Note: MDPI stays neutral with regard to jurisdictional claims in published maps and institutional affiliations.

1. Introduction

Many metal ions have a general tendency to interact with biomolecules, changing and/or modulating their properties and functions, therefore several of them are incorporated to perform crucial roles in organisms carrying out a wide variety of tasks [1,2]. Vanadium is a transition metal that is widely distributed on earth's crust, in soil, crude oil, water and air, so it is not surprising that it found roles in biological systems, being an essential element for many living beings.

Vanadium compounds may have oxidation states ranging from $-$III to +V, but V^{III}, V^{IV} and V^V are those of biological relevance. Vanadium ions bind to a broad range of biological compounds such as proteins, metabolites, membranes or other structures, and as with many other metal ions, a particular oxidation state may be stabilized by forming complexes with suitable ligands.

There are several enzymatic systems using vanadium in their active sites as relevant components for their function [3], and the complexes formed with several bio-ligands are important for the bio-distribution of vanadium [3–7]. For example, it is well established that vanadium binds to transferrin [8–15], this being relevant for its transport and bioavailability in blood. It is also well known that vanadium undergoes redox chemistry and other types of chemical transformations after administration [7,16–19], and understanding these processes is crucial to understanding the mechanisms of action.

In living systems V(IV) is normally in the form of $V^{IV}O$ [oxidovanadium(IV)], and V(V) as V^VO [oxidovanadium(V)] or V^VO_2 [dioxidovanadium(V)], each of these moieties undergoing complex hydrolytic reactions which depend on pH, concentration and ionic strength [20–22]. Additionally, vanadium complexes, which we may designate as

[VO$_n$(L)$_m$], may also be involved in a wide range of reactions, interacting or forming complexes with metabolites, proteins and bio-ligands such as DNA, and these reactions depend on pH (typically in the range 6–8), and nature and concentrations of all species that may bind to vanadium. These include ligand L, H$_2$O, OH$^-$, H$_n$PO$_4^{-(3+n)}$ ions and any bio-ligand present. The actual speciation of vanadium in a system at a fixed pH depends on the total concentration of [VO$_n$(L)$_m$] present, on the type and concentrations of all species that may bind to vanadium and on the formation constants of the vanadium complexes formed.

If a vanadium compound, for example, V^{IV}OSO$_4$ or [V^{IV}O(acac)$_2$] (acac$^-$ = acetylacetonato), is added to a particular biological system, for example, if it is placed in contact with cells, or administered to an animal, and if it exerts some biological effect, it is important to disclose which is/are the particular vanadium species that is responsible for the activity detected. Often the biological effect has been simply, and often wrongly, assigned to either V^{IV}O^{2+} or [V^{IV}O(acac)$_2$], not taking into account the hydrolytic and/or other transformations that these species might have had once added to the biological system.

In other simpler approaches, binding constants of vanadium complexes to bio-ligands have been determined assuming that the complex maintains its integrity once added to the aqueous solution containing the bio-ligand, and often that is not the case. In fact, misinterpretations in evaluating interactions of vanadium complexes with proteins, bio-ligands and/or other biological targets are quite common in the scientific literature, and this text aims to emphasize that care must be taken when interpreting spectroscopic or other analytic information, particularly when examining data involving low concentrations of the vanadium complexes.

2. Discussion

2.1. Hydrolytic Behavior of Oxidovanadium(IV) Ions

Vanadium(IV) normally exists as V^{IV}O-species and is quite susceptible to hydrolysis and oxidation for pH > 3. The hydrolytic behavior of V^{IV}O^{2+} ions in water has been studied [9,22–27], being well understood up to pH ca. 4. To quantify the several complex V^{IV}O-species that may form, we use the usual definition of formation constants:

$$\beta_{pqr}: \quad p\text{V}^{IV}\text{O}^{2+} + q\text{L} + r\text{H}^+ \leftrightarrows \left[\left(\text{V}^{IV}\text{O}\right)_p(\text{L})_q(\text{H})_r\right] \tag{1}$$

When addressing formation constants of hydrolytic species, the coefficient $q = 0$.

At low pH in aqueous solution oxidovanadium(IV) ions exist as [V^{IV}O(H$_2$O)$_5$]$^{2+}$. As the pH is increased, the V^{IV}O^{2+} ions are progressively partly transformed into [V^{IV}O(OH)(H$_2$O)$_4$]$^+$ and [(V^{IV}O)$_2$(OH)$_2$(H$_2$O)$_n$]$^{2+}$, these normally represented as [V^{IV}O(OH)]$^+$ and [(V^{IV}O)$_2$(OH)$_2$]$^{2+}$. For pH > 3.5–4, assuming oxidation to V^V is avoided, other oxidovanadium(IV) hydroxides form and, depending on the total vanadium concentration (C$_V$), V^{IV}O(OH)$_2$ may precipitate (solubility product ~10^{-23}) [9,23,24]. For pH >12 it is clear that the predominant species is [V^{IV}O(OH)$_3$]$^-$ (abbreviation of [V^{IV}O(OH)$_3$(H$_2$O)$_2$]$^-$) [25–27], but what happens in the pH range 4–12 is not well established, and depends on C$_V$ [9,22]. For pH > 5, the formation of a species with a V^{IV}O:OH$^-$ ratio of 2:5 was established [28], but clearly the vanadium species that form are oligomeric, [(V^{IV}O)$_2$(OH)$_5$]$_m^-$, the value of m depending on C$_V$. It was never clarified if oligomeric [(V^{IV}O)$_2$(OH)$_6$]$_n^{2-}$ forms or not at higher pH values, but as mentioned, for pH >12 the formation of monomeric [V^{IV}O(OH)$_3$]$^-$ was established [25–27]. Costa Pessoa, from several calculations based on visible and circular dichroism spectra of solutions containing oxidovanadium(IV) and amino acids (L-Ala [29], L-Ser, L-Thr [30], L-Cys, D-Pen [31], and L-Asp [32]) for pH > 5 determined the values of the formation constants of [(V^{IV}O)$_2$(OH)$_5$]$^-$ and [V^{IV}O(OH)$_3$]$^-$ as log $\beta_{20\text{-}5} \approx -22.3 \pm 0.3$ and log $\beta_{10\text{-}3} \approx -18.2 \pm 0.2$, respectively.

Figure 1A,B depicts species distribution diagrams for the oxidovanadium(IV) system at two different C$_V$ values. It must be kept in mind that V^{IV} forms the insoluble V^{IV}O(OH)$_2$ for pH > 4 if C$_V$ is higher than ca. 10^{-4} M. At pH > 5 oligomeric vanadium species become relevant, and for pH > 7–8 the hydroxide dissolves, but this may take time, oxidation to V^V taking place if dioxygen is not carefully removed. It should be highlighted that for pH > 5–6 $V^{IV}O^{2+}$ ($[V^{IV}O(H_2O)_5]^{2+}$) does not exist as such, and that for $C_V < 10^{-5}$ M oxidovanadium(IV) ions are soluble, being mainly in the form of $[V^{IV}O(OH)_3]^-$ and $[(V^{IV}O)_2(OH)_5]^-$ (Figure 1).

At pH = 7, in C_V conditions that may occur in biological media (from 10^{-7} to 10^{-4} M), the main species present are $[(V^{IV}O)_2(OH)_5]^-$ and $[V^{IV}O(OH)_3]^-$ (Figure 1C); therefore, in any type of calculations, e.g., determination of binding constants, the concentration of $V^{IV}O^{2+}$ cannot be taken as equal to C_V. Additionally, in determination of cytotoxicity, or other similar type of parameters, where an oxidovanadium(IV) salt is added to incubation media of cells, oxidovanadium(IV) ions will be partially or totally oxidized to V^V ions, therefore the biological activity determined will not be due only to V^{IV}-species, but also to the V^V-species formed. The longer the incubation time, the more probable the participation of V^V-species. When testing $V^{IV}O$-salts using C_V values higher than ca. 100 μM, if no precipitation of $V^{IV}O$-hydroxide is detected, this is because either a significant amount of V^{IV} oxidized, or V^{IV} is bound to ligands present in the incubation media, hence protected from oxidation. This should be highlighted when reporting the data to avoid misunderstandings about the identity of the active species.

2.2. Hydrolytic Behavior of Oxidovanadium(V) Ions

The hydrolytic behavior of oxidovanadium(V) ions in water has been studied mainly by using pH potentiometric and ^{51}V NMR spectroscopy measurements [20,21]. The system is complex and the equilibria (and values of formation constants) somewhat depend on the ionic strength and the salt employed to set it [20]. Here, we will consider the physiologically relevant (for blood serum) 0.150 M NaCl medium, and Figure 2 depicts species distribution diagrams of V^V hydrolysis in several distinct conditions.

At the pH values relevant in common biological conditions, and at low concentrations, V^V exists mainly as $H_2VO_4^-$ and HVO_4^{2-}, often referred as VO_3^- or as mono-vanadate (V_1). At pH 7 and low C_V values, (e.g., 10 μM, Figure 2B,C) V^V divanadates (V_2) or tetra-vanadates (V_4) almost do not form, but they become important at higher V concentrations (e.g., 1 mM, Figure 2A,C). In cells, usual physiological vanadate concentrations are also too low to allow the formation of oligovanadates, but in the pH range 3–5.5 or in certain confined cell compartments, decavanadates ($[H_nV_{10}O_{28}]^{(6-n)-}$, V_{10}) may be relevant species [34–38]. If a ligand L, other than OH^-, is present in solution, V^V-L complexes may form, but the fraction of V^V that is not bound to L will be involved in speciation similar to the exemplified in Figure 2.

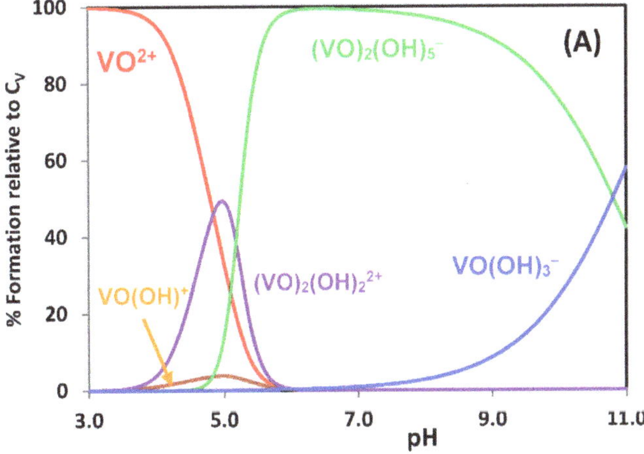

Figure 1. *Cont.*

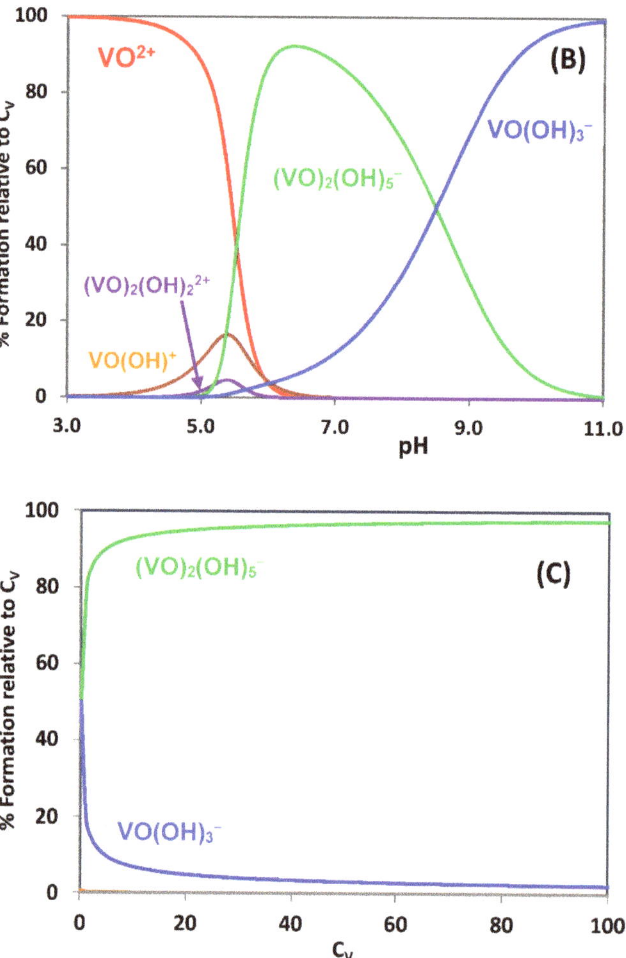

Figure 1. Species distribution diagrams, calculated with the computer program HySS [33], of $V^{IV}O^{2+}$ hydrolysis: (**A**) at $C_V = 2 \times 10^{-3}$ M, (**B**) at $C_V = 1 \times 10^{-5}$ M, in the pH range 3–11; (**C**) at pH = 7 with the total vanadium(IV) concentration (C_V) varying in the range 0.1 to 100 μM. The formation constants of the hydrolytic $V^{IV}O$-species were taken from [9,22]. In the diagram shown in (**A**), in the pH range 4–8 the product $[V^{IV}O^{2+}][OH^-]^2$ is higher than the solubility product of the hydroxide ($\sim 10^{-23}$), thus $V^{IV}O(OH)_2$ will precipitate.

2.3. Evaluation of Binding Constants of Metal Complexes with Bio-Macromolecules

To understand the biological activity of a metal complex or its transport in blood, it is important to evaluate how strong are its interactions with biological macromolecules such as proteins or DNA. Considering a [M(L)$_2$] complex such as $[V^{IV}O(acac)_2]$ or $[V^{IV}O(phen)_2]^{2+}$ (phen = 1,10-phenanthroline), this may correspond to the determination of the equilibrium constant of the following reaction at a particular pH value, e.g., pH = 7:

$$n\left[V^{IV}O(L)_2\right] \; + \; \text{biomolecule} \; \leftrightarrows \; \left[\left[V^{IV}O(L)_2\right]_n(\text{biomolecule})\right] \qquad (2)$$

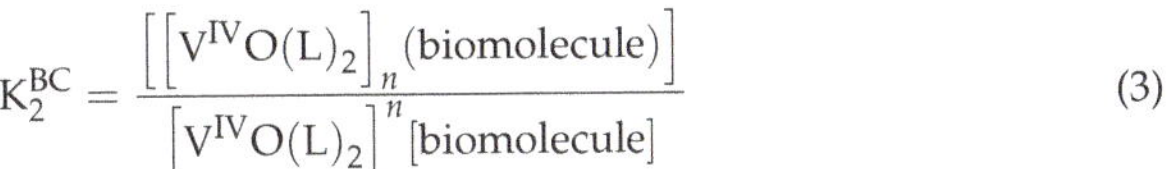

$$K_2^{BC} = \frac{\left[\left[V^{IV}O(L)_2\right]_n(\text{biomolecule})\right]}{\left[V^{IV}O(L)_2\right]^n[\text{biomolecule}]} \tag{3}$$

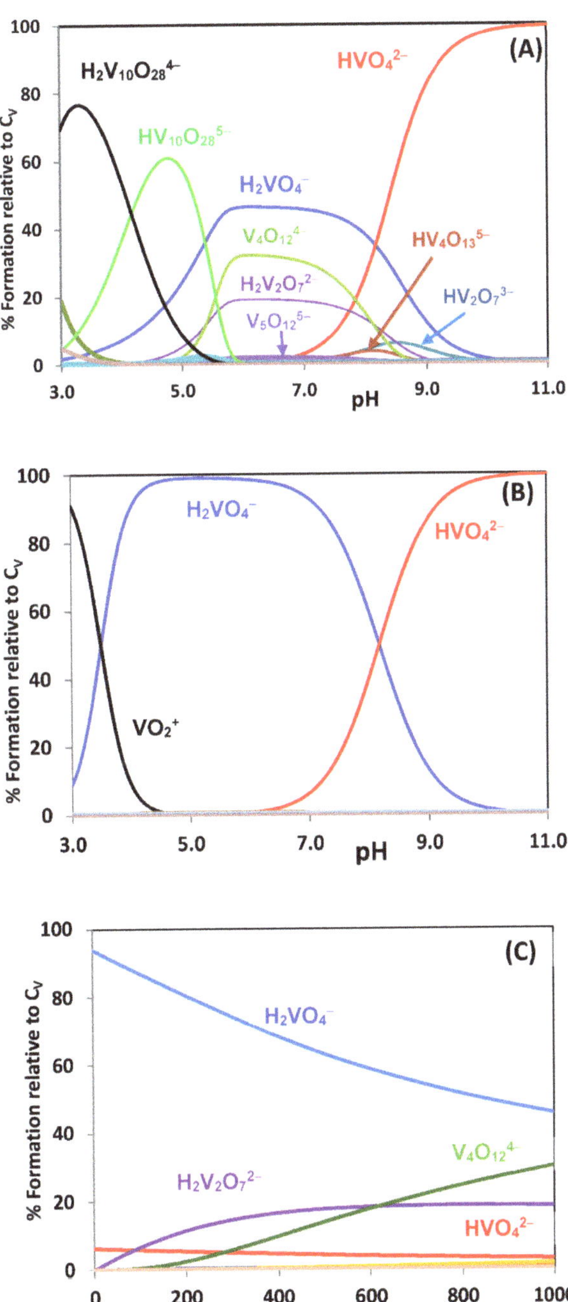

Figure 2. *Cont.*

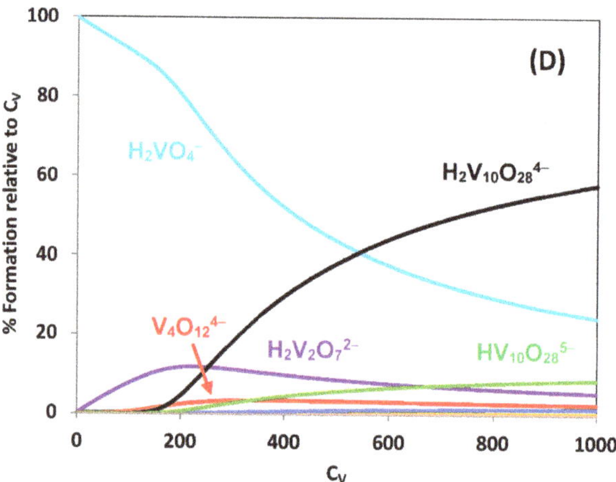

Figure 2. Species distribution diagrams, calculated with the computer program HySS [33], of V^V hydrolysis: (**A**) at $C_V = 1 \times 10^{-3}$ M, (**B**) at $C_V = 1 \times 10^{-5}$ M, in the pH range 3–11; at pH = 7 (**C**) and pH = 5 (**D**) with the total vanadium(V) concentration varying in the range 1 to 1000 μM. The formation constants of the hydrolytic V^V-species were taken from [21].

Often this binding (or association) constant is designated by K_a (or K_b) but these abbreviations should be avoided, as they are the symbols normally used to designate the dissociation constant of acids (or bases). We use K_2^{BC} (and not K^{BC}) because we are assuming the binding of a $[M(L)_2]$ complex (metal:ligand with 1:2 molar ratio); when considering a $[M(L)]$ complex we will use K_1^{BC} (metal:ligand with 1:1 molar ratio). The requirement of these distinct designations is clarified below. We also emphasize that the constants K_1^{BC} and K_2^{BC} correspond to the so called 'conditional' binding constants; their values depend not only on the pH of the solution but also on the type of buffer used. These may have components that have affinity for the metal ion, and thus may act as competitive ligands.

Most methods used to determine parameters of chemical interactions, the procedures to obtain binding constants may involve direct or indirect approaches. Using direct methods such as circular dichroism (CD) and/or UV–Vis electronic absorption spectrophotometry, complex-biomolecule binding constants may be determined accurately, but this typically involves the measurement of a great number of adequate data and the use of suitable computer programs. Indirect approaches such as fluorescence spectroscopy are normally less accurate but are much easier to use and require lower concentrations and less experimental data. In fact, fluorescence titrations must be performed at very low concentrations or preferably at concentrations in which the absorbance at the excitation wavelength is less than 0.05 (A < 0.05). Furthermore, the fluorescence response may no longer be linearly dependent on the light absorbed. Additionally, inner filter effects may also cause deviations from linearity at higher concentrations [39].

In the case of proteins, such as albumin, for which abundant studies are reported in the literature, the value of K_2^{BC} has been determined mainly using fluorescence quenching measurements. The interaction of the metal complex and the protein may involve the quenching of the protein intrinsic fluorescence, due to Trp, Tyr or Phe residues; the interaction often gives rise to significant quenching of the protein's fluorescence and a high analytical signal. These methodologies involve several steps so that it might be confirmed that the fluorescence quenching is due to binding of a compound to the macromolecule and a static quenching process is operating, which is assumed to be due to the non-emitting complex $[V^{IV}O(L)_2]$ associated to the macromolecule.

Titration experiments are usually carried out by fixing the concentration of one component, the biomolecule, while the concentration of the complex is varied. During the course of the experiment, changes in the system are monitored, which are then plotted as a function of complex added. The resulting titration curve, known as a binding isotherm, is then fitted with a mathematical model derived from the expected equilibria to obtain the binding constant (K^{BC}). This model is usually developed from recognizing that the changes observed (e.g., quenching of fluorescence, ΔI) are correlated to the concentration of the species $[([V^{IV}O(L)_2])_n(\text{biomolecule})]$.

When $n = 1$ in Equation (3), the total concentration of biomolecule and of the V-complex are the sum of the concentrations of free and associated forms, respectively:

$$[\text{biomolecule}]_T = [\text{biomolecule}]_{\text{free}} + \left[\left[V^{IV}O(L)_2\right]\right)(\text{biomolecule})]] \tag{4}$$

$$\left[V^{IV}O(L)_2\right]_T = [([V^{IV}O(L)_2])(\text{biomolecule})]] + [V^{IV}O(L)_2]_{\text{free}} \tag{5}$$

It is not easy to measure the concentration of $[[V^{IV}O(L)_2])(\text{biomolecule})]$ (or $[V^{IV}O(L)_2]_{\text{free}}$ and $[\text{biomolecule}]_{\text{free}}$) but the knowledge of these is required to determine K^{BC}. However, these can be used to rewrite the concentration of associated form $[[V^{IV}O(L)_2])(\text{biomolecule})]$ (abbreviated as PC) as a function of the total concentrations, $[\text{biomolecule}]_T$ (C_P) and $[V^{IV}O(L)_2]_T$, (C_C) and of the binding constant, K^{BC} [40].

$$[\text{PC}] = \frac{1}{2}\left(C_P + C_C + \frac{1}{K^{BC}}\right) - \frac{1}{2}\sqrt{\left(C_P + C_C + \frac{1}{K^{BC}}\right)^2 - 4\,C_P C_C} \tag{6}$$

The fluorescence intensity at a given wavelength (I_λ) of a given solution of bio-macromolecule and V-complex may be obtained from the molar fraction average of the fluorescence intensity of the individual species,

$$I_\lambda = \frac{[P]}{C_P}\,I_P + \frac{[PC]}{C_P}\,I_{PC} \tag{7}$$

where I_P and I_{PC} are the fluorescence intensities of the biomolecule in the absence of and presence of the V-complex, respectively. However, most reported K^{BC} constants in the literature are obtained from linearizations, (linear regression methods) which are used due to its simplicity but with the power of modern computational methods should no longer be needed. There are at least two main problems connected with the use of these linearizations: (i) distortion of the experimental errors and (ii) assuming that $[C] \approx C_C$ (only valid when the biomolecule is in large excess) [38]. However, due to its prevalence in the literature and its simplicity we will use them as basis for the current discussion. Another problem with this methodology is that the binding may involve the formation of both a 1:1 and 1:2 ($V^{IV}O{:}L$) complexes, the treatment described below not being valid in these cases.

Conventionally, when molecules bind independently to a set of equivalent sites on a macromolecule, the equilibrium between free and bound molecules may be given by the following equation [41,42],

$$\log\left[(I_0 - I)/I\right] = \log K_2{}^{BC} + n\log\left[Q\right] \tag{8}$$

where I_0 and I are the fluorescence intensities in the absence and presence of the quencher, respectively, $[Q]$ is the quencher concentration (in our case $Q = [V^{IV}O(L)_2]$), $K_2{}^{BC}$ is the binding constant defined according to Equations (2) and (3) and n is the number of binding sites per macromolecule. Using this methodology, if a linear relation between $\log((I_0 - I)/I)$ vs. $\log[Q]$ is obtained, values may be determined for $K_2{}^{BC}$ and n. As stated above, the value for $[Q]$ typically is taken as the total metal complex concentration (here $[V^{IV}O(L)_2]$), this being an approximation of the concentration of $[V^{IV}O(L)_2]_{\text{free}}$. This may be a wrong approximation, but is assumed in most publications. Moreover, as we will show, the fact that a high analytical signal (e.g., a strong quenching of fluorescence) is measured, does not necessarily means a high accuracy for the calculated binding constant.

2.3.1. Binding to Proteins; Human Apo-Transferrin and $[V^{IV}O(acac)_2]$ as an Example

As an example we will consider the case of $[V^{IV}O(acac)_2]$ binding to apo-transferrin (apoHTF) [43]. Fluorescence quenching measurements were done at T = 298 K, pH = 7.4 and λ_{ex} = 295 nm to study the binding of $[V^{IV}O(acac)_2]$ to apo-transferrin. A solution of apoHTF with concentration 1.02×10^{-6} M was prepared, and a solution of $[V^{IV}O(acac)_2]$ was progressively added to get solutions with C_V from 0 to 1.8×10^{-5} M, i.e., with $[V^{IV}O(acac)_2]$:HTF ratios from 1 to ~18. Following the usual linearization methodologies (as described above) [41,42,44], it was assumed that the fluorescence quenching observed should be due to binding of the $[V^{IV}O(acac)_2]$ complex to apoHTF, thus a static quenching is operating which is due to the non-emitting $[V^{IV}O(acac)_2]$ bound to the protein. Using Equation (8), the values of $K_2^{BC} = 1.0 \times 10^4$ and $n = 1.15$ were obtained.

The apparent simplicity of this methodology led to its broad application in the scientific literature. However, fluorescence and its quenching involve an indirect measurement of the interaction of compounds with proteins such as apoHTF and serum albumins, as the Trp residues may not be close to the binding site, and binding at sites that do not significantly affect the fluorescence emission is also possible. Importantly, there are several requirements and possible pitfalls for the validity of use of fluorescence emission data to calculate binding constants, discussed in [41,44,45], emphasizing the most common errors made in their interpretation that should be considered.

Besides these requirements associated to the doubtful validity of the equations used, there are further aspects/issues related to the speciation of systems involving labile metal complexes at low concentrations which deserve further attention. In fact, not considering them has led to errors and misunderstandings. As shown below for the $[V^{IV}O(acac)_2]$–apoHTF system, although it is clear that there is binding of $V^{IV}O$-species to apoHTF that causes the fluorescence quenching, the methodology described, is not valid or reliable to calculate the binding constant of the equilibrium described by Equation (2).

In aqueous solutions containing Hacac and $V^{IV}O$ salts in 2:1 molar ratios and at mM concentrations, as the pH increases the following V-species predominate up to pH = 8: $[V^{IV}O(H_2O)_5]^{2+} \rightarrow [V^{IV}O(acac)]^+ \rightarrow [V^{IV}O(acac)_2]$. Taking the formation constants β_{pqr} for the $V^{IV}O^{2+}$+acac system from [46], defined by Equation (1) with HL = Hacac, species distribution diagrams may be obtained for this system. Figure 3A depicts the speciation diagram calculated for total $V^{IV}O$ and acetylacetone concentrations of 3 and 6 mM, respectively. It is clear that in the pH range 6–8 $[V^{IV}O(acac)_2]$ is the main V^{IV}-complex present in solution. However, if the concentration of $[V^{IV}O(acac)_2]$ is lowered to ca. 2 μM, close to the concentrations used in the fluorescence quenching measurements in the system $[V^{IV}O(acac)_2]$ + apoHTF, the relative importance of most V^{IV}-species totally differs (see Figure 3B).

It is clear from Figure 3B that in the pH range 6–8, the relative concentrations of $[V^{IV}O(acac)]^+$ and $[V^{IV}O(acac)_2]$ in solution are low. At these low complex concentrations most of V^{IV} is in the form of $[(V^{IV}O)_2(OH)_5]^-$ and $[V^{IV}O(OH)_3]^-$; therefore, in the conditions used to measure the fluorescence quenching, Equation (3) cannot be used with $[Q] = [V^{IV}O(acac)_2]$, and the fluorescence quenching observed may be due to several distinct species other than $[V^{IV}O(acac)_2]$.

To further understand how wrong it is to apply the methodology described above, we will obtain the speciation diagram for the system $V^{IV}O^{2+}$ + acetylacetone + apoHTF, assuming the formation of $[((V^{IV}O(acac)_2])_1(apoHTF)]$ with the value of K_2^{BC} (= 1.0×10^4) obtained at pH = 7.4 based on data of fluorescence measurements. The value of log β_{120} = 16.27 corresponds to the formation of $[V^{IV}O(acac)_2]$; therefore, the formation constant of $[((V^{IV}O(acac)_2])_1(apoHTF)]$, corresponding to the reaction of Equation (9) at pH = 7.4, is: $\beta_2^{BC} = K_2^{BC} \times \beta_{120} = 10^{20.27}$.

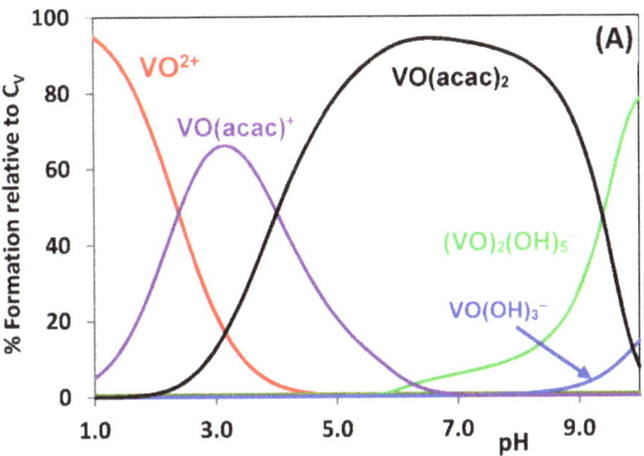

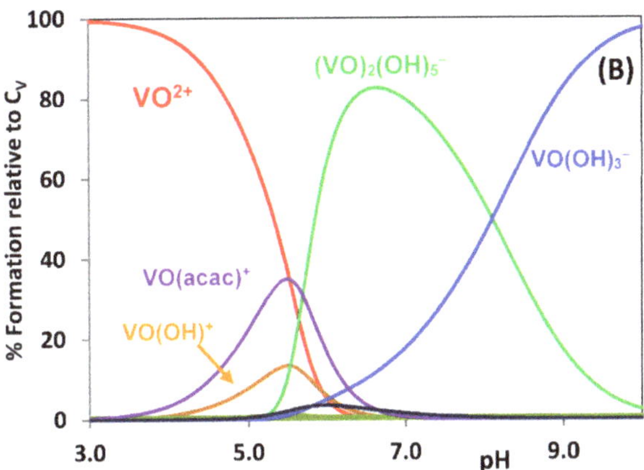

Figure 3. Species distribution diagrams, calculated with the computer program HySS [33], for total $V^{IV}O$ and acetylacetone concentrations of (**A**) 3 and 6 mM, respectively, and (**B**) 2 μM and 4 μM, respectively, considering the formation constants log β_{110} = 8.73 and log β_{120} = 16.27, reported in [46] and the $V^{IV}O$-hydrolytic constants: $[VO(OH)]^+$, $[(VO)_2(OH)_2]^{2+}$, $[(VO)_2(OH)_5]^-$ and $[VO(OH)_3]^-$ in refs. [9,22].

$$V^{IV}O^{2+} + 2acac^- + apoHTF \rightleftharpoons \left[\left(\left[V^{IV}O(acac)_2\right]\right)_1 (apoHTF)\right] \tag{9}$$

It is known that $V^{IV}O^{2+}$ binds to proteins, and the binding constants, at a particular pH value, may be defined as:

$$n\,V^{IV}O^{2+} + protein \rightleftharpoons \left(V^{IV}O\right)_n (protein) \tag{10}$$

$$\beta_n = \frac{\left[(V^{IV}O)_n(protein)\right]}{\left[V^{IV}O^{2+}\right]^n [protein]} \tag{11}$$

In the case of apo-transferrin $V^{IV}O^{2+}$ may form $[V^{IV}O(apoHTF)]$ and $[(V^{IV}O)_2(apoHTF)]$, and their formation constants were previously determined at pH 7.4 [15]: log β_1 = 13.4 and log β_2 = 25.2. The speciation may be calculated at pH = 7.4, see Figure 4, taking the concentration of apoHTF of 2×10^{-6} M and varying the concentration of [$V^{IV}O(acac)_2$] from 1×10^{-6} M up to 20×10^{-6} M, those used in the fluorescence quenching measurements.

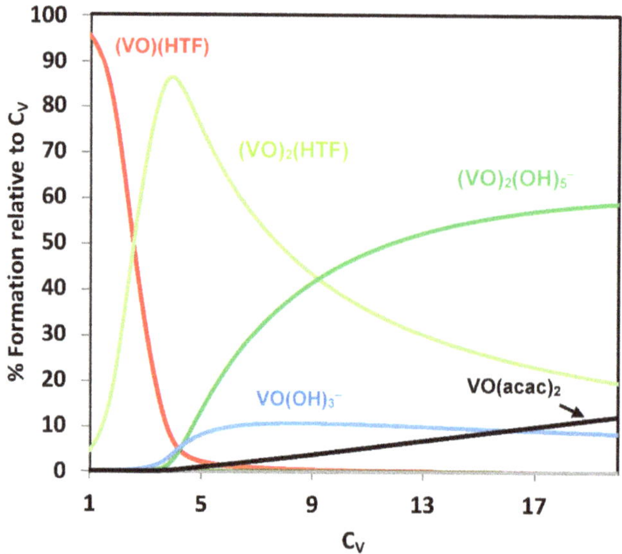

Figure 4. Species distribution diagram for the system $V^{IV}O^{2+}$ + acetylacetone + apoHTF, for [apoHTF]= 2×10^{-6} M and [$V^{IV}O(acac)_2$] in the range of 1 to 20 μM, at pH = 7.4, considering the formation constants reported in [9,15,22,46], and the formation constant of ([$V^{IV}O(acac)_2$])$_1$(apoHTF)] ($=10^{20.27}$) as obtained from fluorescence quenching methods. The calculations were carried out with the computer program HySS [33].

It is clear from Figure 4 that with a binding constant $K_2^{BC} = 10^4$ (K_a in many publications) the amount of [([$V^{IV}O(acac)_2$])$_1$(apoHTF)] formed is too low to have any relevant concentration at pH 7.4. Thus, the observed quenching of fluorescence cannot be solely due to the formation of [([$V^{IV}O(acac)_2$])$_1$(apoHTF)], the value of K_2^{BC} being clearly wrong.

Researchers should be aware that with values of association constants (log K_1^{BC} or log K_2^{BC}) in the range of 4–8, typically found in the literature at pH 6–8, the relevance of binding of [M(L)$_n$] (M = $V^{IV}O$) complexes to proteins such as apoHTF, BSA or HSA will be negligible in most cases. Similar conclusions may be extended to several other complexes of labile metal ions such as e.g., Cu(II), Zn(II) or Fe(III).

In the particular case of the $V^{IV}O^{2+}$ + acetylacetone + apoHTF system, only for values of log K_2^{BC} > ~11, which correspond to logβ_2^{BC} > 27, the amount of [([$V^{IV}O(acac)_2$])$_1$(apoHTF)] formed starts being visible in speciation diagrams at pH 7.4 (Figure 5). However, the system is much more complex as besides [$V^{IV}O(acac)_2$(apoHTF)], other species probably form, e.g., [$V^{IV}O(acac)(apoHTF)$], [($V^{IV}O)_2(acac)(apoHTF)$] and [($V^{IV}O)_2(acac)_2(apoHTF)$], and to properly address it and determine formation constants, spectroscopic techniques other than fluorescence should be used.

That was done for example in the $V^{IV}O^{2+}$ + maltol + apoHTF system at pH = 7.4, in which the formation constants were determined based on circular dichroism and EPR spectroscopy data, and calculated log β values of [$V^{IV}O(mal)(apoHTF)$], [($V^{IV}O)_2(mal)(apoHTF)$] and [($V^{IV}O)_2(mal)_2(apoHTF)$] (mal = maltolato) were 17.7, 30.3 and 34.8, respectively. Similarly, in the $V^{IV}O^{2+}$ + dhp + apoHTF system (dhp = 1,2-dimethyl-3-hydroxy-4(1H)-pyridinone), the calculated log β values of [$V^{IV}O(dhp)(apoHTF)$], [($V^{IV}O)_2(dhp)(apoHTF)$] and [($V^{IV}O)_2(dhp)_2(apoHTF)$] were 21.3, 33.0 and 40.3, respectively [15]. The systems $V^{IV}O^{2+}$ + picolinato + apoHTF, $V^{IV}O^{2+}$ + lactate + apoHTF, $V^{IV}O^{2+}$ + HSA and $V^{IV}O^{2+}$+

dhp + HSA and the corresponding formation constants were also determined based on EPR data [47].

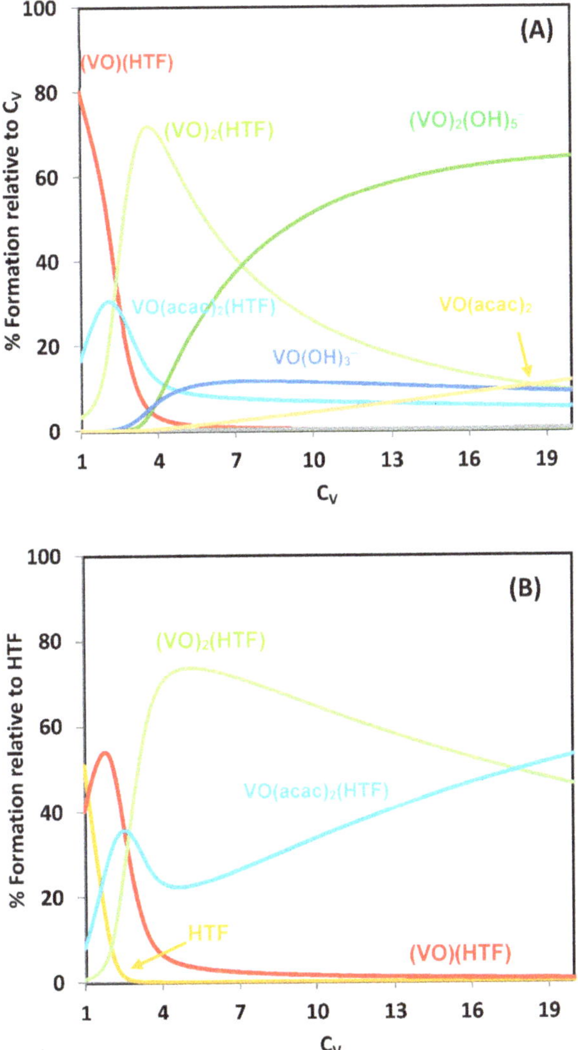

Figure 5. Species distribution diagrams for the system $V^{IV}O^{2+}$ + acetylacetone + apoHTF, for [apoHTF]= 2×10^{-6} M and [$V^{IV}O(acac)_2$] in the range of 1 to 20 μM, at pH = 7.4, considering the formation constants reported in [9,15,22,46], and the formation constant of [($V^{IV}O(acac)_2)_1$(apoHTF)] (=10^{27}). (**A**) Representation of vanadium containing species; (**B**) Representation of apoHTF containing species. The calculations were carried out with the computer program HySS [33].

2.3.2. Binding to DNA

It is known that vanadate is possibly carcinogenic to humans. In vitro vanadate can induce DNA strand breaks in human fibroblasts [48], but it was also found that distinct cell lines may respond differently to $NaVO_3$, and that its carcinogenic role at low concentrations may result from stimulation of proliferation of tumorigenic cells [49]. $V^{IV}OSO_4$ was found genotoxic for normal and cancer cells being able to induce DNA damage in lymphocytes (producing DNA single- and double-strand breaks) and in HeLa cancer cells (imposing only single strand breaks). Reactive oxygen species seem to be involved in the formation of DNA lesions [50].

Many studies reported cleavage/damage of DNA by either vanadium salts [48–52] or complexes [51,53–62]. It is known that several transition metals increase the oxidative stress in cells, which for vanadium has been mainly attributed to the generation of hydroxyl radicals by Fenton-like reactions promoted by physiological hydrogen peroxide [54,55]. Typically it is reported that vanadium(IV) reacts with H_2O_2 generating $HO^{\bullet}$ radicals, and these radicals may hydroxylate nucleobases and/or yield DNA strand breaks [51,52]. The involvement of singlet oxygen as the reactive species has also been suggested in some studies [56,63]. However, when reporting these reactions, the actual mechanisms of action of vanadium complexes is rarely accessed, most studies only stating if these processes are or not inhibited or activated by the presence of the ligand or of other chemical agents. In a study that included $V^{IV}OSO_4$, as well as various V-complexes (with picolinic acid and maltolate) and different oxidants, only $V^{IV}OSO_4$ showed ability to modify a double stranded 167-bp DNA fragment in the presence of $KHSO_5$. The mechanism proposed by the authors involved the oxidation of V^{IV} to V^V and the formation of a "caged" sulfate radical in the presence of $KHSO_5$. DNA cleavage at the guanine base was also proposed [64].

Notwithstanding, when reporting the action of metal complexes it is important to fully understand the role of the metal, of the ligand and of other species present. For example, the nuclease activity of $[V^{IV}O(acac)_2]$ and of several derivatives was studied by several techniques [54,55]; the mechanism was shown to be oxidative and the DNA cleavage mainly due to the formation of reactive oxygen species (ROS). Hydrolytic cleavage of the phosphodiester bond was also observed, but at much slower rate and did not compete with the oxidative route. Noteworthy, it was shown that the generation of ROS was much higher in experiments where phosphate was used as buffer, this being attributed to the formation of a mixed-ligand/mixed-valence complex containing phosphate, $[(V^{IV}O)(V^VO)(acac)_2(H_nPO_4{}^{n-3})]$ [54]. These studies focused on $[V^{IV}O(acac)_2]$ but in fact could/should be applied to other vanadium compounds, emphasizing the requirement of carefully accessing the speciation of the systems, which in this particular case is modified by the presence of phosphate. As phosphate is present at variable concentrations in most biological systems, this may have important implications in the interpretation of the biological activity of vanadium compounds. The extent of this effect depends on the phosphate:vanadium ratio and the formation of mixed-ligand species such as VO-L-phosphate (L = acac), which strongly enhances the oxidative stress, allegedly caused by $[V^{IV}O(acac)_2]$, most probably is not restricted to this particular vanadium system.

Bernier et al. investigated the nuclease activity of V(IV)-complexes of hydroxysalen derivatives under oxidative or reducing conditions. In the absence of activating agents, none of the complexes induced DNA cleavage; while in the presence of mercaptopropionic acid (MPA) or oxone all induced DNA modifications. The complexes reacted with DNA at guanine residues in the presence of oxone, and the mechanism proposed, based on spin-trapping EPR experiments, involved the oxidation of V^{IV} to V^V, the production of $SO_4{}^{-\bullet}$ and $SO_5{}^{-\bullet}$ radicals via a redox reaction and the trapping of the sulfate radicals by the metal through the formation of a "caged" radical [58].

Besides the ability to cleave/damage DNA, the binding interactions of vanadium complexes to DNA have also been studied by different techniques; for example electronic absorption [56,59–62], circular dichroism [60–62], and fluorescence emission have been used frequently [42,60–62,65]; other techniques such as atomic force microscopy [66–69], viscosity [56,59–61,69] and have also been applied. ^{51}V NMR, [53,54] capillary electrophoresis and Fourier transform infrared difference spectroscopy [70] has also been used, but much less frequently, as well as DNA melting, applied to evaluate e.g., binding and/or DNA crosslink formation by the complexes [56,59,60,71].

These studies are normally done at pH ~7 and, as emphasized above, in most cases extensive hydrolysis of the complexes is expected. While many studies have confirmed the binding of $V^{IV}O^{2+}$ and $V^{V}O_2{}^+$ to proteins such as HSA, BSA, apoHTF and immunoglobulin G (IgG), and formation constants determined [12,15,72–75], not much is known about the binding of these ions to DNA.

To the best of our knowledge, the only literature study on the binding $V^{IV}O^{2+}$ and $V^{V}O_3{}^-$ ions to *calf thymus* DNA in aqueous solutions at physiological pH, used capillary electrophoresis and FTIR spectroscopy. The authors reported that $V^{IV}O^{2+}$ binds DNA through guanine and adenine N-7 atoms and the phosphate groups with apparent binding constants of 8.8×10^5 M^{-1} and 3.4×10^5 M^{-1} for guanine and adenine respectively. The $V^{V}O_3{}^-$ ion shows weaker binding through thymine, adenine, and guanine bases, with $K^{BC} = 1.9 \times 10^4$ M^{-1} and no interaction with the backbone phosphate moieties was found. The authors also reported a partial B-to-A DNA transition upon $V^{IV}O$–DNA binding. Again, the speciation occurring at physiological pH for V(IV) and V(V) species was not taken into account [70]. For DNA this is particularly relevant since it contains negatively charged phosphate groups that may participate in electrostatic binding to vanadium. Upon dissolution of $V^{IV}O^{2+}$ at physiological pH negatively charged species are formed— $[(V^{IV}O)_2(OH)_5]^-$ and $[V^{IV}O(OH)_3]^-$ -and these have to be taken into account.

However, even if we assume that vanadium(IV or V) ions do not bind to DNA, when evaluating the binding of $[V^{IV/V}O_n(L)_m]$ complexes to DNA, to assume that the complexes do not hydrolyze at low concentration in aqueous solvents, is an oversimplification that normally is not acceptable.

2.4. Behavior of Metal Complexes when Added to Incubation Media of Cells

Recent studies regarding applications of vanadium compounds in therapeutics have been mainly focused on their anti-cancer and anti-parasitic potential and many vanadium complexes have been tested, mainly by in vitro studies [8,76]. Within these fields several polypyridyl complexes of V(IV), as well as mixed-ligand complexes containing polypyridyl ligands have been prepared and their anti-proliferative activity, cytotoxicity and ability to induce apoptosis tested in vitro, normally displaying high activities [8,77–96]. Possible biological targets and mechanisms of action of V-phen compounds have been discussed, and interaction with DNA has been typically considered as relevant [8,79].

$V^{IV}O$-complexes with polypyridyl ligands also hydrolyse and/or change their composition at low μM concentration when compared with mM concentrations [95,96], and this is clearly demonstrated in Figure 6 for the $V^{IV}O^{2+}$ + phen system. Decomposition of the $V^{IV}O$-complexes with polypyridyl ligands leads to the release of the free ligands, which by themselves are biologically active and cytotoxic [95–103].

Metal complexes may react with cell culture media components. This is important for their in vitro biological activity, [104] particularly for labile metal ions, but has been normally neglected in studies reporting cytotoxicity of vanadium compounds, as well as of other metal complexes. The role of the ligands in the cytotoxicity of $[V^{IV}O(OSO_3)(Me_2phen)_2]$ (Me_2phen = 4,7-dimethyl-1,10-phenanthroline), and of a few related $V^{IV}O$-complexes, was investigated by the group of Lay and co-workers [95]. These researchers reported cytotoxicity assays with human lung cancer A549 cells and tested the stability of the complexes in aqueous solutions as well as in cell culture media. They concluded that the high cytotoxicity at 72 h incubation of the free ligands and corresponding $V^{IV}O$-complexes is due to the free ligands upon decomposition of the complexes in cell culture medium.

More recently, Nunes et al. [96] reported cytotoxicity studies of a group of $V^{IV}O$-polypyridyl complexes at several incubation times against three different types of cancer cells. The compounds addressed were $[V^{IV}O(OSO_3)(phen)_2]$ $[V^{IV}O(OSO_3)(Me_2phen)_2]$, and $[V^{IV}O(OSO_3)(amphen)_2]$ (amphen = 5-amino-1,10-phenanthroline). We will globally designate these 1,10-phenanthroline compounds by Xphen. Upon incubation for 72 h with several types of cells, the cytotoxicity of all compounds was approximately equal, while at incubation times of 3 and 24 h, where the IC_{50} values measured are much higher, the complexes show significantly higher activity than the free ligands. This difference in behavior is probably due to the distinct type of speciation taking place in cell media. Next, we shortly discuss these observations.

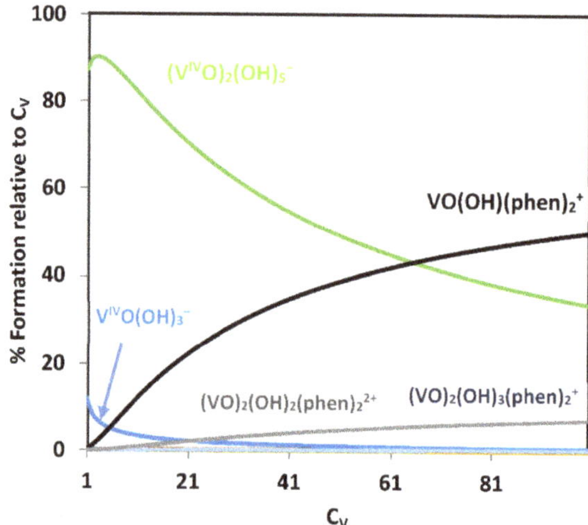

Figure 6. Species distribution diagrams in water at pH 7.0 for the system $V^{IV}O$-phen [97] in the concentration range C_V = 1 to 100 μM (C_V is the total $V^{IV}O$ concentration), with the molar ratio $V^{IV}O$:phen of 1:2. The concentrations of the dinuclear species $[(V^{IV}O)_2(phen)_2(OH)_2]^{2+}$ and $[(V^{IV}O)_2(phen)_2(OH)_3]^+$ almost coincide. The diagrams were calculated using the HySS computer program [33] including the hydrolytic species $[V^{IV}O(OH)]^+$, $[(V^{IV}O)_2(OH)_2]^{2+}$, $[(V^{IV}O)_2(OH)_5]_n^-$ and $[V^{IV}O(OH)_3]^-$ [9,22,29].

Cell culture media used in cytotoxicity studies with mammalian cells contain many potential ligands for $V^{IV}O^{2+}$, namely relatively high amounts of bovine serum albumin (BSA), mainly due to the addition of fetal bovine serum (FBS) [104,105]. In the typical case of addition of 10% FBS, the concentration of BSA is ca. 40 μM. Binding of vanadium salts and complexes to proteins has been studied, particularly aspects associated with their transport in blood [3,9,13,106–111]. Binding of vanadium compounds to serum proteins [73,108–112], including complexes containing polypyridyl co-ligands [96,97,107], have also been reported. It is important to understand that as soon as each of the $V^{IV}O$-Xphen complexes is added to the cell incubation media, they decompose, no longer being present as $[V^{IV}O(Xphen)_2]^{2+}$. This was demonstrated in [96], and we will now highlight the main points addressed.

For this purpose, we must clarify the definitions of formation constants to be used. The stability constants of complexes of $V^{IV}O^{2+}$ with phen and derivatives are defined in the usual way, considering the reaction in Equation (1) with L = phen, and the conditional binding constants of $V^{IV}O^{2+}$ to apoHTF, HSA, BSA or any other protein, at a set pH, may be defined according to Equations (10) and (11).

Regarding conditional binding (or association) constants of $[V^{IV}O(phen)_n]^{2+}$ complexes to a protein (e.g., BSA) at a set pH, these may be defined similarly to Equations (2) and (3), and in the case of phen (omitting charges) as:

$$K_1^{BC}: \quad \left[V^{IV}O(phen)\right] + BSA \rightleftharpoons \left[V^{IV}O(phen)(BSA)\right] \tag{12}$$

$$K_2^{BC}: \quad \left[V^{IV}O(phen)_2\right] + BSA \rightleftharpoons \left[V^{IV}O(phen)_2(BSA)\right] \tag{13}$$

The formation constants for $[V^{IV}O(phen)(BSA)]$ and $[V^{IV}O(phen)_2(BSA)]$ correspond to:

$$\beta_1^{BC}: \quad V^{IV}O^{2+} + phen + BSA \leftrightharpoons \left[V^{IV}O(phen)(BSA)\right] \tag{14}$$

$$\beta_2^{BC}: \quad V^{IV}O^{2+} + 2phen + BSA \leftrightharpoons \left[V^{IV}O(phen)_2(BSA)\right] \tag{15}$$

The constants defined in Equations (10)–(15) may be determined (or estimated) at predefined pH values; these are required for the speciation calculations at set pH values. The values of K_1^{BC} and K_2^{BC} may be related to β_1^{BC} and β_2^{BC} using Equations (1) and (12), to get Equation (14):

$$\log K_1^{BC} = \log \beta_1^{BC} - \log \beta_{110} \qquad \log K_2^{BC} = \log \beta_2^{BC} - \log \beta_{120} \tag{16}$$

β_{110} and β_{120} being the formation constants of $[V^{IV}O(phen)]^{2+}$ and $[V^{IV}O(phen)_2]^{2+}$ as defined in Equation (1).

In Figure 6, we depict speciation diagrams for the $V^{IV}O$-phen system at pH 7.0 in the range of total oxidovanadium(IV) concentrations (C_V) from 1 to 100 μM. It is clear from Figure 6 that at pH 7.0, for $C_{VO} < {\sim}10$ μM most V^{IV} is in the form of hydrolytic $V^{IV}O$-species (mainly $[(V^{IV}O)_2(OH)_5]^-$ and $[V^{IV}O(OH)_3]^-$), and that the most relevant $V^{IV}O$-phen complex at this pH is $[V^{IV}O(OH)(phen)_2]^+$, its relative importance increasing with C_V.

The binding constants of $V^{IV}O^{2+}$ to BSA were not determined, but as previously done [96], we will make the approximation that these are equal to those determined for HSA at pH 7.4 [47]. Assuming that [BSA] = 40 μM, addition of $[V^{IV}O(phen)_2]^{2+}$ in the concentration range 1–100 μM and the formation of $V^{IV}O$-BSA species, but at this stage not considering the binding of $V^{IV}O$-phen complexes to BSA, the calculated species distribution diagram (Figure S1), totally differs from that shown in Figure 6. It may be observed that $V^{IV}O^{2+}$ predominantly binds to BSA, and only for higher vanadium concentrations, when there is not enough BSA to bind the oxidovanadium(IV) ions present (the total BSA concentration is 40 μM), $V^{IV}O$-phen species start being relevant. We also highlight that in the same conditions most of phen is either bound to BSA or free in solution.

$V^{IV}O$-phen complexes bind to BSA but it was not feasible to adequately calculate the formation constant for the $[V^{IV}O(phen)(BSA)]$ species by CD spectroscopy; the value of the constant, K_1^{BC}, for the binding of $[V^{IV}O(phen)]$ to BSA could only be approximately estimated with this technique [96]. In order that the VO:phen:BSA species (1:1:1) might have a reasonable concentration on speciation diagrams the corresponding stability constant must be at least $\log\beta_1((VO(phen)(BSA)) = \log \beta_1^{BC} {\sim}15$, which corresponds to a binding constant of $[V^{IV}O(phen)]$ to BSA of ca. $K_1^{BC} = 10^9$. The speciation diagrams included in Figure 7 are obtained for conditions modelling the amount of BSA present in cell incubation media.

As the binding constants of all vanadium-BSA-containing species are only an approximation of their true values, the diagrams depicted in Figure 7 may only be used for semi-quantitative purposes. The conditions assumed are not equal to those of cell media used in the cytotoxicity determinations, but provide a reasonable model and allow stating that upon addition of $[V^{IV}O(phen)_2]^{2+}$ to cell culture media containing 10% FBS (where [BSA] $\approx$ 40 μM), the complex loses its integrity yielding several different species. Moreover, the type of species that have relevant concentrations depends on the amount of complex added. Additionally, as V^{IV} may oxidize to V^V in incubation media of cells, also forming V^V–phen and V^V–hydrolytic compounds, the complexity of the speciation of V-phen species is much higher than what Figure 7 suggests.

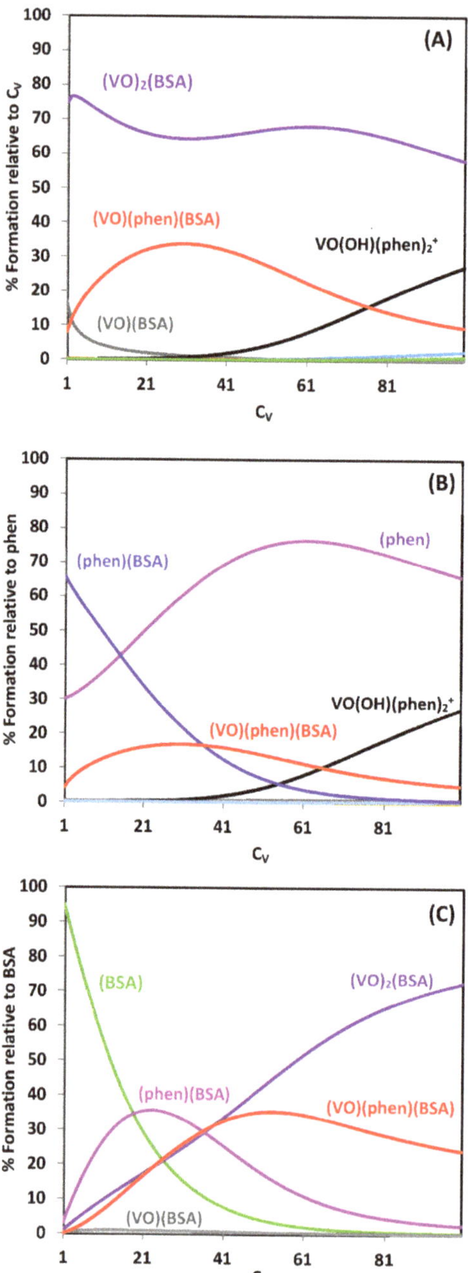

Figure 7. Species distribution diagrams of the system $V^{IV}O^{2+}$+phen+BSA at pH = 7 (range of C_V values: 1 to 100 μM) calculated using the HySS program [33], taking a molar ratio $V^{IV}O$:phen of 1:2 and [BSA] = 40 μM. The binding of phen to BSA is taken into account with a binding constant of 5.7×10^4, and the formation of $V^{IV}O$-phen-BSA species with a stability constant of 10^{15} (which corresponds to $K_1^{BC} \sim 10^9$, see also text). (**A**) Representation of vanadium containing species; (**B**) Representation of 1,10-phenanthroline containing species; (**C**) Representation of BSA containing species.

Overall, it is clear that the interpretation of the obtained cytotoxicity data must take into account the speciation that the V-complexes undergo in the medium, and we cannot simply assume the existence of $[V^{IV}O(phen)_2]^{2+}$ or $[V^{IV}O(phen)]^{2+}$ either outside or inside the cells to discuss the mechanism of cytotoxic action.

Another interesting example was reported by Tshuva and co-workers [113] on a highly promising diaminotris(phenolato) V^V-complex that showed high hydrolytic stability in water [114] due to its six-coordination and lack of labile ligands, along with promising and broad in vitro and in vivo cytotoxic activity. The resemblance of the biological activity of the complexes and their corresponding free ligands, implying their participation as active species, inspired an investigation into the complex interactions in the cellular environment [113]. They observed: (i) the presence of free ligand inside cells, as well a V^V-species, which was not the initial complex; (ii) the ligand uptake seemed to be slower than the medium-formed species of the V^VO complex, suggesting that vanadium promoted the cell penetration and (iii) no significant impact was found for the vanadium center on the apoptotic process, which was similar for free ligand and V-complex. Overall, the studies indicated the free ligand as the active species, despite the high stability of the complexes in water, suggesting that some cellular components promote complex dissociation and that vanadium facilitates cell uptake. Like the $V^{IV}O$-Xphen system, higher activity is observed at shorter incubation periods with the complexes than with free ligands alone, highlight the role of the vanadium complex as a pro-drug.

2.5. Interactions of Metal Complexes with Biological Targets

Many studies reported effects of vanadium salts and complexes with biological targets such as proteins or DNA. While effects are indeed observed, misunderstandings about possible structural effects of complexes and which are the active species operating are frequent.

For example, in a study of $V^{IV}OSO_4$ and several oxidovanadium(IV) complexes with a set of six β-diketonato ligands, it was reported that $V^{IV}O$ and the complexes showed significant inhibitory activity of snake venom phosphodiesterase I enzyme. [115] All seven compounds exhibited a non-competitive type inhibition, with one complex exhibiting $IC_{50} \approx 9.8$ μM, the other six compounds (including $V^{IV}OSO_4$) exhibiting IC_{50} values in the 17–33 μM range. The authors discussed the inhibitory activity mainly based on structural characteristics of the compounds, these mostly measured in organic solvents. However, the $V^{IV}O$-β-diketonato complexes certainly undergo hydrolysis and/or oxidation in aqueous solvents, as discussed above for $[V^{IV}O(acac)_2]$, and this must be taken into account in the discussion of their activity.

Another example is a study of $V^{IV}OSO_4$, $[V^{IV}O(mal)_2]$ and $[V^{IV}O(alx)_2]$ (alx = 3-hydroxy-5-methoxy-6-methyl-2-pentyl-4-pyrone), reported to act as antidiabetic agents. An in vitro Akt kinase assay was carried out using the GSK3 fusion protein, and the experiments were carried out with the complexes in 'kinase buffer' (25 mM Tris–HCl, 10 mM $MgCl_2$, 1 mM EGTA, 2 mM dithiothreitol, 0.1% BSA, pH 7.5), also containing 0.2 mM ATP and 0.5 μg of the GSK3 fusion protein substrate (EGTA = ethylene glycol bis(2-aminoethyl ether)tetraacetate, ATP = adenosine triphosphate) [116]. $[V^{IV}O(alx)_2]$ phosphorylated the Akt kinase and GSK of 3T3-L1 and GSK of adipocytes, but not $V^{IV}OSO_4$ or $[V^{IV}O(mal)_2]$. It was also reported that the vanadium uptake by the adipocytes upon 10-min stimulus with the $[V^{IV}O(alx)_2]$ and $[V^{IV}O(mal)_2]$, was 11 ± 1.4 and 2.5 ± 0.2 nmol of vanadium per 10^6 cells, respectively [116].

These data are interesting and important, but care should be taken on elaborations of further conclusions. Besides the ligands of the complexes and the possible V^{IV}-oxidation, the mixture used for the Akt kinase assay contains at least two other relevant $V^{IV}O^{2+}$ binders: EGTA and BSA in quite high concentrations, therefore, a significant amount of vanadium will not be bound to allixinato or maltolato. Moreover, the vanadium uptaken and inside the adipocytes is possibly not bound to these ligands, but to other bio-ligands of the cells.

We will mention a last example. Complex N,N'-ethylenebis(pyridoxylideneiminato) vanadium(IV), [V^{IV}O(pyren)], originally prepared by Correia et al. [117], was reported to have relevant anticancer properties, demonstrated in in vitro tests with several human cancer cell lines such as A375 (human melanoma) and A549 (human lung carcinoma) cells. The mechanism of the effect of [V^{IV}O(pyren)] was assigned to apoptosis induction, triggered by ROS increase, followed by mitochondrial membrane depolarization [118].

The experiments were carried out with different concentrations of H_2pyren and [V^{IV}O(pyren)] at 24, 48, and 72 h of incubation; pyren showed no relevant cytotoxic affect. The data obtained seems promising, but again the assignment of the active species to [V^{IV}O(pyren)] is elusive. It was previously reported that the ligand precursor H_2pyren as well as [V^{IV}O(pyren)] are more soluble in water than similar salen and [V^{IV}O(salen)] complexes, and this allowed the determination of the formation constants in the $V^{IV}O^{2+}$ + pyren system up to pH = 5, where the neutral [V^{IV}O(pyren)] starts precipitating [117]. Speciations are depicted in Figure 8.

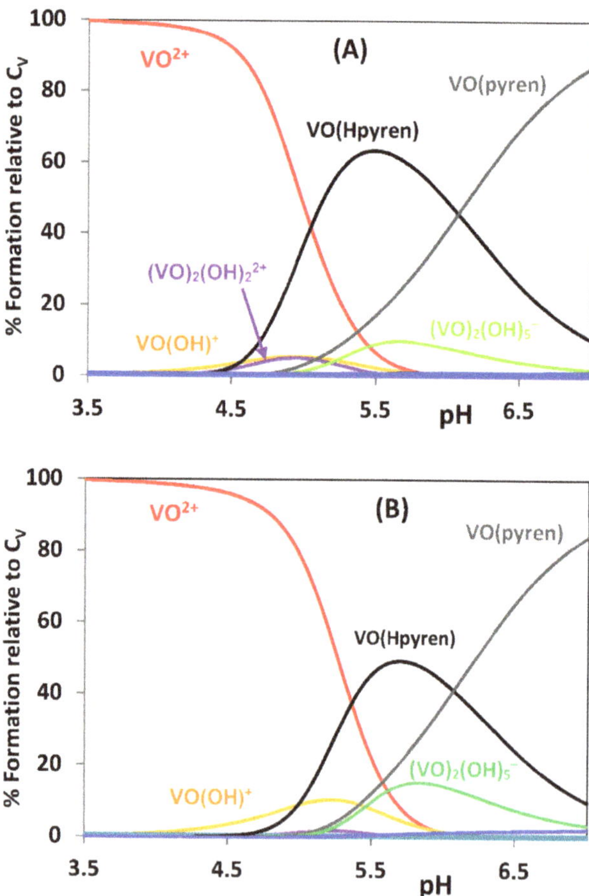

Figure 8. Species distribution diagrams of V^{IV}O-complexes formed in solutions containing $V^{IV}O^{2+}$ and H_2pyren, with $C_V = C_{pyren} = 100\ \mu M$ (**A**) and $10\ \mu M$ (**B**), assuming the formation constants determined are valid up to pH = 7. [V^{IV}O(Hpyren)]$^+$ corresponds to [V^{IV}O(pyren)] protonated at one of the pyridine N-atoms. In both complexes the pyren ligand binds in a tetradentate fashion to V^{IV}.

The calculated concentration distribution curves of Figure 8 indicate that even at 10 μM the [$V^{IV}O$(pyren)] complex maintains its integrity. However, in solutions containing [$V^{IV}O$(pyren)] the metal progressively oxidizes and the ligand undergoes hydrolysis; after standing at 4 °C for two weeks, crystals of [(V^VO)$_2$(pyr$_{1/2}$en)$_2$] were isolated (pyr$_{1/2}$en is the half Schiff base of pyren, where one of the C=N bonds hydrolysed, see Scheme 1). This means that at lower complex concentration, both these processes may proceed much faster. Therefore, the biological properties cannot be unambiguously simply assigned to [$V^{IV}O$(pyren)], as several other V^{IV} and V^V species may be present.

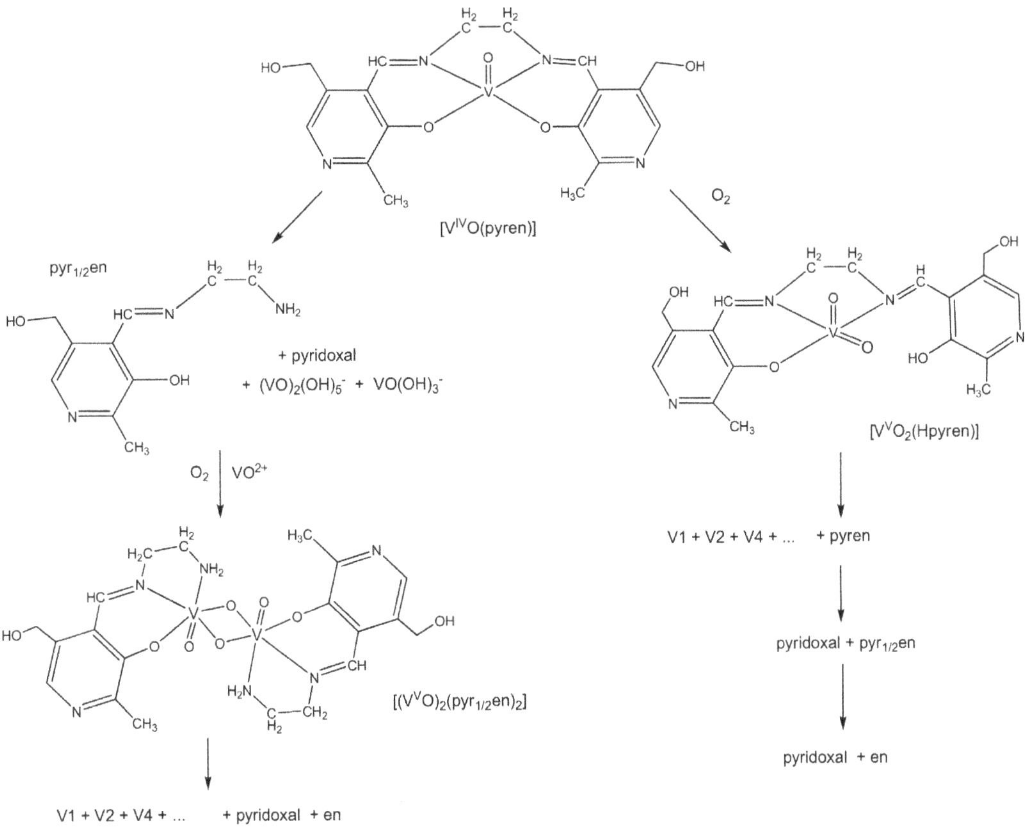

Scheme 1. Fate of [$V^{IV}O$(pyren)] in aqueous solution and outline of some possible pathways during aging. V_1, V_2 and V_4 are V^V species—see Section 2.2-and en is ethylenediamine [117].

3. Conclusions

It is known that the behavior of vanadium and of most first-row transition metal complexes in solution, namely in biological fluids, depend on their environment, namely on the presence of other potential ligands. In aqueous media most metal complexes undergo hydrolysis, ligand exchange and redox reactions, as well as chemical changes that depend on pH and concentration, and on the presence and nature of the several biological components present in the media. However, when reporting the biological action of metal complexes, often the possibility of chemical changes is not taken into account.

Vanadium(IV) ions are very susceptible to oxidation, and both V^{IV} and V^V ions are also very prone to hydrolysis. In this work we highlight that in the particular case of most vanadium compounds, besides oxidation and/or hydrolysis, as soon as they are dissolved in aqueous media they undergo several other types of chemical transformations and they no longer exist in their initial form. These changes are particularly extensive at the low concentrations normally used in biological experiments. If a biological effect is observed, it is clear that in order to determine which is the active species and/or propose mechanisms of action, it is essential to evaluate the speciation of the metal-complex in the media where it is acting, but this has been neglected in most of the studies published.

Besides concentration of the metal ion, ionic strength, nature of buffer used, etc., other factors that are relevant for the hydrolytic behavior of vanadium(V) ions are time and temperature of experiments. These may influence the type and the concentration of the V-species present. Typically, chemical and spectroscopic studies are done at ca. 25 °C, while biological experiments at ~37 °C. Both these factors are relevant e.g., for the formation/decomposition of polyoxidovanadates. Moreover, we did not discuss issues associated to non-oxido vanadium(IV) complexes. Some of these complexes are quite resistant to hydrolysis, but with time they may produce oxidovanadium(IV or V) complexes, and these processes should be accounted for when discussing their biological action.

It should also be understood that in in vitro experiments, besides transformations in the incubation media, which will be relevant for the uptake process, once taken up by cells, most probably vanadium will no longer be bound to the original ligand, but to bio-ligands such as proteins of the cellular system.

If a vanadium compound is administered in vivo, either orally or by injection, the situation is much more complex. The ligand bound to the metal certainly changes during its bioprocessing, new complexes form, also with a distinct speciation profile, which may have (or not) a beneficial biological activity. The use of any metal-complex in the clinic requires the evaluation of its speciation chemistry associated with its pharmacokinetic and pharmacodynamic properties. This once more requires the evaluation of speciation of the vanadium species formed in each biological compartment and tissue involved. All these processes correspond to huge tasks and facilitate experiments and interpretations may yield misunderstandings and pseudo-conclusions that do not contribute to the advance of knowledge of the systems.

Supplementary Materials: The following are available online at https://www.mdpi.com/2304-6740/9/2/17/s1. Figure S1: Species distribution diagrams of the system VIVO^{2+} + phen + BSA at pH = 7 (range of CV values: 1 to 100 μM) calculated using the HySS program, taking a molar ratio VIVO:phen of 1:2 and [BSA] = 40 μM. The binding of phen to BSA is taken into account with a binding constant of 5.7 × 10^4, but in this figure the formation of VIVO-phen-BSA species is not con-sidered.

Author Contributions: Conceptualization, J.C.P.; Methodology, J.C.P. and I.C.; Software (use), J.C.P. and I.C.; Validation, J.C.P. and I.C.; Formal Analysis, J.C.P. and I.C.; Investigation, J.C.P. and I.C.; Resources, J.C.P. and I.C.; Data Curation, J.C.P. and I.C.; Writing—Original Draft Preparation, J.C.P.; Writing—Review & Editing, J.C.P. and I.C.; Visualization, J.C.P. and I.C.; Supervision, J.C.P. and I.C.; Project Administration, J.C.P.; Funding Acquisition, J.C.P. and I.C. All authors have read and agreed to the published version of the manuscript.

Funding: This research received funding from Fundação para a Ciência e Tecnologia (FCT) projects UIDB/00100/2020), Programa Operacional Regional de Lisboa 2020. I.C. thanks program FCT Investigator (IF/00841/2012).

Institutional Review Board Statement: Not applicable.

Informed Consent Statement: Not applicable.

Data Availability Statement: Not applicable.

Acknowledgments: We thank Centro de Química Estrutural, financed by Fundação para a Ciência e Tecnologia.

Conflicts of Interest: The authors declare no conflict of interest.

Abbreviations

Acac	acetylacetonate
alx	3-hydroxy-5-methoxy-6-methyl-2-pentyl-4-pyrone
Amphen	5-amino-1,10-phenanthroline
ATP	Adenosine triphosphate
Bipy	2,2′-bipyridine
BSA	Bovine serum albumin
CD	Circular dichroism
C_V	total vanadium concentration
dhp	1,2-dimethyl-3-hydroxy-4(1H)-pyridinone
DMEM	Dulbecco's Modified Eagle Medium
DMSO	Dimethyl sulfoxide
EGTA	ethylene glycol bis(2-aminoethyl ether)tetraacetate
EPR	Electronic paramagnetic spectroscopy
FBS	Fetal bovine serum
HSA	Human serum albumin
IC_{50}	Minimum inhibitory concentration
ICP-MS	Inductively coupled plasma mass spectrometry
Mal	maltolato
MEM	Minimum Essential Medium Eagle
Me$_2$phen	4,7-dimethyl-1,10-phenanthroline
MeOH	Methanol
NMR	Nuclear magnetic resonance
Phen	1,10-phenanthroline
ROS	Reactive oxygen species
RPMI medium	Roswell Park Memorial Institute medium
SDS	Sodium dodecyl sulfate
Tris	Tris(hydroxymethyl)aminomethane

References

1. Williams, R.J.P.; Fraústo da Silva, J.J.R. *The Natural Selection of the Chemical Elements—The Environment and Life's Chemistry*; Oxford University Press: Oxford, UK, 1996.
2. Williams, R.J.P.; Fraústo da Silva, J.J.R. *The Chemistry of Evolution—The Development of our Ecosystem*; Elsevier: Amsterdam, The Netherlands, 2006.
3. Pessoa, J.C.; Garribba, E.; Santos, M.F.A.; Santos-Silva, T. Vanadium and proteins: Uptake, transport, structure, activity and function. *Coord. Chem. Rev.* **2015**, *301*, 49–86. [CrossRef]
4. Willsky, G.R.; Chi, L.H.; Godzala, M., III; Kostyniak, P.J.; Smee, J.J.; Trujillo, A.M.; Alfano, J.A.; Ding, W.; Hu, Z.; Crans, D.C. Anti-diabetic effects of a series of vanadium dipicolinate complexes in rats with streptozotocin-induced diabetes. *Coord. Chem. Rev.* **2011**, *255*, 2258–2269. [CrossRef] [PubMed]
5. Crans, D.C.; Trujillo, A.M.; Pharazyn, P.S.; Cohen, M.D. How environment affects drug activity: Localization, compartmentalization and reactions of a vanadium insulin-enhancing compound, dipicolinatooxovanadium (V). *Coord. Chem. Rev.* **2011**, *255*, 2178–2192. [CrossRef]
6. Metelo, A.M.; Pérez-Carro, R.; Castro, M.M.C.A.; López-Larrubia, P. VO(dmpp)$_2$ normalizes pre-diabetic parameters as assessed by in vivo magnetic resonance imaging and spectroscopy. *J. Inorg. Biochem.* **2012**, *115*, 44–49. [CrossRef]
7. Trevino, S.; Díaz, A.; Sánchez-Lara, E.; Sanchez-Gaytan, B.L.; Perez-Aguilar, J.M.; González-Vergara, E. Vanadium in Biological Action: Chemical, Pharmacological Aspects, and Metabolic Implications in Diabetes Mellitus. *Biol. Trace Elem. Res.* **2019**, *188*, 68–98. [CrossRef]
8. Pessoa, J.C.; Etcheverry, S.; Gambino, D. Vanadium compounds in medicine. *Coord. Chem. Rev.* **2015**, *301*, 24–48. [CrossRef] [PubMed]
9. Pessoa, J.C. Thirty years through vanadium chemistry. *J. Inorg. Biochem.* **2015**, *147*, 4–24. [CrossRef]
10. Pessoa, J.C.; Goncalves, G.; Roy, S.; Correia, I.; Mehtab, S.; Santos, M.F.A.; Santos-Silva, T. New insights on vanadium binding to human serum transferrin. *Inorg. Chim. Acta* **2014**, *420*, 60–68. [CrossRef]
11. Mehtab, S.; Goncalves, G.; Roy, S.; Tomaz, A.I.; Santos-Silva, T.; Santos, M.F.A.; Romão, M.J.; Jakusch, T.; Kiss, T.; Pessoa, J.C. Interaction of vanadium(IV) with human serum apo-transferrin. *J. Inorg. Biochem.* **2013**, *121*, 187–195. [CrossRef]
12. Sanna, D.; Biro, L.; Buglyo, P.; Micera, G.; Garribba, E. Biotransformation of BMOV in the presence of blood serum proteins. *Metallomics* **2012**, *4*, 33–36. [CrossRef] [PubMed]

13. Sanna, D.; Micera, G.; Garribba, E. New Developments in the Comprehension of the Biotransformation and Transport of Insulin-Enhancing Vanadium Compounds in the Blood Serum. *Inorg. Chem.* **2010**, *49*, 174–187. [CrossRef]

14. Pessoa, J.C.; Tomaz, I. Transport of Therapeutic Vanadium and Ruthenium Complexes by Blood Plasma Components. *Curr. Med. Chem.* **2010**, *17*, 3701–3738. [CrossRef]

15. Jakusch, T.; Hollender, D.; Enyedy, E.A.; Gonzalez, C.S.; Montes-Bayon, M.; Sanz-Medel, A.; Pessoa, J.C.; Tomaz, I.; Kiss, T. Biospeciation of various antidiabetic (VO)-O-IV compounds in serum. *Dalton Trans.* **2009**, 2428–2437. [CrossRef]

16. Chasteen, N.D.; Grady, J.K.; Holloway, C.E. Characterization of the binding, kinetics, and redox stability of vanadium(IV) and vanadium(V) protein complexes in serum. *Inorg. Chem.* **1986**, *25*, 2754–2760. [CrossRef]

17. Thompson, K.H.; Lichter, J.; LeBel, C.; Scaife, M.C.; McNeill, J.H.; Orvig, C. Vanadium treatment of type 2 diabetes: A view to the future. *J. Inorg. Biochem.* **2009**, *103*, 554–558. [CrossRef] [PubMed]

18. Crans, D.C.; Smee, J.J.; Gaidamauskas, E.; Yang, L. The Chemistry and Biochemistry of Vanadium and the Biological Activities Exerted by Vanadium Compounds. *Chem. Rev.* **2004**, *104*, 849–902. [CrossRef] [PubMed]

19. Kremer, L.E.; McLeod, A.I.; Aitken, J.B.; Levina, A.; Lay, P.A. Vanadium(V) and -(IV) complexes of anionic polysaccharides: Controlled release pharmaceutical formulations and models of vanadium biotransformation products. *J. Inorg. Biochem.* **2015**, *147*, 227–234. [CrossRef]

20. Pettersson, L.; Andersson, I.; Gorzsas, A. Speciation in peroxovanadate systems. *Coord. Chem. Rev.* **2003**, *237*, 77–87. [CrossRef]

21. Elvingson, K.; Gonzalez Baro, A.; Pettersson, L. Speciation in Vanadium Bioinorganic Systems. 2. An NMR, ESR, and Potentiometric Study of the Aqueous H+-Vanadate-Maltol System. *Inorg. Chem.* **1996**, *35*, 3388–3393. [CrossRef]

22. Vilas Boas, L.F.; Pessoa, J.C. Vanadium. In *Comprehensive Coordination Chemistry*; Wilkinson, G., Gillard, R.D., McCleverty, J.A., Eds.; Pergamon Press: Oxford, UK, 1987; Volume 3, pp. 453–583.

23. Ducret, L.P. Vanadium Tétravalent. *Ann. Chim. S12 Chp. III* **1951**, *6*, 723–737.

24. Rossotti, F.J.C.; Rossotti, H.S. Studies on the Hydrolysis of Metal Ions. XII. The Hydrolysis of the Vanadium(IV) ion. *Acta Chem. Scand.* **1955**, *9*, 1177–1192. [CrossRef]

25. Iannuzzi, M.M.; Kubiak, C.P.; Rieger, P.H. Electron spin resonance line width studies of vanadium(IV) in acidic and basic aqueous solutions. *J. Phys. Chem.* **1976**, *80*, 541–545. [CrossRef]

26. Iannuzzi, M.M.; Rieger, P.H. Nature of vanadium(IV) in basic aqueous solution. *Inorg. Chem.* **1975**, *14*, 2895–2899. [CrossRef]

27. Copenhafer, W.C.; Rieger, P.H. Proton, deuteron, oxygen-17, and electron spin magnetic resonance of the trihydroxovanadyl(IV) ion. Kinetics and relaxation. *Inorg. Chem.* **1977**, *16*, 2431–2437. [CrossRef]

28. Komura, K.; Hayashi, M.; Imanaga, H. Hydrolytic Behavior of Oxovanadium(IV) Ions. *Bull. Chem. Soc. Jpn.* **1977**, *50*, 2927–2931. [CrossRef]

29. Pessoa, J.C.; Boas, L.F.V.; Gillard, R.D.; Lancashire, R.J. Oxovanadium(Iv) and Amino Acids—I. The System L-Alanine+VO^{2+}—A Potentiometric and Spectroscopic Study. *Polyhedron* **1988**, *7*, 1245–1262. [CrossRef]

30. Costa Pessoa, J.; Vilas Boas, L.F.; Gillard, R.D. Oxovanadium (IV) and Aminoacids—II. The systems L-Serine and L-Threonine +VO^{2+}; A Potentiometric and Spectroscopic Study. *Polyhedron* **1989**, *8*, 1173–1199. [CrossRef]

31. Costa Pessoa, J.; Vilas Boas, L.F.; Gillard, R.D. Oxovanadium(IV) and Aminoacids—IV. The system L-Cysteine and D-Penicillamine +VO^{2+}; A Potentiometric and Spectroscopic Study. *Polyhedron* **1990**, *9*, 2101–2125. [CrossRef]

32. Costa Pessoa, J.; Marques, R.L.; Vilas Boas, L.F.; Gillard, R.D. Oxovanadium(IV) and Aminoacids—III. The system L-Aspartic Acid +VO^{2+}; A Potentiometric and Spectroscopic Study. *Polyhedron* **1990**, *9*, 81–99. [CrossRef]

33. Alderighi, L.; Gans, P.; Ienco, A.; Peters, D.; Sabatini, A.; Vacca, A. Hyperquad simulation and speciation (HySS): A utility program for the investigation of equilibria involving soluble and partially soluble species. *Coord. Chem. Rev.* **1999**, *184*, 311–318. [CrossRef]

34. Aureliano, M.; Crans, D.C. Decavanadate (V$_{10}$O$_{28}$$^{6-}$) and oxovanadates: Oxometalates with many biological activities. *J. Inorg. Biochem.* **2009**, *103*, 536–546. [CrossRef]

35. Turner, T.L.; Nguyen, V.H.; McLauchlan, C.C.; Dymon, Z.; Dorsey, B.M.; Hooker, J.D.; Jones, M.A. Inhibitory effects of decavanadate on several enzymes and Leishmania tarentolae In Vitro. *J. Inorg. Biochem.* **2012**, *108*, 96–104. [CrossRef] [PubMed]

36. Aureliano, M.; Ohlin, C.A. Decavanadate in vitro and in vivo effects: Facts and opinions. *J. Inorg. Biochem.* **2014**, *137*, 123–130. [CrossRef]

37. Crans, D.C.; Peters, B.J.; Wu, X.; McLauchlan, C.C. Does anion-cation organization in Na+-containing X-ray crystal structures relate to solution interactions in inhomogeneous nanoscale environments: Sodium-decavanadate in solid state materials, minerals, and microemulsions. *Coord. Chem. Rev.* **2017**, *344*, 115–130. [CrossRef]

38. Samart, N.; Althumairy, D.; Zhang, D.; Roess, D.A.; Crans, D.C. Initiation of a novel mode of membrane signaling: Vanadium facilitated signal transduction. *Coord. Chem. Rev.* **2020**, *416*, 213286. [CrossRef]

39. Coutinho, A.A.; Prieto, M. Ribonuclease TI and Alcohol Dehydrogenase Fluorescence Quenching by Acrylamide. *J. Chem. Ed.* **1993**, *70*, 425–428. [CrossRef]

40. Thordarson, P. Determining association constants from titration experiments in supramolecular chemistry. *Chem. Soc. Rev.* **2011**, *40*, 1305–1323. [CrossRef] [PubMed]

41. Van de Weert, M.; Stella, L. Fluorescence quenching and ligand binding: A critical discussion of a popular methodology. *J. Mol. Struct.* **2011**, *998*, 144–150. [CrossRef]
42. Galkina, P.A.; Proskurnin, M.A. Supramolecular interaction of transition metal complexes with albumins and DNA: Spectroscopic methods of estimation of binding parameters. *Appl. Organomet. Chem.* **2018**, *32*, e4150. [CrossRef]
43. Correia, I.; Chorna, I.; Cavaco, I.; Roy, S.; Kuznetsov, M.L.; Ribeiro, N.; Justino, G.; Marques, F.; Santos-Silva, T.; Santos, M.F.A.; et al. Interaction of [(VO)-O-IV(acac)(2)] with Human Serum Transferrin and Albumin. *Chem. Asian J.* **2017**, *12*, 2062–2084. [CrossRef] [PubMed]
44. Macii, F.; Biver, T. Spectrofluorimetric analysis of the binding of a target molecule to serum albumin: Tricky aspects and tips. *J. Inorg. Biochem.* **2021**. [CrossRef] [PubMed]
45. Van de Weert, M. Fluorescence Quenching to Study Protein-ligand Binding: Common Errors. *J. Fluoresc.* **2010**, *20*, 625–629. [CrossRef] [PubMed]
46. Crans, D.C.; Khan, A.R.; Mahroof-Tahir, M.; Mondal, S.; Miller, S.M.; la Cour, A.; Anderson, O.P.; Jakusch, T.; Kiss, T. Bis(acetylamido)oxovanadium(IV) complexes: Solid state and solution studies. *J. Chem. Soc. Dalton Trans.* **2001**, 3337–3345. [CrossRef]
47. Sanna, D.; Buglyo, P.; Micera, G.; Garribba, E. A quantitative study of the biotransformation of insulin-enhancing VO^{2+} compounds. *J. Biol. Inorg. Chem.* **2010**, *15*, 825–839. [CrossRef]
48. Ivancsits, S.; Pilger, A.; Diem, E.; Schaffer, A.; Rüdiger, H.W. Vanadate induces DNA strand breaks in cultured human fibroblasts at doses relevant to occupational exposure. *Mutation Res.* **2002**, *519*, 25–35. [CrossRef]
49. Desaulniers, D.; Cummings-Lorbetskie, C.; Leingartner, K.; Xiao, G.-H.; Zhou, G.; Parfett, C. Effects of vanadium (sodium metavanadate) and aflatoxin-B1 on cytochrome p450 activities, DNA damage and DNA methylation in human liver cell lines. *Toxicol. In Vivo* **2021**, *70*, 105036. [CrossRef] [PubMed]
50. Wozniak, K.; Blasiak, J. Vanadyl sulfate can differentially damage DNA in human lymphocytes and HeLa cells. *Arch. Toxicol.* **2004**, *78*, 7–15. [CrossRef]
51. Shi, X.; Jiang, H.; Mao, Y.; Ye, J.; Saffiotti, U. Vanadium(IV)-mediated free radical generation and related 2′-deoxyguanosine hydroxylation and DNA damage. *Toxicology* **1996**, *106*, 27–38. [CrossRef]
52. Sakurai, H.; Nakai, M.; Miki, T.; Tsuchiya, K.; Takada, J.; Matsushit, R. DNA cleavage by hydroxyl radicals generated in a vanadyl ion-hydrogen peroxide system. *Biochem. Biophys. Res. Commun.* **1992**, *189*, 1090–1095. [CrossRef]
53. Patra, D.; Biswas, N.; Kumari, B.; Das, P.; Sepay, N.; Chatterjee, S.; Drew, M.G.B.; Ghosh, T. A family of mixed-ligand oxidovanadium(V) complexes with aroylhydrazone ligands: A combined experimental and computational study on the electronic effects of para substituents of hydrazone ligands on the electronic properties, DNA binding and nuclease activities. *RSC Adv.* **2015**, *5*, 92456–92472.
54. Butenko, N.; Pinheiro, J.P.; Da Silva, J.P.; Tomaz, A.I.; Correia, I.; Ribeiro, V.; Pessoa, J.C.; Cavaco, I. The effect of phosphate on the nuclease activity of vanadium compounds. *J. Inorg. Biochem.* **2015**, *147*, 165–176. [CrossRef]
55. Butenko, N.; Tomaz, A.I.; Nouri, O.; Escribano, E.; Moreno, V.; Gama, S.; Ribeiro, V.; Telo, J.P.; Pessoa, J.C.; Cavaco, I. DNA cleavage activity of (VO)-O-IV(acac)(2) and derivatives. *J. Inorg. Biochem.* **2009**, *103*, 622–632. [CrossRef]
56. Sasmal, P.K.; Patra, A.K.; Nethaji, M.; Chakravarty, A.R. DNA Cleavage by New Oxovanadium(IV) Complexes of N-Salicylidene alfa-Amino Acids and Phenanthroline Bases in the Photodynamic Therapy Window. *Inorg. Chem.* **2007**, *46*, 11112–11121. [CrossRef]
57. Palmajumder, E.; Sepay, N.; Mukherjea, K.K. Development of oxidovanadium and oxido-peroxido vanadium-based artificial DNA nucleases via multi spectroscopic investigations and theoretical simulation of DNA binding. *J. Biomol. Struct. Dyn.* **2018**, *36*, 919–927. [CrossRef]
58. Verquin, G.; Fontaine, G.; Bria, M.; Zhilinskaya, E.Z.; Abi-Aad, E.; Aboukaïs, A.; Baldeyrou, B.; Bailly, C.; Bernier, J.-L. DNA modification by oxovanadium(IV) complexes of salen derivatives. *J. Biol. Inorg. Chem.* **2004**, *9*, 345–353. [CrossRef]
59. Sasmal, P.K.; Patra, A.K.; Chakravarty, A.R. Synthesis, structure, DNA binding and DNA cleavage activity of oxovanadium(IV) N-salicylidene-S-methyldithiocarbazate complexes of phenanthroline bases. *J. Inorg. Biochem.* **2008**, *102*, 1463–1472. [CrossRef] [PubMed]
60. Khan, N.H.; Pandya, N.; Maity, N.C.; Kumar, M.; Patel, R.M.; Kureshy, R.I.; Abdi, S.H.R.; Mishra, S.; Das, S.; Bajaj, H.C. Influence of chirality of V(V) Schiff base complexes on DNA, BSA binding and cleavage activity. *Eur. J. Med. Chem.* **2011**, *46*, 5074–5085. [CrossRef]
61. Saha, U.; Mukherjea, K.K. DNA binding and nuclease activity of an oxovanadium valinato-Schiff base complex. *Intern. J. Biol. Macromol.* **2014**, *66*, 166–171. [CrossRef]
62. Correia, I.; Roy, S.; Matos, C.P.; Borovic, S.; Butenko, N.; Cavaco, I.; Marques, F.; Lorenzo, J.; Rodriguez, A.; Moreno, V.; et al. Vanadium(IV) and copper(II) complexes of salicylaldimines and aromatic heterocycles: Cytotoxicity: DNA binding and DNA cleavage properties. *J. Inorg. Biochem.* **2015**, *147*, 134–146. [CrossRef] [PubMed]
63. Kwong, D.W.J.; Chan, O.Y.; Wong, R.N.S.; Musser, S.M.; Vaca, L.; Chan, S.I. DNA-Photocleavage Activities of Vanadium(V)−Peroxo Complexes. *Inorg. Chem.* **1997**, *36*, 1276–1277. [CrossRef]
64. Stemmler, A.; Burrows, C. Guanine versus deoxyribose damage in DNA oxidation mediated by vanadium(IV) and vanadium(V) complexes. *J. Biol. Inorg. Chem.* **2001**, *6*, 100–106. [CrossRef] [PubMed]
65. Scalese, G.; Machado, I.; Correia, I.; Pessoa, J.C.; Bilbao, L.; Pérez-Diaz, L.; Gambino, D. Exploring oxidovanadium(IV) homoleptic complexes with 8-hydroxyquinoline derivatives as prospective antitrypanosomal agents. *New J. Chem.* **2019**, *43*, 17756–17773. [CrossRef]

66. Fernández, M.; Varela, J.; Correia, I.; Birriel, E.; Castiglioni, J.; Moreno, V.; Pessoa, J.C.; Cerecetto, H.; González, M.; Gambino, D. A new series of heteroleptic oxidovanadium(IV) compounds with phenanthroline-derived co-ligands: Selective Trypanosoma cruzi growth inhibitors. *Dalton Trans.* **2013**, *42*, 11900–11911. [CrossRef]

67. Fernández, M.; Becco, L.; Correia, I.; Benítez, J.; Piro, O.E.; Echeverria, G.A.; Medeiros, A.; Comini, M.; Lavaggi, M.L.; González, M.; et al. Oxidovanadium(IV) and dioxidovanadium(V) complexes of tridentate salicylaldehyde semicarbazones: Searching for prospective antitrypanosomal agents. *J. Inorg. Biochem.* **2013**, *127*, 150–160. [CrossRef]

68. Machado, I.; Fernández, M.; Becco, L.; Garat, B.; Brissos, R.F.; Zabarska, N.; Gamez, P.; Marques, F.; Correia, I.; Pessoa, J.C.; et al. New metal complexes of NNO tridentate ligands: Effect of metal center and co-ligand on biological activity. *Inorg. Chim. Acta* **2014**, *420*, 39–46. [CrossRef]

69. Benítez, J.; Cavalcanti de Queiroz, A.; Correia, I.; Amaral Alves, M.; Alexandre-Moreira, M.S.; Barreiro, E.J.; Moreira Lima, L.J.; Varela, J.; González, M.H.; Cerecetto, H.H.; et al. New oxovanadium(IV) N-acylhydrazone complexes: Promising antileishmanial and antitrypanosomal agents. *Eur. J. Med. Chem.* **2013**, *62*, 20–27. [CrossRef]

70. Ouameur, A.A.; Arakawa, H.; Tajmir-Riahi, H.A. Binding of oxovanadium ions to the major and minor grooves of DNA duplex: Stability and structural models. *Biochem. Cell Biol.* **2006**, *84*, 677–683. [CrossRef] [PubMed]

71. Kumar, A.; Dixit, A.; Sahoo, S.; Banerjee, S.; Bhattacharyya, A.; Garaia, A.; Karandeb, A.A.; Chakravarty, A.R. Crystal structure, DNA crosslinking and photo-induced cytotoxicity of oxovanadium(IV) conjugates of boron-dipyrromethene. *J. Inorg. Biochem.* **2020**, *202*, 110817. [CrossRef]

72. Kiss, T.; Jakusch, T.; Hollender, D.; Dornyei, A.; Enyedy, E.A.; Pessoa, J.C.; Sakurai, H.; Sanz-Medel, A. Biospeciation of antidiabetic VO(IV) complexes. *Coord. Chem. Rev.* **2008**, *252*, 1153–1162. [CrossRef]

73. Sanna, D.; Garribba, E.; Micera, G. Interaction of VO^{2+} ion with human serum transferrin and albumin. *J. Inorg. Biochem.* **2009**, *103*, 648–655. [CrossRef]

74. Jakusch, T.; Pessoa, J.C.; Kiss, T. The speciation of vanadium in human serum. *Coord. Chem. Rev.* **2011**, *255*, 2218–2226. [CrossRef]

75. Sanna, D.; Bíró, L.; Buglyó, P.; Micera, G.; Garribba, E. Transport of the anti-diabetic VO^{2+} complexes formed by pyrone derivatives in the blood serum. *J. Inorg. Biochem.* **2012**, *115*, 87–99. [CrossRef] [PubMed]

76. Gambino, D. Potentiality of vanadium compounds as antiparasitic agents. *Coord. Chem. Rev.* **2011**, *255*, 2193–2203. [CrossRef]

77. Benítez, J.; Correia, I.; Becco, L.; Fernández, M.; Garat, B.; Gallardo, H.; Conte, G.; Kuznetsov, M.L.; Neves, A.; Moreno, V.; et al. Searching for vanadium-based prospective agents against Trypanosoma cruzi: Oxidovanadium(IV) compounds with phenanthroline derivatives as ligands. *Z. Anorg. Allg. Chem.* **2013**, *639*, 1417–1425. [CrossRef]

78. Rehder, D. Perspectives for vanadium in health issues. *Future Med. Chem.* **2016**, *8*, 325–338. [CrossRef]

79. Kioseoglou, E.; Petanidis, S.; Gabriel, C.; Salifoglou, A. The chemistry and biology of vanadium compounds in cancer therapeutics. *Coord. Chem. Rev.* **2015**, *301*, 87–105. [CrossRef]

80. Benitez, J.; Guggeri, L.; Tomaz, I.; Pessoa, J.C.; Moreno, V.; Lorenzo, J.; Aviles, F.X.; Garat, B.; Gambino, D. A novel vanadyl complex with a polypyridyl DNA intercalator as ligand: A potential anti-protozoa and anti-tumor agent. *J. Inorg. Biochem.* **2009**, *103*, 1386–1394. [CrossRef]

81. Benitez, J.; Becco, L.; Correia, I.; Leal, S.M.; Guiset, H.; Pessoa, J.C.; Lorenzo, J.; Tanco, S.; Escobar, P.; Moreno, V.; et al. Vanadium polypyridyl compounds as potential antiparasitic and antitumoral agents: New achievements. *J. Inorg. Biochem.* **2011**, *105*, 303–312. [CrossRef]

82. Sakurai, H.; Tamura, H.; Okatami, K. Mechanism for a new antitumor vanadium complex hydroxyl radical-dependent DNA-cleavage by 1,10-phenanthroline-vanadyl complex in the presence of hydrogen peroxide. *Biochem. Biophys. Res. Commun.* **1995**, *206*, 133–137. [CrossRef] [PubMed]

83. Narla, R.K.; Dong, Y.; D'Cruz, O.J.; Navara, C.; Uckun, F.M. Bis(4,7-dimethyl-1,10-phenanthroline) Sulfatooxovanadium(IV) as a Novel Apoptosis-inducing Anticancer Agent. *Clin. Cancer Res.* **2000**, *6*, 1546–1556.

84. Rehder, D. Implications of vanadium in technical applications and pharmaceutical issues. *Inorg. Chim. Acta* **2017**, *455*, 378–389. [CrossRef]

85. Benitez, J.; Guggeri, L.; Tomaz, I.; Arrambide, G.; Navarro, M.; Pessoa, J.C.; Garat, B.; Garnbino, D. Design of vanadium mixed-ligand complexes as potential anti-protozoa agents. *J. Inorg. Biochem.* **2009**, *103*, 609–616. [CrossRef]

86. Maurya, M.R.; Khan, A.A.; Azam, A.; Kumar, A.; Ranjan, S.; Mondal, N.; Pessoa, J.C. Dinuclear Oxidovanadium(IV) and Dioxidovanadium(V) Complexes of 5,5′-Methylenebis(dibasic tridentate) Ligands: Synthesis, Spectral Characterisation, Reactivity, and Catalytic and Antiamoebic Activities. *Eur. J. Inorg. Chem.* **2009**, *2009*, 5377–5390. [CrossRef]

87. Maurya, M.R.; Khan, A.A.; Azam, A.; Ranjan, S.; Mondal, N.; Kumar, A.; Avecilla, F.; Pessoa, J.C. Vanadium complexes having [(VO)-O-IV](2+) and [(VO2)-O-V](+) cores with binucleating dibasic tetradentate ligands: Synthesis, characterization, catalytic and antiamoebic activities. *Dalton Trans.* **2010**, *39*, 1345–1360. [CrossRef]

88. Leon, I.E.; Porro, V.; Di Virgilio, A.L.; Naso, L.G.; Williams, P.A.M.; Bollati-Fogolin, M.; Etcheverry, S.B. Antiproliferative and apoptosis-inducing activity of an oxidovanadium(IV) complex with the flavonoid silibinin against osteosarcoma cells. *J. Biol. Inorg. Chem.* **2014**, *19*, 59–74. [CrossRef]

89. Leon, I.E.; Di Virgilio, A.L.; Porro, V.; Muglia, C.I.; Naso, L.G.; Williams, P.A.M.; Bollati-Fogolin, M.; Etcheverry, S.B. Antitumor properties of a vanadyl(IV) complex with the flavonoid chrysin [VO(chrysin)(2)EtOH](2) in a human osteosarcoma model: The role of oxidative stress and apoptosis. *Dalton Trans.* **2013**, *42*, 11868–11880. [CrossRef]

90. Tasiopoulos, A.J.; Tolis, E.J.; Tsangaris, J.M.; Evangelou, A.; Woollins, J.D.; Slawin, A.M.; Pessoa, J.C.; Correia, I.; Kabanos, T.A. Model investigations for vanadium-protein interactions: Vanadium(III) compounds with dipeptides and their oxovanadium(IV) analogues. *J. Biol. Inorg. Chem.* **2002**, *7*, 363–374. [CrossRef]

91. Dong, Y.H.; Narla, R.K.; Sudbeck, E.; Uckun, F.M. Synthesis, X-ray structure, and anti-leukemic activity of oxovanadium(IV) complexes. *J. Inorg. Biochem.* **2000**, *78*, 321–330. [CrossRef]

92. Scalese, G.; Correia, I.; Benítez, J.; Rostán, S.; Marques, F.; Mendes, F.; Matos, A.P.; Pessoa, J.C.; Gambino, D. Evaluation of cellular uptake, cytotoxicity and cellular ultrastructural effects of heteroleptic oxidovanadium(IV) complexes of salicylaldimines and polypyridyl ligands. *J. Inorg. Biochem.* **2017**, *166*, 162–172. [CrossRef] [PubMed]

93. Narla, R.K.; Chen, C.L.; Dong, Y.H.; Uckun, F.M. In vivo antitumor activity of bis(4,7-dimethyl-1,10-phenanthroline) sulfatooxo-vanadium(IV) {METVAN [VO(SO$_4$)(Me-2-Phen)(2)]}. *Clin. Cancer Res.* **2001**, *7*, 2124–2133. [PubMed]

94. Narla, R.K.; Dong, Y.H.; Uckun, F.M. Apoptosis inducing novel anti-leukemic agent, bis(4,7-dimethyl-1,10 phenanthroline) sulfa-tooxovanadium(IV) [VO(SO$_4$)(Me$_2$-Phen)(2)] depolarizes mitochondrial membranes. *Leuk. Lymphoma* **2001**, *41*, 625. [CrossRef]

95. Sanna, D.; Buglyo, P.; Tomaz, A.I.; Pessoa, J.C.; Borovic, S.; Micera, G.; Garribba, E. (VO)-O-IV and Cu-II complexation by ligands based on pyridine nitrogen donors. *Dalton Trans.* **2012**, *41*, 12824–12838. [CrossRef]

96. Le, M.; Rathje, O.; Levina, A.; Lay, P.A. High cytotoxicity of vanadium(IV) complexes with 1,10-phenanthroline and related ligands is due to decomposition in cell culture medium. *J. Biol. Inorg. Chem.* **2017**, *22*, 663–672. [CrossRef]

97. Nunes, P.; Correia, I.; Cavaco, I.; Marques, F.; Pinheiro, T.; Avecilla, F.; Costa Pessoa, J. Therapeutic potential of vanadium complexes with 1,10-phenanthroline ligands, quo vadis? Fate of complexes in cell media and cancer cells. *J. Inorg. Biochem.* **2021**, *217*, 111350. [CrossRef]

98. Coyle, B.; Kavanagh, K.; McCann, M.; Devereux, M.; Geraghty, M. Mode of anti-fungal activity of 1,10-phenanthroline and its Cu(II), Mn(II) and Ag(I) complexes. *Biometals* **2003**, *16*, 321–329. [CrossRef] [PubMed]

99. Deegan, C.; Coyle, B.; McCann, M.; Devereux, M.; Egan, D.A. In Vitro anti-tumour effect of 1,10-phenanthroline-5,6-dione (phendione), [Cu(phendione)$_3$](ClO$_4$)$_2$·4H$_2$O and [Ag(phendione)$_2$]ClO$_4$ using human epithelial cell lines. *Chem. Biol. Interact.* **2006**, *164*, 115–125. [CrossRef] [PubMed]

100. Kellett, A.; O'Connor, M.; McCann, M.; Howe, O.; Casey, A.; McCarron, P.; Kavanagh, K.; McNamara, M.; Kennedy, S.; May, D.D.; et al. Water-soluble bis(1,10-phenanthroline) octanedioate Cu^{2+} and Mn^{2+} complexes with unprecedented nano and picomolar in vitro cytotoxicity: Promising leads for chemotherapeutic drug development. *MedChemComm* **2011**, *2*, 579–584. [CrossRef]

101. McCann, M.; Santos, A.L.S.; da Silva, B.A.; Romanos, M.T.V.; Pyrrho, A.S.; Devereux, M.; Kavanagh, K.; Fichtner, I.; Kellett, A. In vitro and in vivo studies into the biological activities of 1,10-phenanthroline, 1,10-phenanthroline-5,6-dione and its copper(II) and silver(I) complexes. *Toxicol. Res. UK* **2012**, *1*, 47–54. [CrossRef]

102. Stavrianopoulos, J.G.; Karkas, J.D.; Chargaff, E. DNA Polymerase of Chicken Embryo: Purification and Properties. *Proc. Nat. Acad. Sci. USA* **1972**, *69*, 1781–1785. [CrossRef]

103. Falchuk, K.H.; Krishan, A. 1,10-Phenanthroline Inhibition of Lymphoblast Cell Cycle. *Cancer Res.* **1977**, *37*, 2050–2056. [PubMed]

104. Levina, A.; Crans, D.C.; Lay, P.A. Speciation of metal drugs, supplements and toxins in media and bodily fluids controls in vitro activities. *Coord. Chem. Rev.* **2017**, *352*, 473–498. [CrossRef]

105. *Advanced DMEM (Dulbecco's Modified Eagle's Medium)*; Catalog Number 12491023; Thermo Fisher Scientific: Waltham, MA, USA, 2017.

106. Sanna, D.; Ugone, V.; Serra, M.; Garribba, E. Speciation of potential anti-diabetic vanadium complexes in real serum samples. *J. Inorg. Biochem.* **2017**, *173*, 52–65. [CrossRef]

107. Sanna, D.; Ugone, V.; Micera, G.; Buglyo, P.; Biro, L.; Garribba, E. Speciation in human blood of Metvan, a vanadium based potential anti-tumor drug. *Dalton Trans.* **2017**, *46*, 8950–8967. [CrossRef] [PubMed]

108. Sanna, D.; Serra, M.; Micera, G.; Garribba, E. Interaction of Antidiabetic Vanadium Compounds with Hemoglobin and Red Blood Cells and Their Distribution between Plasma and Erythrocytes. *Inorg. Chem.* **2014**, *53*, 1449–1464. [CrossRef]

109. Sanna, D.; Micera, G.; Garribba, E. Interaction of Insulin-Enhancing Vanadium Compounds with Human Serum holo-Transferrin. *Inorg. Chem.* **2013**, *52*, 11975–11985. [CrossRef]

110. Levina, A.; McLeod, A.I.; Gasparini, S.J.; Nguyen, A.; De Silva, W.G.M.; Aitken, J.B.; Harris, H.H.; Glover, C.; Johannessen, B.; Lay, P.A. Reactivity and Speciation of Anti-Diabetic Vanadium Complexes in Whole Blood and Its Components: The Important Role of Red Blood Cells. *Inorg. Chem.* **2015**, *54*, 7753–7766. [CrossRef] [PubMed]

111. Sciortino, G.; Sanna, D.; Lubinu, G.; Maréchal, J.-D.; Garribba, E. Unveiling VIVO^{2+} binding modes to human serum albumins by an integrated spectroscopic-computational approach. *Chem. Eur. J.* **2020**. [CrossRef]

112. Cobbina, E.; Mehtab, S.; Correia, I.; Gonçalves, G.; Tomaz, I.; Cavaco, I.; Jakusch, T.; Enyedi, E.; Kiss, T.; Pessoa, J.C. Binding of Oxovanadium(IV) Complexes to Blood Serum Albumins. *J. Mex. Chem. Soc.* **2013**, *57*, 180–191. [CrossRef]

113. Reytman, L.; Hochman, J.; Tshuva, E.Y. Anticancer diaminotris(phenolato) vanadium(V) complexes: Ligand-metal interplay. *J. Coord. Chem.* **2018**, *71*, 2003–2011. [CrossRef]

114. Reytman, L.; Braitbard, O.; Hochman, J.; Tshuva, E.Y. Highly Effective and Hydrolytically Stable Vanadium(V) Amino Phenolato Antitumor Agents. *Inorg. Chem.* **2016**, *55*, 610–618. [CrossRef] [PubMed]

115. Mahroof-Tahir, M.; Brezina, D.; Fatima, N.; Choudhary, M.I. Atta-ur-Rahman Synthesis and characterization of mononuclear oxovanadium(IV) complexes and their enzyme inhibition studies with a carbohydrate metabolic enzyme, phosphodiesterase I. *J. Inorg. Biochem.* **2005**, *99*, 589–599. [CrossRef] [PubMed]

116. Hiromura, M.; Nakayama, A.; Adachi, Y.; Doi, M.; Sakurai, H. Action mechanism of bis(allixinato)oxovanadium(IV) as a novel potent insulin-mimetic complex: Regulation of GLUT4 translocation and FoxO1 transcription factor. *J. Biol. Inorg. Chem.* **2007**, *12*, 1275–1287. [CrossRef] [PubMed]
117. Correia, I.; Pessoa, J.C.; Duarte, M.T.; Henriques, R.T.; Piedade, M.F.M.; Veiros, L.F.; Jakusch, T.; Kiss, T.; Dörnyei, A.; Castro, M.M.C.A.; et al. *N,N'*-ethylenebis(pyridoxylideneiminato) and *N,N'*-ethylenebis(pyridoxylaminato): Synthesis, characterisation, potentiometric, spectroscopic and DFT study of their vanadium(IV) and vanadium(V) complexes. *Chem. Eur. J.* **2004**, *10*, 2301–2317. [CrossRef] [PubMed]
118. Strianese, M.; Basile, A.; Mazzone, A.; Morello, S.; Turco, M.C.; Pellecchia, C. Therapeutic Potential of a Pyridoxal-Based Vanadium(IV) Complex Showing Selective Cytotoxicity for Cancer Versus Healthy Cells. *J. Cell Physiol.* **2013**, *228*, 2202–2209. [CrossRef] [PubMed]

Article

Kinetic Studies of Sodium and Metforminium Decavanadates Decomposition and In Vitro Cytotoxicity and Insulin- Like Activity

Aniela M. Silva-Nolasco [1,2], Luz Camacho [2], Rafael Omar Saavedra-Díaz [1], Oswaldo Hernández-Abreu [1], Ignacio E. León [3] and Irma Sánchez-Lombardo [1,*]

[1] Centro de Investigación de Ciencia y Tecnología Aplicada de Tabasco, División Académica de Ciencias Básicas (CICTAT), Universidad Juárez Autónoma de Tabasco, Carretera Cunduacán-Jalpa km. 1 Col. La Esmeralda, Cunduacán 86690, Tabasco, Mexico; animonsino@gmail.com (A.M.S.-N.); rafael.saavedra@ujat.mx (R.O.S.-D.); oswaldo.hernandez@ujat.mx (O.H.-A.)

[2] Laboratorio de Nutrición Experimental, Instituto Nacional de Pediatría, Ciudad de Mexico 04530, Mexico; camacho.luz@gmail.com

[3] Centro de Química Inorgánica CEQUINOR (CONICET, UNLP), Bv 120 1465, La Plata 1900, Argentina; iel86@yahoo.com.ar

* Correspondence: irma.sanchez@ujat.mx

Received: 22 October 2020; Accepted: 2 December 2020; Published: 8 December 2020

Abstract: The kinetics of the decomposition of 0.5 and 1.0 mM sodium decavanadate (NaDeca) and metforminium decavanadate (MetfDeca) solutions were studied by ^{51}V NMR in Dulbecco's modified Eagle's medium (DMEM) medium (pH 7.4) at 25 °C. The results showed that decomposition products are orthovanadate $[H_2VO_4]^-$ (V_1) and metavanadate species like $[H_2V_2O_7]^{2-}$ (V_2), $[V_4O_{12}]^{4-}$ (V_4) and $[V_5O_{15}]^{5-}$ (V_5) for both compounds. The calculated half-life times of the decomposition reaction were 9 and 11 h for NaDeca and MetfDeca, respectively, at 1 mM concentration. The hydrolysis products that presented the highest rate constants were V_1 and V_4 for both compounds. Cytotoxic activity studies using non-tumorigenic HEK293 cell line and human liver cancer HEPG2 cells showed that decavanadates compounds exhibit selectivity action toward HEPG2 cells after 24 h. The effect of vanadium compounds (8–30 μM concentration) on the protein expression of AKT and AMPK were investigated in HEPG2 cell lines, showing that NaDeca and MetfDeca compounds exhibit a dose-dependence increase in phosphorylated AKT. Additionally, NaDeca at 30 μM concentration stimulated the glucose cell uptake moderately (62%) in 3T3-L1 adipocytes. Finally, an insulin release assay in βTC-6 cells (30 μM concentration) showed that sodium orthovanadate (MetV) and MetfDeca enhanced insulin release by 0.7 and 1-fold, respectively.

Keywords: polyoxometalates; decavanadate; cytotoxicity; insulin-like activity; diabetes therapy; vanadium biochemistry; vanadium speciation

---

1. Introduction

Polyoxometalates (POMs) have several applications in biology and medicine. Interactions between the highly charged POM molecules and biological molecules frequently occur through hydrogen-bonding and electrostatic interactions [1]. Moreover, POMs have shown pharmacological activities in vitro and in vivo, such as antitumor, antimicrobial, and antidiabetic [2,3]. Their roles in biological systems are non-functional or functional kind of interactions with biomolecules [4], like the tungstate cluster that helps to solve the X-ray structure of ribosome [5] or the insulin-like properties of the decavanadates [6].

In recent years, several organic and inorganic decavanadate compounds have been synthesized, exhibiting a wide structural supramolecular diversity in one, two or three dimensions [7–9]. However, the interaction of decavanadates with biological targets under physiological conditions are scarce reported since the decavanadate anion can be formed at vanadium concentrations up to 0.1 mM and in the pH range of 2–6 [10], and some organic decavanadates compounds are water-insoluble [11].

In biological studies, buffer solutions are extensively used, although just a few studies have addressed the speciation of the decomposition products of the decavanadate compounds in such reaction media. The decomposition of the decameric species at neutral pH can be followed by ^{51}V NMR showing a decrease in the peaks associated with the three magnetic independent vanadium nuclei of the decavanadate V_A, V_B and V_C (Figure 1), albeit an increase of the signals for the metavanadate peak $[H_2VO_4]^-$ (V_1) and the appearance of the orthovanadates species signals like $[H_2V_2O_7]^{2-}$ (V_2), $[V_4O_{12}]^{4-}$ (V_4) and $[V_5O_{15}]^{5-}$ (V_5) [12–14]. Moreover, monomeric vanadate is always present in decavanadate solutions at neutral pH [15]. The decavanadate decomposition rate is faster in acid than in basic solutions [16,17]. In the latter, the reaction proceeds via base-dependent or base-independent paths, and it depends on the counterions present in the solution [17].

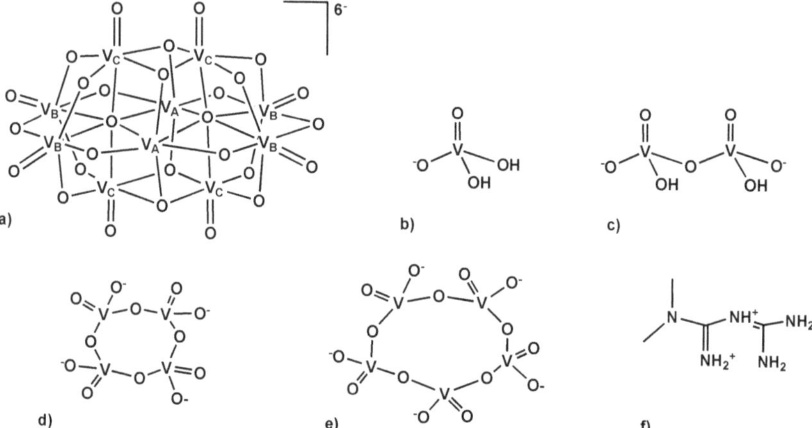

Figure 1. Schematic structure of (**a**) decavanadate anion $[V_{10}O_{28}]^{6-}$, (**b**) orthovanadate $[H_2VO_4]^-$ (V_1), metavanadate species like (**c**) $[H_2V_2O_7]^{2-}$ (V_2), (**d**) $[V_4O_{12}]^{4-}$ (V_4), (**e**) $[V_5O_{15}]^{5-}$ (V_5) and (**f**) diprotonated metformin (Metf).

Vanadium speciation is complicated under physiological conditions, many known forms of vanadium V^{4+} and V^{5+} species have been shown to readily interconvert through redox and hydrolytic reactions, and it is, therefore, difficult to determine which are the active species [18]. Additionally, in biological studies, the active vanadium species will depend on the sample preparation and handling, that is, whether the compounds were dissolved in media or buffer before addition to the cell culture and for how long the complexes have been in solution before adding aliquots to the medium [19].

Metabolic diseases like diabetes mellitus type 2 (DM2) and cancer are non-communicable diseases (NCD) that have become one of the major health hazards of the modern world [20]. Carcinogenesis occurs when normal cells receive genetic "hits", after which a full neoplastic phenotype of growth, invasion, and metastasis develops. Diabetes may influence this process through chronic inflammation, endogenous or exogenous hyperinsulinemia, or hyperglycemia, but potential biologic links between the two diseases are incompletely understood [21]. The development of innovative therapeutic modalities [22] that increase the effectiveness of clinical drugs like *cis*-platin or metformin hydrochloride and arrest their chemoresistance or side effects is a topic trend for scientists. In this context, AMP-activated kinase (AMPK) signaling has become a promising therapeutic target in hepatocellular carcinoma [23]. Another interesting target is the identification of exploitable vulnerabilities for the treatment of hyperactive phosphatidylinositol 3-kinase (PI3K/AKT) tumors [24], and combining inhibitors of the pentose phosphate pathway (PPP) may represent a promising approach for selectively causing oxidative stress-induced cell killing in ovarian and lung cancer cells [25].

The medicinal potentiality of vanadium compounds is a challenging task that demands investigation [26] and in general few groups have pursued it. The insulin-like effects of vanadium have been tested in vitro and in vivo [27,28]; however, the applied necessary dose of vanadium still was close to the levels at which side effects are observed [29]. In fact, there is only one vanadium compound that has been tested in humans, the bis(ethylmaltolato)oxovanadium(IV) (BEOV). In general, 20 mg of vanadium compound was well tolerated [30], but at the end of Phase IIa clinical trial, the trial was abandoned due to renal problems of some patients [31]. However, several questions about the transport and mode of action of the vanadium compounds need to be addressed [28] due to the distinct action mechanism that regulates glucose metabolism by vanadium [32].

In this work, we have studied the kinetics of the decomposition of 0.5 and 1 mM sodium decavanadate (NaDeca) and metforminium decavanadate (MetfDeca) in Dulbecco's modified Eagle's medium (DMEM) solution at pH 7.4 by ^{51}V NMR, with the aim to understand the medium and the vanadium concentration effects in both, the decomposition rate and the influence in the ratio of the final products, namely V_1, V_2, V_4 and V_5. To our knowledge, the ammonium decavanadate compound decomposition in MES; MES = 2-(*N*-morpholino)ethanesulfonic acid by ^{51}V NMR is the only report that describes the decomposition reaction [14]. Thus, NaDeca stability has not been extensively studied in buffer solutions. NaDeca and MetfDeca compounds are composed of the highly negative charged decavanadate and the positive counter ions. The counter ions bonding with the decameric moiety are ionic [33]. In that regard, the same biological activity of both compounds was expected if metformin hydrochloride (Metf) was pharmacologically an inactive molecule. Nevertheless, due to Metf antidiabetic properties, different results were expected in the biological activity of the NaDeca and MetfDeca compounds. MetfDeca compound in vivo exhibited hypoglycemic and lipid-lowering properties in type 1 diabetes mellitus (T1DM) [34] and type 2 diabetes mellitus (T2DM) models [35]. However, some questions were not addressed in those studies, like if MetV and MetfDeca regulated hyperglycemia and oxidative stress with the same action mechanisms, MetfDeca stability and toxicological effects [35].

With the aim to address some of the former questions and to estimate if two different counter ions could play a role as activators or inhibitors in the biological activity of decavanadates, we investigated how the decomposition products in DMEM medium at pH 7.4 can promote damage on the cell viability of HEK293 human embryonic kidney cells and HEPG2 human liver cancer cells. A comparison of these results with the cytotoxic effect of sodium orthovanadate and metformin hydrochloride was also performed. In addition, the activation of AKT and AMPK pathways for the HEPG2 cell line by the vanadium compounds were studied in order to establish if the hydrolysis products promote the same activation mechanism in the metabolic pathways. Finally, glucose uptake in 3T3L-1 differentiated adipocytes study is presented along with an insulin release assay in βTC-6 cells at 30 μM concentration of the vanadium compounds, with the purpose of identifying if the same active species are promoting the desirable effects in each case.

2. Results and Discussion

2.1. Characterization of the Sodium and Metforminium Decavanadate Solutions

The metforminium decavanadate (MetfDeca) $(C_4H_{13}N_5)_3V_{10}O_{28}\cdot8H_2O$ and the sodium decavanadate (NaDeca) $Na_6V_{10}O_{28}\cdot18H_2O$ were prepared according to previously reported procedures [33,36].

The ^{51}V NMR spectra for 1 mM concentration of NaDeca and MetfDeca compounds were recorded at pH 4 in 10% DMSO-*d6* and 90% H_2O (*v/v*), showing three signals at −420, −494, −510 ppm that were assigned to decameric species $[V_{10}O_{28}]^{6-}$ (V_{10}), attributed to the three different vanadium atoms of the decavanadate structure V_{10A}, V_{10B} and V_{10C} respectively and one signal at −556 ppm assigned to the diprotonated monomeric species $[H_2VO_4]^-$ (V_1) (Figure 2a) [37]. The ^{51}V NMR spectra for both 0.5 and 1.0 mM concentration samples for NaDeca and MetfDeca complexes show the same species present in the solution; additionally, the complexes are stable through time. These results are in agreement with the reported V_{10A}, V_{10B} and V_{10C} peaks that were observed for 10 mM NaDeca solution in D_2O at pH 3.1 and in Middlebrook 7H9 broth medium supplemented with 10% ADC enrichment (5% BSA, 2% dextrose, 5% catalase), glycerol (0.2%, *v/v*) and Tween-80 (0.05%, *v/v*) at pH 6.5. [13] In contrast, NaDeca and MetfDeca are not stable in the DMEM medium at pH 7. Their hydrolysis products are orthovanadate $[H_2VO_4]^-$ (V_1) and metavanadate species like $[H_2V_2O_7]^{2-}$ (V_2), $[V_4O_{12}]^{4-}$ (V_4) and $[V_5O_{15}]^{5-}$ (V_5) that are observed at −556, −570, −578 and −586 ppm, respectively (Figure 2b) like previously reported for 1 mM solution of $(NH_4)_6 [V_{10}O_{28}]\cdot6H_2O$ in MES buffer (0.1 M), NaCl (0.5 M) at pH 8 [14].

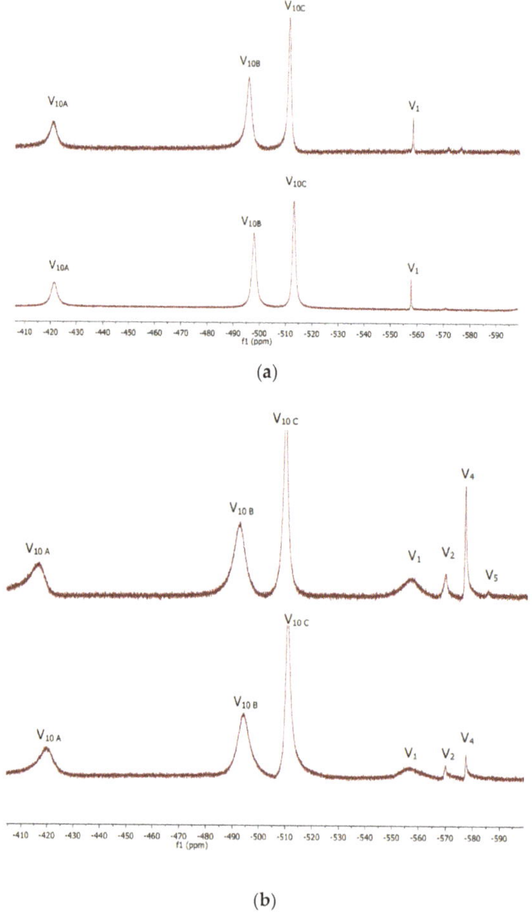

Figure 2. ^{51}V NMR spectra of (**a**) 1 mM NaDeca (top) and MetfDeca (bottom) in 10% DMSO-*d6* and 90% H_2O (*v/v*) DMSO at pH 4, (**b**) 1 mM NaDeca (top) and MetfDeca (bottom) in DMEM medium at pH 7.4.

2.2. Kinetic Studies by ^{51}V NMR

In vanadium(V) solutions, different oligomeric vanadate species can occur simultaneously, depends on several factors such as vanadate concentration, pH and ionic strength [12], so at 0.5 and

1 mM of NaDeca and MetfDeca, the V_{10} and V_1 species were present at pH 4, but the hydrolysis of both compounds in DMEM medium allowed us to follow by ^{51}V NMR the formation and the increment in the concentration over time at 25 °C of the orthovanadate, V_1 and metavanadate species V_2, V_4 and V_5 at −556, −570, −578 and −586 ppm, respectively. The kinetics of the decomposition of 1.0 mM NaDeca and MetfDeca (10 mM total vanadium) are plotted in Figure 3a, where the vanadium concentration for V_{10} species was calculated by integration of the V_{10A} (2 vanadium atoms), V_{10B} (4 vanadium atoms) and V_{10C} (4 vanadium atoms) resonances at −420, −494 and −510 ppm, respectively, and the rate constants for the three decavanadate signals V_A, V_B and V_C are shown with a negative sign by convention in Table 1. For comparison, the increase in concentration of the V_1 and V_4 vanadate species as a function of time are plotted in Figure 3b. Interestingly, the reaction is faster at 0.5 mM concentration of decavanadate than at 1 mM for NaDeca and MetfDeca compounds (Table 1). The rate constants of 0.5 mM NaDeca $(2.28 \pm 0.08) \times 10^{-3}$ and $(1.72 \pm 0.07) \times 10^{-4}$ for the appearance of V_4 and V_5 species, respectively, are three and four times higher than the ones calculated for 0.5 mM MetfDeca compound $(7.63 \pm 0.8) \times 10^{-4}$ and $(4.09 \pm 0.3) \times 10^{-5}$ for V_4 and V_5 species, respectively. Surprisingly, the rate constants for the appearance of the V_4 and V_5 species (Table 1) do not differ significantly for 1 mM NaDeca compared with 1 mM MetfDeca.

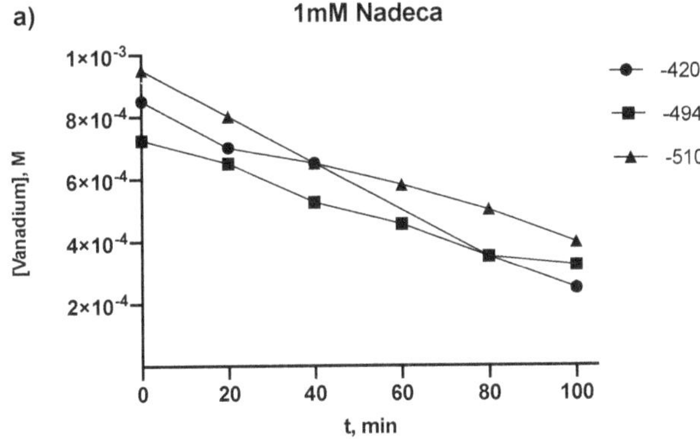

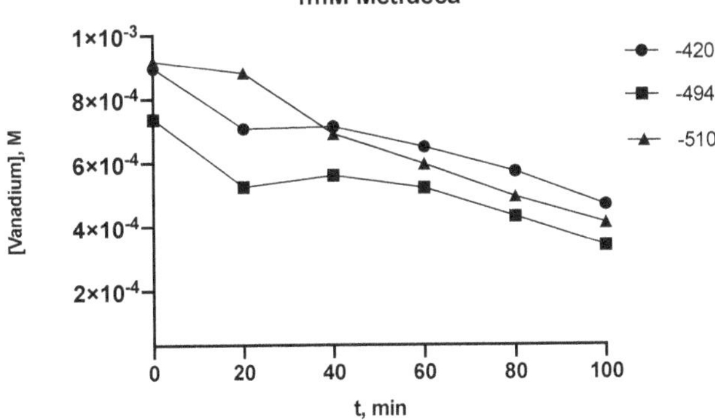

Figure 3. *Cont.*

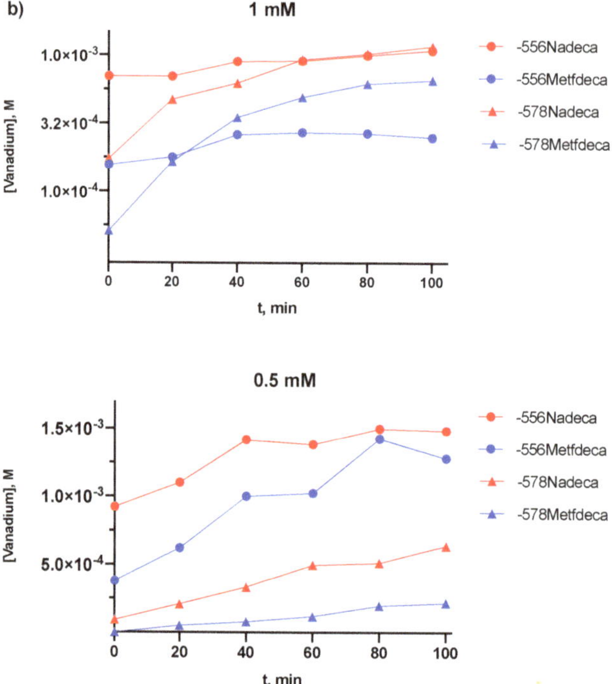

Figure 3. (**a**) ^{51}V NMR decomposition of 1 mM NaDeca and MetfDeca in Dulbecco's modified Eagle's medium (DMEM) medium, plotted as vanadium concentration (10 mM total vanadium concentration) associated with the decameric species V_{10A} (circle), V_{10B} (square), V_{10C} (triangle) over time at 25 °C. (**b**) Hydrolysis of (left) 1 mM and (right) 0.5 mM NaDeca and MetfDeca in DMEM medium followed by the formation of the orthovanadate V_1 at −556 ppm (red circle for NaDeca and blue circle for MetfDeca) and metavanadate species V_4 at −578 ppm (red triangle for NaDeca and blue triangle for MetfDeca) over time at 25 °C.

The decomposition of NaDeca and MetfDeca show first-order dependence versus time. In the case of NaDeca at 0.5 and 1 mM concentration, the calculated half-life time of the decomposition in DMEM medium at 25 °C is 9 h. In contrast, the calculated lifetime for MetfDeca is 9 h and 11 h for 0.5 mM and 1 mM concentration, respectively (Table 1). These results are in line with the half-life time for the decomposition of decameric species found by Ramos et al., where for 10 μM decavanadate concentration in different buffers pH 7–7.5, the half-life time is between 5 to 10 h. In that study, the authors performed a stabilization study of the decavanadate species with the G-actin protein, and due to the coordination of the protein with the decameric species, its half-life time was increased five times from 5 to 27 h at 10 μM of decavanadate concentration, however, in the same study the addition of 200 μM of ATP to the medium prevented the actin polymerization by V_{10} and the half-life time decreased from 27 to 10 h [12].

The decomposition rate of the decavanadate moiety is sensitive to the cations present in solution [16], the fast reaction in acid media can be accelerated by alkali metal cations and slowed down by large cations such as tetra-alkylammonium ions due to the formation of ionic-pairs with the protonated decavanadate to form $[VO_2]^+$ in seconds [16]. In basic media, the reaction is slower than in acid media, but the decomposition reaction proceeds via base independent ($k_{1'}$) and base dependent (k_2) paths (Equation (1)). In the absence of sodium ions, the rate of reaction is independent of $[OH^-]$ [17]. In this work, it seems that the base-dependent decomposition path is active as well, because for

NaDeca and MetfDeca, the observed rate of decomposition is not increasing with decavanadate concentration in both cases (Table 1), and the presence of a high sodium concentration in the DMEM media ($\mu = 0.1$ M NaCl), produces an increase in the decomposition rate via a reactive alkali-metal decavanadate species (k_2) Equation (1) [17,38].

Table 1. Summary of rate constants for the decomposition of 0.5 and 1.0 mM NaDeca and MetfDeca compound hydrolysis in DMEM medium at 25 °C and pH 7.4.

Decavanadate Compound Concentration	^{51}V NMR Signal [a]	k_{obs}, min^{-1} NaDeca	k_{obs}, min^{-1} MetfDeca
0.0005 M	-420 (V_{10A})	$(-1.25 \pm 0.03) \times 10^{-3}$	$(-1.40 \pm 0.03) \times 10^{-3}$
	-494 (V_{10B})	$(-1.26 \pm 0.2) \times 10^{-3}$	$(-1.48 \pm 0.7) \times 10^{-3}$
	-510 (V_{10c})	$(-1.82 \pm 0.4) \times 10^{-3}$	$(-1.44 \pm 0.4) \times 10^{-3}$
	-556 (V_1)	$(4.59 \pm 0.7) \times 10^{-3}$	$(2.97 \pm 0.4) \times 10^{-3}$
	-570 (V_2)	$(7.07 \pm 0.7) \times 10^{-4}$	$(6.17 \pm 0.8) \times 10^{-4}$
	-578 (V_4)	$(2.28 \pm 0.08) \times 10^{-3}$	$(7.63 \pm 0.8) \times 10^{-4}$
	-586 (V_5)	$(1.72 \pm 0.07) \times 10^{-4}$	$(4.09 \pm 0.3) \times 10^{-5}$
0.001 M	-420 (V_{10A})	$(-1.46 \pm 0.7) \times 10^{-3}$	$(-1.36 \pm 0.2) \times 10^{-3}$
	-494 (V_{10B})	$(-1.13 \pm 0.1) \times 10^{-3}$	$(-8.39 \pm 0.2) \times 10^{-4}$
	-510 (V_{10c})	$(-1.16 \pm 0.2) \times 10^{-3}$	$(-9.49 \pm 0.5) \times 10^{-4}$
	-556 (V_1)	$(2.62 \pm 0.6) \times 10^{-3}$	$(2.00 \pm 0.2) \times 10^{-3}$
	-570 (V_2)	$(6.81 \pm 0.4) \times 10^{-4}$	$(5.34 \pm 0.4) \times 10^{-4}$
	-578 (V_4)	$(1.96 \pm 0.2) \times 10^{-3}$	$(1.85 \pm 0.09) \times 10^{-3}$
	-586 (V_5)	$(1.89 \pm 0.1) \times 10^{-4}$	$(1.38 \pm 0.8) \times 10^{-4}$

[a] For calculating the rate of consumption of the decavanadate complexes, three different resonances were used, -420, -494 and -510, whereas for calculating the rate of appearance for the V_1, V_2, V_4 and V_5, only one resonance was used.

$$-d[V_{10}O_{28}{}^{6-}]_{tot}/dt = [k_{1'} + k_2[OH^-]][V_{10}O_{28}{}^{6-}]_{tot} \tag{1}$$

$$d[V_4O_{12}{}^{4-}]/dt = [k_{1'} + k_2[OH^-]][V_{10}O_{28}{}^{6-}]_{tot} \tag{2}$$

Goddard and Druskovich's [17,38] decomposition experiments were followed by UV-Vis techniques, although metavanadate species formation was not reported. Decavanadate ^{51}V NMR signals are wide, and the spectrum acquisition takes longer than the UV-Vis one. However, metavanadate species formation can be followed by ^{51}V NMR. In Table 1, NaDeca hydrolysis products formation rates are moderately faster than the ones calculated for MetfDeca, and the reaction rate is not increasing with the decavanadate concentration, so Equation (1) for the decomposition reaction was rewritten as Equation (2), where the reaction rate was expressed in terms of the metavanadate species formation. Based on the literature and our results, we proposed that in high alkali metal concentration, like in DMEM medium, the sodium ions form an ionic aggregate with the V_{10} species (Scheme 1), which then reacts with the hydroxide ion [17]. In this work, M^+ is the sodium ion, and the M' is the metformin cation ($C_4H_{12}N_5{}^+$), which at pH 7 is monoprotonated [33].

$$[(C_4H_{12}N_5)V_{10}O_{28}]^{5-} + Na^+ \underset{}{\overset{K_{IP}}{\rightleftharpoons}} [NaV_{10}O_{28}]^{5-} + (C_4H_{12}N_5)^+$$

$$[NaV_{10}O_{28}]^{5-} + Na^+ \underset{}{\overset{K'}{\rightleftharpoons}} [Na_2V_{10}O_{28}]^{4-}$$

$$[Na_2V_{10}O_{28}]^{4-} + OH^- \overset{k_2{}^M}{\longrightarrow} PRODUCTS$$

Scheme 1. Putative reaction mechanism for decavanadate decomposition reaction in DMEM medium at pH 7.4.

The base dependent equation can be rewritten as:

$$Rate = k_2[OH^-][V_{10}O_{28}{}^{6-}]_{tot} = k_2{}^M[OH^-][V_{10}O_{28}{}^{6-}]_{tot} \tag{3}$$

where

$$[V_{10}O_{28}{}^{6-}]_{tot} = [(C_4H_{12}N_5)V_{10}O_{28}{}^{5-}] + [NaV_{10}O_{28}{}^{5-}] + [Na_2V_{10}O_{28}{}^{4-}] \tag{4}$$

In DMEM medium, it seems likely that the ion-pair association is stronger with the metformin cation $(C_4H_{12}N_5)V_{10}O_{28}^{5-}$ at 1 mM MetfDeca concentration, which does not form at the same rate as the $Na_2V_{10}O_{28}^{4-}$ active species to react with the free OH^- anion (Scheme 1) like 1 mM NaDeca, where the total V_{10} anion concentration (Equation (4)) is almost in the higher ionic aggregate $Na_2V_{10}O_{28}^{4-}$ species, and on that way can follow the base-dependent path (Equation (3)). However, the same calculated values for the decomposition rate of 0.5 mM and 1 mM NaDeca (Table 1) suggest that the concentration of $Na_2V_{10}O_{28}^{4-}$ species remains the same under the buffer conditions; several ion-pairs can be proposed by the combination of monovalent cation and hexavalent anion. Nevertheless, Schwarzenbach and Geier [39] showed that the alkali metal cations formed the ion-pair complexes $MHV_{10}O_{28}^{4-}$, $MV_{10}O_{28}^{5-}$, and $M_2V_{10}O_{28}^{4-}$ base on their formation constants 91% of the decavanadate is in the ion-pair form $M_2V_{10}O_{28}^{4-}$ and 9% in the form $MV_{10}O_{28}^{5-}$ for M = Li or Na [39].

In vanadium speciation diagrams, at total vanadium concentration lower than 5 µM, the decavanadate anion is not formed [10], but some meta and orthovanadate species are present in solution at neutral pH. In that regard, this kinetic study was performed to have an approximate of the constant rate values at which the oligomer vanadium species were formed and, with some cautions in the interpretations of the data, would allow us to compare the biological activity of MetV (V_1) and Metf versus NaDeca and MetfDeca to show if the hydrolysis products produce a different biological response than the orthovanadate (V_1) and to quantify if MetfDeca compound promotes a synergistic effect between its components that increase the decavanadate antidiabetic properties. In that regard, the biological experiments that are shown in the next sections were performed in DMEM solution at pH 7.4, and the cells were incubated with the compounds for 24 h, with the exception of the insulin release assay, where the cells were incubated with the compounds for one hour.

2.3. Cell Viability

To investigate the cytotoxicity of vanadium compounds against non-tumoral and tumoral human cells and potential anticancer activity, the compounds NaDeca, MetfDeca, MetV and Metf were tested against HEK293 human embryonic kidney cells and HEPG2 human liver cancer cells. In Figure 4a, the percentage of cell viability vs. compound concentration for the four compounds against HEK293 is shown. The IC_{50} value found for NaDeca was 40 ± 4 µM, for MetfDeca was 85 ± 5 µM, for sodium MetV was 181 ± 7 µM and for Metf was 420 ± 11 µM. In the case of the HEPG2, the cytotoxicity dose dependence is shown in Figure 4b. The highest cytotoxic activity was observed for NaDeca, with an IC_{50} value of 9.0 ± 0.7 µM, follow by the MetfDeca with an IC_{50} of 29 ± 0.7 µM, and IC_{50} values of 93 ± 5 and 540 ± 4 µM for MetV and Metf, respectively.

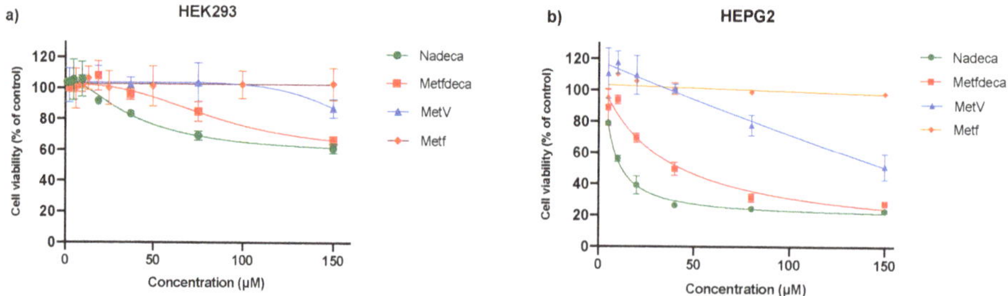

Figure 4. Cell viability assay at different vanadium compounds concentrations after treatment for 24 h (**a**) HEK293 and (**b**) HEPG2 cells. The cell viability of each treatment group was compared with the corresponding untreated control, which was normalized to 100% of cell viability. Error bars represent the standard deviation for triplicate runs ($n = 3$).

As can be seen in Figure 4a,b, the cell viability decreases in a dose manner response. The IC_{50} of the vanadium compounds against HEPG2 cells is around the same value as other compounds previously reported, like cis-platin (15.9 μM) [40] and monomeric V^{4+} compounds [41–43].

The metformin hydrochloride does not reduce the cellular viability in the range of concentrations that the decavanadate compounds do; the NaDeca compound exhibits more activity than MetfDeca and MetV regardless of the cell line after 24 h. The cytotoxicity of the tested compounds against HEPG2 is different for the non-tumorigenic HEK293 cells indicating that the toxicity of the compounds exhibits a good correlation on selectivity toward HEPG2 cancer cells in 24 h (see Table 2). The three vanadium and Metf compounds do not affect the viability of the HEK293 cells; this is an important result from this work, which may have an impact due to the new strategies intended to reduce the renal toxicity induced by cisplatin [44,45].

Table 2. Cytotoxic activity (IC_{50}) and selectivity index (SI) of compounds against HEK293 and HEPG2 cells after 24 h.

Compound	HEK293 Lower Cell Viability (%)	HEK293 IC_{50} (μM)	HEPG2 Lower Cell Viability (%)	HEPG2 IC_{50} (μM)	SI (IC_{50} HEK293/ IC_{50} HEPG2)
NaDeca	61	40 ± 4	22	9 ± 0.7	4.4
MetfDeca	67	85 ± 5	28	29 ± 0.7	2.9
MetV	88	181 ± 7	51	93 ± 5	1.9
Metf	100	420 ± 11	98	540 ± 4	0.77

In the case of decavanadate compounds, the IC_{50} seems strongly dependent on the type of cell line and the counter ion; for example, the IC_{50} of the $Na_4[(HOCH_2CH_2)_3NH]_2[V_{10}O_{28}]\cdot6H_2O$ towards HEPG2 cell line is 16.4 ± 3 μg/mL while for human cervical cancer cell line (Hela cells) is 53.1 ± 12.1 μg/mL [46], the compounds $[(H_2tmen)_3V_{10}O_{28}]\cdot6H_2O$ and $[(H_2en)_3V_{10}O_{28}]\cdot2H_2O$ were tested in human normal hepatocytes L02, and their IC_{50} values are 6.5 ± 0.6 and 7.2 ± 0.7 μM, respectively indicating that are cytotoxic for the L02 human cell line [47]. In 2018 Nunes and coworkers studied the cytotoxicity effect of three decavanadates compounds in African green monkey kidney (Vero) cells, and the three compounds exhibit low effect; 200 μM of the compounds reduced 50% of the Vero cells viability in 96 h. The compounds tested were the decavanadate complexes of sodium, nicotinamidium $[(3\text{-Hpca})_4H_2V_{10}O_{28}]\cdot2H_2O\cdot2(3\text{-pca})$ and isonicotinamidium $[(4\text{-Hpca})_4H_2V_{10}O_{28}]\cdot2(4\text{-pca})$ [48]. However, in the three studies presented before [46–48] for decavanadate compounds, the effect of the counter ion in the cytotoxic studies was not studied.

It seems that the decavanadate compounds—or their decomposition products V_1, V_2 and V_4—decreased the viability of hepatocarcinoma HEPG2 cells faster than the normal HEK293 cells (Table 2) after 24 h. The cytotoxicity of cancer and normal cells can be attributed to a different mechanism like Wang and coworkers reported in 2010 [49] that for 100 μM of MetV in MEM (minimum essential medium) in normal hepatocytes L02, the cell arrest mechanism is ROS-dependent and for HEPG2 is ROS-independent to mediated ERK (extracellular signal-regulated protein kinase activation) after 72 h. In the present study, the Metf cation association with the decavanadate moiety promotes some kind of protection against the normal HEK293 cells. However, the dissociation of the ion-pairs $NaV_{10}O_{28}{}^{5-}$, $(C_4H_{12}N_5)V_{10}O_{28}{}^{5-}$ and the further hydrolysis to V_1 and other products will not protect the vanadium atoms for the reduction into V^{4+} that could significantly increase the ROS levels and the apoptosis for the normal cells.

2.4. Proteins Expression

Protein kinase B (AKT) is a crucial mediator of insulin-resistant glucose and lipid digestion [50]. To evaluate the effect that decavanadate compounds have in phosphatidylinositol 3-kinase (PI3K/AKT) and AMPK pathways in HEPG2 cells, a Western blot examination was performed.

The cells were cultured with various concentrations of the compounds for 24 h without insulin. In Figure 5a we can see that NaDeca highly phosphorylates AKTα while the MetfDeca, MetV and Metformin show moderate activity. Thus, NaDeca and MetfDeca compounds exhibit a dose-dependence increase in phosphorylated AKT (p-AKT) as shown in Figure 5c, where 8, 16, and 30 μM of NaDeca induced a 2, 4 and 6-fold-increase in the phosphorylation, respectively. In contrast, the expression of the AMPK, a cellular metabolism energy sensor, by its phosphorylation p-AMPKα is not significantly elevated by the compounds (Figure 5d). However, NaDeca in 8 μM concentration exhibits around 33% of the increase in the AMPK phosphorylation. The low percentage of phosphorylation in AMPK by the vanadium compounds and metformin (Figure 5d) can be explained as follows: in hepatocellular carcinoma (HCC), the pathway function is downregulated [51], it seems like a low level of AMPK is required to maintain viability during the metabolic stress of tumor cells by different mechanisms [52].

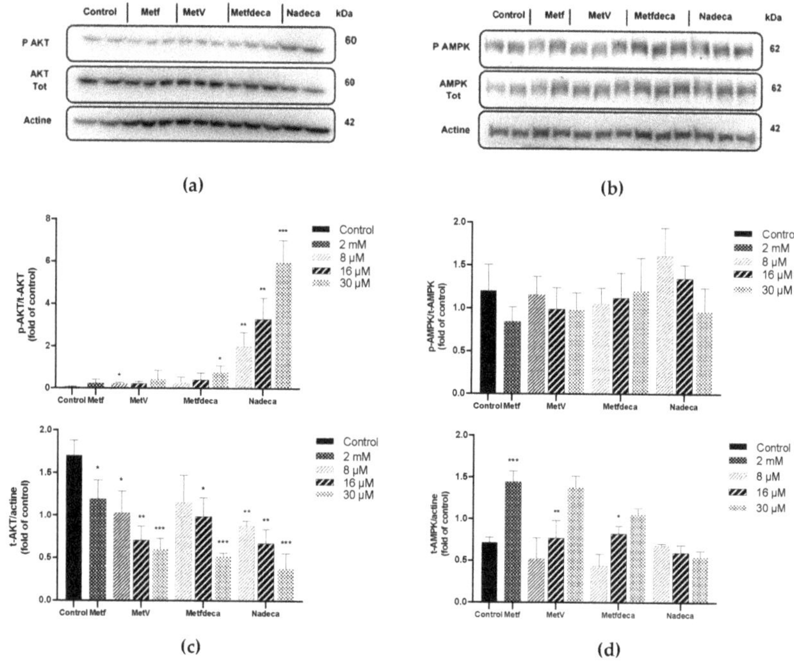

Figure 5. Effect of vanadium compounds on the protein expression of (**a**) protein kinase B (AKT) representative Western blot and (**b**) AMP-activated kinase (AMPK) representative Western blot in HEPG2 human hepatocarcinoma cells. Metf 2 mM and vanadium compounds 8, 16 and 30 μM concentrations, respectively, from left to right. Quantitative data of (**c**) p-AKT/t-AKT and t-AKT/actine (**d**) p-AMPK/t-AMPK and t-AMPK/actine. All the values are the mean ± SD. * $p < 0.05$, ** $p < 0.01$ and *** $p < 0.001$ vs. untreated control cells.

In the present work, the activation of AKT by 2 mM of Metf and by 8–30 μM of MetfDeca is moderate; in the case of Metf, p-AKT is increased by 21% while for 30 μM of MetfDeca, the increase is 80%; however, we observed that the NaDeca formation rate of metavanadate species is moderately faster than MetfDeca under the same experimental conditions (Table 1), due to the weaker ionic pairs for NaDeca than for MetfDeca (Scheme 1), so if all the 8 μM NaDeca decomposition product is V_1 the p-AKT fold should be ten times 0.21, the fold value that we found experimentally for the MetV is 2.1 (Figure 5c), this clearly indicates that 8 μM of NaDeca is decomposed to 80 μM of V_1. Nevertheless, the decomposition of NaDeca at higher concentration solutions shows lower amounts of V_1 produced, based on the p-AKT fold activity. If we double NaDeca concentration to 16 μM, the experimental fold value is 3.3 for the decameric compound, and for MetV is 0.21, while for 30 µM of MetV, the p-AKT fold value is 0.45 and for NaDeca is 6. It seems that higher vanadium concentrations induce the formation of larger oligomers like V_2, V_4 and V_5. Thus the active species could be a combination of the orthovanadate and the metavanadate units, with differing AKT signaling activation mechanisms promoted by different vanadium species.

Activation of the IR kinases by vanadium compounds exhibit different mechanisms, which depend on the type of cell and also the oxidation state of the metal [53,54]. Recently, several lines of evidence suggest that cancer cells upregulate the oxidative pentose phosphate pathway (PPP) to support cell growth and survival, by consequence exhibited increased PPP flux, NADPH/NADPC ratio, and ROS [25], in the liver 30% of the glucose oxidation occurs via PPP, so, it is not surprising that the AKT activation could be in a phosphatidylinositol 3-kinase (PI3K)-dependent manner by ROS [24]. However, in our control experiments the AKT signal in HEPG2 cells has not been activated (Figure 5c), suggesting that AKT phosphorylation by MetV, MetfDeca and NaDeca can be attributed to the activation of PTB-1B by orthovanadate (V_1) [32]. On the other hand, for NaDeca compound, the tetramer species is formed at the same speed than V_1 (Table 1), V_4 could be the one that is reduced [55] and the vanadium (IV) species VO^{2+} is activating the AKT pathway in a PI3K-dependent manner by ROS, like in the case of $VOSO_4$ that exhibited a 17-fold increase in the phosphorylation of AKT in HEPG2 cells at 25 µM concentration [56]. In 2015 Levina and coworkers performed a speciation study by XANES spectroscopy, where for 1 mM of orthovanadate in HEPG2 cells with DMEM medium after 24 h, 50% of the initial vanadium was found as tetrahedral species of V^{5+} (V_1, V_2, V_4 and V_5 are tetrahedral), 30% as V^{4+} moieties with a coordination number of six and 20% as V^{4+} with a coordination number of five [19], this study supports our observation that after 24 h not more decavanadate species are present in solution. It also supports our hypothesis that not all the vanadium in solution is present in the highest oxidation state (V^{5+}) and some has been reduced to V^{4+} promoting different mechanism of AKT activation, particularly for the NaDeca compound.

2.5. Glucose Uptake Assay

To establish whether MetV, MetfDeca, NaDeca and Metf compounds stimulate glucose uptake on adipocytes, the effect on the 2-NDBG cell uptake in 3T3-L1 differentiated adipocytes was evaluated. The experiments were performed at 16, 30 µM concentration for vanadium compounds and 2 mM for Metf in the absence of insulin. Insulin (100 nM) was used as a positive control. As it can be seen in Figure 6, NaDeca (30 µM) stimulates the glucose cell uptake on 62%, MetfDeca on 52%, MetV on 37% and Metf (2 mM) on 33%, while control conditions stimulate around to 20%. At 16 µM, NaDeca stimulates 29% and is the only compound that shows a notable difference between both concentrations. Our results suggest that the uptake is moderate due to the low concentration of the compounds; it has been shown that elevated concentrations of decavanadate 100 µM [6] and vanadate 325 µM were required for stimulation of glucose uptake in rat adipocytes, the later associated with IR Tyr auto-phosphorylation [53]. The activation of the insulin receptor substrates (IRS) has been demonstrated to occur in a dose-dependent manner in cardiomyocytes for MetV [54] and in 3T3-L1 cells for $VOSO_4$ [57] due to different mechanisms of actions. Our results indicate that the PI3K pathway was activated due to the activation of IRS-1 by PTPB1 phosphorylation for MetfDeca, MetV and Metf by a combination of different mechanisms that includes PTPB1 phosphorylation, and for NaDeca by a ROS production, where V_1 and the metavanadate species are involved, the ROS production by a decavanadate compound and the activation of the semicarbazide-sensitive amine oxidase (SSAO)/vascular adhesion protein-1 (VAP-1) was reported by Ybarola [58] the compound hexakis(benzylammonium) decavanadate showed that can stimulate glucose uptake in rat adipocytes in a dose dependent manner EC_{50} 150 µM, an in vitro assay they confirmed that hexakis(benzylammonium) decavanadate is oxidized in the same extension by SSAO enzyme as benzylamine and vanadate, using ^{51}V NMR the authors also found that for 10 mM of the compound in the presence of 2.5 mM of H_2O_2 at pH 7.4, the major products of the decavanadate decomposition were V_1, V_4 and $[V(OH)_2(OO)_2(OH)_2]^{2-}$. The decomposition products promoted the inhibition of PTP and the activation of SSAO that regulates the translocation of the GLUT4 transport and stimulates glucose transport [58], like in the case of the vanadium compounds tested in this work, where the GLUT4 transport is translocated and the glucose is transported by the cell.

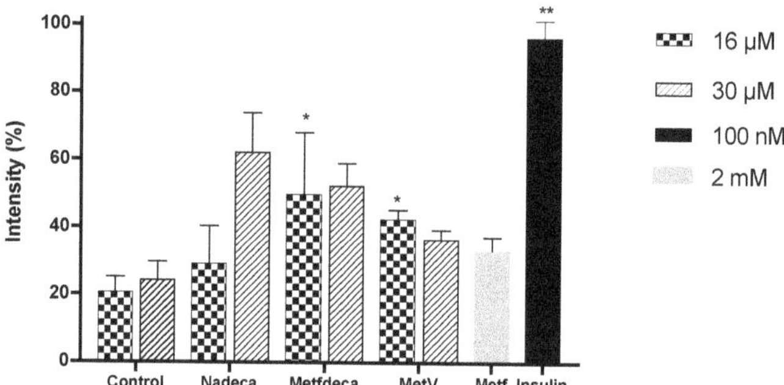

Figure 6. Effect of vanadium compounds on the glucose uptake in 3T3-L1 differentiated adipocytes at 16, 30 µM for vanadium compounds and 2 mM for Metf in the absence of insulin. All the values are the mean ± SD. * $p < 0.05$ and ** $p < 0.01$ vs. untreated control cells.

2.6. Insulin Release Assay

The effect on insulin release of NaDeca, MetfDeca, MetV at 30 µM concentration was studied in βTC-6 cells. Glucose 10 mM and repaglinide 30 µM concentration were used as control. The latter was used due to the pharmacological activity, such as blocking ATP-dependent K$^+$ channels and stimulate the release of insulin from the pancreas in a dose-dependent manner [59]. Figure 7 shows that MetV and MetfDeca enhanced insulin release by 0.7 and 1-fold relative to glucose control. In addition, both vanadium compounds showed more activity than the repaglinide at the same concentration, while NaDeca shows lower activity at 30 µM concentration than the glucose and repaglinide controls. βTC-6 cells secrete insulin in response to glucose; however, this cell line derived from transgenic mice develop a high hexokinase activity [60]; in normal pancreatic β-cells isolated from mouse islets, the effects of NaVO$_3$ were studied at 0.1–1 mM concentration [61], the authors found that vanadate did not affect basal insulin release, although, vanadate potentiated the glucose effect by a different mechanism than blocking the sodium pump or affecting the AmpC levels [61].

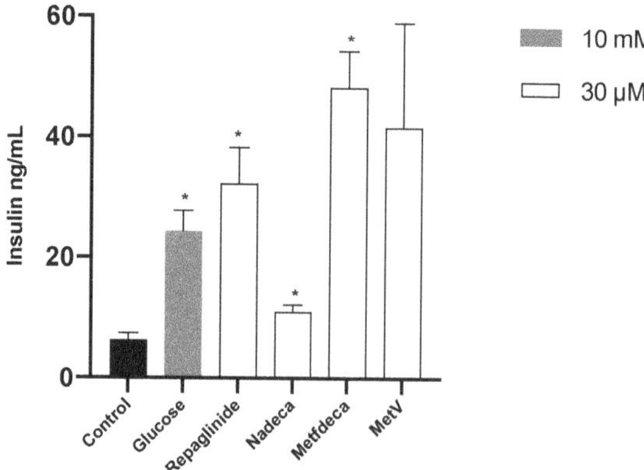

Figure 7. Effect of the vanadium compounds (30 µM concentration) on the insulin release in βTC-6 cells with 10 mM of glucose and 30 µM of repaglinide. All the values are the mean ± SD. * $p < 0.05$ vs. untreated control cells.

In 1999, Proks and coworkers performed an experiment in different types of cloned K_{ATP} channels expressed in *Xenopus oocytes* [62]. Their results showed that sodium decavanadate in 2 mM concentration made by a solution of Na_3VO_4 at pH 7.2 modulated K_{ATP} channel activity via the SUR subunit, the Hill coefficients for both activation and inhibition of K_{ATP} currents suggested that the cooperativity action of more than one vanadate molecule was involved in these effects. They also found that the effects were abolished by boiling the solution where the vanadate polymers where virtually absent [62]. Our results indicate that vanadium species promote more than one insulin release mechanism in βTC-6 cells (Figure 8), MetV, NaDeca, and MetfDeca decomposition in V_1 augment insulin secretion by tyrosine phosphorylation of IRS-1 and IRS-2 [63,64], while in the second mechanism, vanadium oligomers can be active blocking ATP-dependent K^+ channels, however, we propose that the active species in the decavanadate solutions are the vanadium dimers V_2, although, V_2 formation rate is slower (Table 1), it can be present in considerable amounts blocking ATP-dependent K^+ channels [62,65]. In the case of V_4, the higher oligomer formation is promoted by the decomposition reaction of the NaDeca compound (Table 1), and the tetramer V_4 has not followed any of the two mechanisms (Figure 8). It has been shown that vanadium compounds like $VOSO_4$ and $NaVO_3$ (1.6–100 μM) stimulated ROS production in isolated rat liver mitochondria [66]. In 2013 Hosseini and coworkers showed that V^{5+} (25–200 μM) interaction with respiratory complex III is the major source of V^{5+} induced ROS in rat liver mitochondria [67]. Interestingly, the concentration of ROS formation highly increases with 200 μM of sodium metavanadate in 60 min while with just 50 μM, it is not the case [67]. We hypothesized that NaDeca at 30 μM concentration product V_4 has some interaction with the cell mitochondria like its membrane depolarization [68] through ROS production that inhibits the insulin release by NaDeca compound.

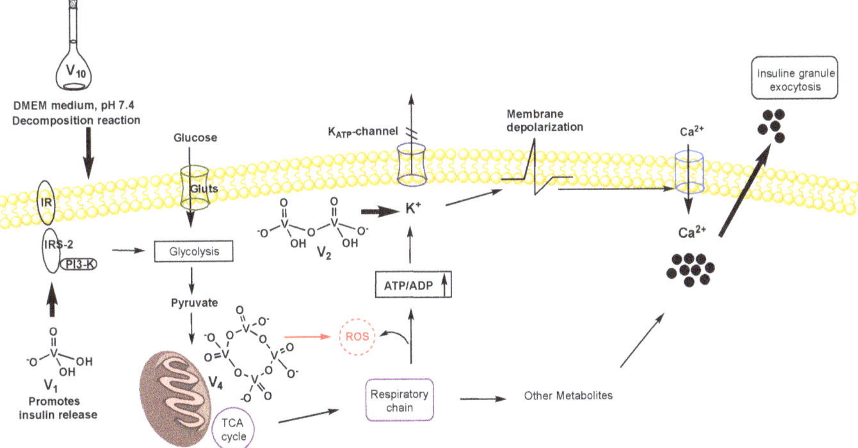

Figure 8. Illustration of the vanadium species mechanisms of action on the insulin release in βTC-6 cells. The total vanadium concentration is 300 μM, NaDeca decomposition reaction is moderately faster than MetfDeca reaction (Table 1), so NaDeca and MetfDeca majority decomposition products are V_1 and V_4; however, some V_2 is present, the putative mechanism of action for V_1 is that enhanced tyrosine phosphorylation, and on that way, V_1 species is able to further augment insulin secretion. A second putative mechanism involves inhibition of the K_{ATP} channel by V_2 species.

3. Experimental

3.1. Chemicals and Reagents

Ammonium metavanadate (NH_4VO_3), sodium metavanadate ($NaVO_3$), hydrochloric acid (HCl 37% *w/v* in H_2O), dimethyl sulfoxide (DMSO), *d*6-DMSO, deuterium oxide (D_2O), 3-(4,5-dimethylthiazol-2-yl)

-2,5-diphenyltetrazolium Bromide (MTT) 98%, 4-(2-hydroxyethyl)-1-piperazineethanesulfonic acid (HEPES), potassium chloride (KCl), sodium chloride (NaCl), ethylenediaminetetraacetic acid disodium salt (EDTA), ethylene glycol-bis(2-aminoethylether)-*N,N,N′,N′*-tetraacetic acid (EGTA), β-glycerol-phosphate, triton X-100, NaF, sodium pyrophosphate dibasic, sodium orthovanadate (Na_3VO_4) and 1,4-dithiothreitol (DDT) were purchased from Sigma-Aldrich (St Louis, MO, USA). Phenylmethylsulfonyl fluoride (PMSF) from Calbiochem (San Diego, CA, USA). COMPLETE (protease inhibitor cocktail) from ROCHE (Mannheim, Germany). Dulbecco's modified Eagle's medium (DMEM) high glucose, fetal bovine serum (FBS) and penicillin/streptomycin from Gibco (Gaithersburg, MD, USA). All the cell lines used were purchased from ATCC (HEP-G2 HB-8065, 3T3-L1 CL-173, Beta-TC-6 RL 11506) (Manassas, VA, USA).

Metformin hydrochloride ($C_4H_{11}N_5{\cdot}HCl$) was isolated directly from commercial brand tablets. The metforminium decavanadate (MetfDeca) $(C_4H_{13}N_5)_3V_{10}O_{28}{\cdot}8H_2O$ was prepared according to the literature [33]. The sodium decavanadate (NaDeca) $Na_6V_{10}O_{28}{\cdot}18H_2O$ was prepared by suspending $NaVO_3$ (0.12 g, 1 mmol) in distilled water (30 mL). After the suspension was stirred at room temperature for 1 h, the pH was adjusted to 4 by the addition of HCl (1 M). The resulting orange solution was filtered, and the filtrate was allowed to evaporate at 4 °C. Orange crystals were obtained after one week, according to a previously reported procedure [36].

The concentrations of the stock solutions for the biological studies in water for metformin hydrochloride (Metf), sodium metavanadate (MetV), metforminium decavanadate (MetfDeca) were 30 mM, whereas for sodium decavanadate (NaDeca) was 15 mM. The metforminium decavanadate crystals are water-insoluble, so it was solubilized in 10% DMSO before the addition of water. For the ^{51}V NMR studies, 10% DMSO-*d6* was used.

3.2. Kinetic Studies

The kinetics of the decomposition reaction of sodium and metforminium decavanadates in DMEM medium at 25 °C was determined by ^{51}V NMR at 0.5 and 1.0 mM of decavanadate concentration. The spectra were acquired using 0.5 mL as a final volume with 10% DMSO-*d6* in a Bruker Ascend 600 MHz spectrometer. ^{51}V spectra were recorded using parameters reported previously [12,69] at 157.85 MHz. The chemical shifts were obtained using an external reference using 100 mM Na_3VO_4 solution in 1.0 M NaOH ($[VO_4]^{3-}$ signal at −541 ppm) [70]. The concentrations of each vanadate species V_x were calculated from the fractions of the total integrated areas using the following equation: $[V_x] = (A_x/A_t) \times ([V_t]/n)$, where A_x corresponds to the area measured for the x vanadate species with n as the oligomer number (number of vanadium atoms), A_t is the sum of the measured areas and $[V_t]$ is the total vanadate concentration [71]. In the case of the decameric species, three signals at −420, −494 and −510 ppm were integrated for 2, 4 and 4 vanadium atoms, respectively [72].

The rate constants were calculated by the initial rates method, where the species concentration V_x was plotted over time (100 min), the ^{51}V NMR spectra were acquired every 20 min, and the reaction was started when the decavanadate compound aliquot was added to the DMEM medium.

3.3. Cell Viability Assay

Cell viability of the three vanadium compounds NaDeca, MetfDeca, MetV and metformin hydrochloride against HEPG2 and HEK293 was tested using MTT assay (Sigma-Aldrich, St Louis, MO, USA). The cells were placed in a 96-well micro-assay culture plate (ULTRACRUZ, Santa Cruz Dallas, TX, USA) at a density of 1×10^5 cells per well in 0.2 mL of DMEM-high glucose culture medium supplemented with fetal bovine serum FBS (10%) and penicillin/streptomycin (1%), and grown at 37 °C in a humidified 5% CO_2 incubator for 24 h. After this, the cells were treated with 0.002 mL of each compound per well by triplicate; sequential dilutions 1:2 were made for each compound, DMSO was used as a blank. The cells were incubated for 24 h. The surviving cells were determined. We added 0.01 mL of MTT (5 mg/mL in phosphate-buffered saline) to each well, and the cells were incubated for 3 h at 37 °C in a humidified 5% CO_2 incubator. After this time, the medium was removed from the cells, and 0.1 mL of DMSO was added to each well, and the cells were incubated for 1 h. The cells viability was determined by measure their ability to reduce MTT (yellow) to formazan product (violet). The absorbance was quantified at 600 nm by a Modulus microplate Luminometer spectrophotometer (Turner BioSystems, Sunnyvale, CA, USA).

3.4. Western Blot Analysis

The cells were placed in 6-well micro-assay culture plates at a density of 5×10^5 cells per well in 3 mL of DMEM-high glucose culture medium supplemented with fetal bovine serum FBS (10%) and penicillin/streptomycin (1%); the cells were treated with 8, 16 and 30 µM of each compound and the cells were grown at 37 °C in a humidified 5% CO_2 incubator for 24 h.

Cultured cells were washed with 1 mL of cold phosphate buffer solution (PBS). For AMPK assays cells were lysed by 0.25 mL of ice-cold HEPES lysis buffer: HEPES (50 mM, pH 7.4), EDTA (1 mM), EGTA (1 mM), KCl (50 mM), glycerol (5 mM), Triton X100 (0.1% w/v), NaF (50 mM), NaPPi (5 mM), Na_3VO_4 (1 mM), DDT (1 mM), PMSF (0.2 mM) and COMPLETE 1X as protease inhibitor. Homogenates were centrifuged at 16,128× g for 20 min at 4 °C in an Eppendorf centrifuge 5804R.

Supernatants were collected for their protein quantitation by Lowry method; 50 µg of protein were separated by 10% SDS-page and transferred to PVDF for blotting using the following antibodies (cell signaling 1:1000) anti-pAKT (Ser473), anti-p-AMPKα (Thr172), anti-AMPKα, anti-AKT (PKBα) and anti-β-actin at 4 °C overnight. Blots were visualized with HRP-conjugated goat anti-rabbit IgG or HRP-conjugated goat anti-mouse IgG at room temperature for one hour. Actin was used as loading controls for the total protein content. Proteins were visualized and quantified in a Bio-Rad ChemiDoc XRS (Bio-Rad, Hercules, CA, USA). with the Quantity One software (Version 4.5, Bio-Rad, Hercules, CA, USA).

3.5. Adipocyte Differentiation

Preadipocytes 3T3-L1 were obtained from ATCC and differentiated, as previously described [73]. Briefly, cells were grown to confluency in a 75 cm flask (CORNING) with DMEM medium supplemented with 10% calf serum (Biowest, Riverside, MO, USA) and standard temperature and CO_2 conditions (37 °C and 5% CO_2). Two days after reaching confluency, media was replaced to induce differentiation (DMEM supplemented with 10% fetal bovine serum (FBS), 1.0 µg/mL human insulin, 0.5 mM 3-isobutyl-1-methylxanthine and 1 µM dexamethasone). After 48 h, media was changed with DMEM supplemented with 10% fetal bovine serum and 1.0 µg/mL human insulin and cells were incubated for 48 h. Finally, the media was replaced with DMEM supplemented with 10% FBS for 4 days, media was refreshed every 2 days.

3.6. Glucose Uptake Assay

3T3-L1 differentiated adipocytes cells were seeded in a 96-well plate (ULTRACRUZ, Santa Cruz) 1×10^5 cells per well. The next day media was changed to starving media (DMEM without supplementation, no glucose), compounds were added at 16 and 30 µM final concentration and incubated 20 h at standard conditions. Cells were incubated with or without 100 nM insulin for 1 h. After this, 300 µM of 2-NBDG (Invitrogen by Thermo Fisher Scientific) were added to each well and incubated 20 min at 37 °C and 5% CO_2, cells were washed once with PBS, and 100 µL/well of fresh PBS were added. Fluorescence was read at 485/535 nm (Modulus Microplate Luminometer).

3.7. Insulin Release Assay

Studies were performed with βTC-6 cells. The cells were placed in 24-well micro-assay culture plates at a density of 2.5×10^5 cells per well in DMEM culture medium; the cells were incubated overnight at 37 °C in a humidified 5% CO_2 incubator. The next day medium was changed, and 10 mM glucose and 30 µM repaglinide were added as controls. Compounds were added at 30 µM final concentration and incubated for one hour at standard conditions. The insulin quantification was made with a mouse insulin ELISA kit (ALPCO, INSMS-E01, ALPCO, Salem NH, USA).

3.8. Statistical Analysis

Data were presented as mean ± SEM of three independent experiments. Statistical significance of data was analyzed by Student's *t*-test and one-way analysis of variance (ANOVA). A probability of the value of $p < 0.05$ was considered as statistically significant. Calculations and figures were made using Grad Pad Prism version 8 (GraphPad Software, San Diego, CA, USA).

4. Conclusions

Vanadium solution chemistry represents a challenge due to its complexity. However, new therapeutic approaches can be explored with decavanadate compounds in biological reaction media, vanadium therapeutic potential in different diseases like DM2, cancer, metabolic syndrome and cardiovascular diseases should be addressed. Decavanadate decomposition products like V_2 and V_4 action mechanisms in cytotoxic activity, AMPK and AKT expression still have open questions; however, V_1 is well known as a glucose uptake promoter and insulin release agent. Nevertheless, the combination of orthovanadate and methavanadate species can increase the desirable therapeutic effects of vanadium, as shown in this work.

Our results show that at least two mechanisms are promoted AKT activation by NaDeca, and MetfDeca hydrolysis products in HEPG2 cells, the first one with the orthovanadate (V_1) species involved in PTP-1B mediated AKT activation, while the second mechanism involves the activation of the AKT pathway in a PI3K-dependent manner by ROS, in this regard, we hypothesized that V_4 could be involved in a vanadium reduction process that promotes the ROS exacerbation in HEPG2 cells in DMEM medium and that ROS production results in a decrement of the cell viability in normal (HEK293) and carcinogenic cells (HEPG2).

In this sense, our results indicate that a combination of at least two mechanisms is associated with the glucose uptake in 3T3L-1 differentiated adipocytes that includes PTP-1B phosphorylation and ROS production in the case of NaDeca.

MetfDeca and MetV at 30 μM concentration enhanced insulin release in βTC-6 cells; surprisingly, the NaDeca compound is almost inactive in the assay. Our results suggest that MetfDeca decomposition products (V_1 and V_2) promote more than one insulin release mechanism in the DMEM medium. The first proposed mechanism is that V_1 augment insulin secretion by tyrosine phosphorylation of the IRS, and in a second putative mechanism, vanadium oligomers like V_2 can be active, blocking ATP-dependent K^+ channels. However, V_4 species that are produced by the decomposition reaction of NaDeca and MetfDeca are not following either mechanism.

The data presented in this paper demonstrate that decavanadate decomposition products are able to promote different biological mechanisms of action, than the ones promoted by orthovandate (MetV) and metformin hydrochloride (Metf). Thus, more chemical and biological experiments are necessary to establish the active species and their composition with the aim to explore new therapies in the treatment of some metabolic diseases.

Author Contributions: A.M.S.-N. and R.O.S.-D. performed the kinetics experiments; A.M.S.-N. and L.C. performed the biological studies; I.E.L. and O.H.-A. reviewed and edited the manuscript; I.S.-L. wrote most of the manuscript and carried out the kinetics analysis. All authors have read and agreed to the published version of the manuscript.

Funding: This research received no external funding. A.M.S.-N. and I.S.-L. would like to thank the UJAT for funding A.M.S.-N. fellowship.

Acknowledgments: We specially thanks to Boris Rodenak Kladniew to participate in the biological data analysis.

Conflicts of Interest: The authors declare no conflict of interest.

References

1. Van Rompuy, L.S.; Parac-Vogt, T.N. Interactions between polyoxometalates and biological systems: From drug design to artificial enzymes. *Curr. Opin. Biotechnol.* **2019**, *58*, 92–99. [CrossRef]
2. Čolović, M.B.; Lacković, M.; Lalatović, J.; Mougharbel, A.S.; Kortz, U.; Krstić, D.Z. Polyoxometalates in Biomedicine: Update and Overview. *Curr. Med. Chem.* **2020**, *27*, 362–379. [CrossRef]

3. Aureliano, M. Decavanadate contribution to vanadium biochemistry: In vitro and in vivo studies. *Inorg. Chim. Acta* **2014**, *420*, 4–7. [CrossRef]

4. Bijelic, A.; Rompel, A. The use of polyoxometalates in protein crystallography—An attempt to widen a well-known bottleneck. *Coord. Chem. Rev.* **2015**, *299*, 22–38. [CrossRef] [PubMed]

5. Crans, D.C.; Sánchez-Lombardo, I.; McLauchlan, C.C. Exploring Wells-Dawson Clusters Associated with the Small Ribosomal Subunit. *Front. Chem.* **2019**, *7*, 462–476. [CrossRef] [PubMed]

6. Aureliano, M.; Crans, D.C. Decavanadate ($V_{10}O_{28}{}^{6-}$) and oxovanadates: Oxometalates with many biological activities. *J. Inorg. Biochem.* **2009**, *103*, 536–546. [CrossRef] [PubMed]

7. Nakamura, S.; Ozeki, T. Hydrogen-bonded aggregates of protonated decavanadate anions in their tetraalkylammonium salts. *J. Chem. Soc. Dalton Trans.* **2001**, 472–480. [CrossRef]

8. Ferreira da Silva, J.L.; Fátima Minas da Piedade, M.; Teresa Duarte, M. Decavanadates: A building-block for supramolecular assemblies. *Inorg. Chim. Acta* **2003**, *356*, 222–242. [CrossRef]

9. Bošnjaković-Pavlović, N.; Prévost, J.; Spasojević-de Biré, A. Crystallographic Statistical Study of Decavanadate Anion Based-Structures: Toward a Prediction of Noncovalent Interactions. *Cryst. Growth Des.* **2011**, *11*, 3778–3789. [CrossRef]

10. Povar, I.; Spinu, O.; Zinicovscaia, I.; Pintilie, B.; Ubaldini, S. Revised Poubaix diagrams for the vanadium–water system. *J. Electrochem. Sci. Eng.* **2019**, *9*, 620. [CrossRef]

11. Kojima, T.; Antonio, M.; Ozeki, T. Solvent-Driven Association and Dissociation of the Hydrogen-Bonded Protonated Decavanadates. *J. Am. Chem. Soc.* **2011**, *133*, 7248–7251. [CrossRef] [PubMed]

12. Ramos, S.; Manuel, M.; Tiago, T.; Duarte, R.; Martins, J.; Gutiérrez-Merino, C.; Moura, J.J.G.; Aureliano, M. Decavanadate interactions with actin: Inhibition of G-actin polymerization and stabilization of decameric vanadate. *J. Inorg. Biochem.* **2006**, *100*, 1734–1743. [CrossRef] [PubMed]

13. Samart, N.; Arhouma, Z.; Kumar, S.; Murakami, H.A.; Crick, D.C.; Crans, D.C. Decavanadate Inhibits Mycobacterial Growth More Potently Than Other Oxovanadates. *Front. Chem.* **2018**, *6*. [CrossRef] [PubMed]

14. Krivosudský, L.; Roller, A.; Rompel, A. Tuning the interactions of decavanadate with thaumatin, lysozyme, proteinase K and human serum proteins by its coordination to a pentaaquacobalt(II) complex cation. *New J. Chem.* **2019**, *43*, 17863–17871. [CrossRef]

15. Soares, S.S.; Martins, H.; Duarte, R.O.; Moura, J.J.G.; Coucelo, J.; Gutiérrez-Merino, C.; Aureliano, M. Vanadium distribution, lipid peroxidation and oxidative stress markers upon decavanadate in vivo administration. *J. Inorg. Biochem.* **2007**, *101*, 80–88. [CrossRef] [PubMed]

16. Clare, B.W.; Kepert, D.L.; Watts, D.W. Acid decomposition of decavanadate: Specific salt effects. *J. Chem. Soc. Dalton Trans.* **1973**, 2481–2487. [CrossRef]

17. Druskovich, D.M.; Kepert, D.L. Base decomposition of decavanadate. *J. Chem. Soc. Dalton Trans.* **1975**, 947–951. [CrossRef]

18. Crans, D.; Rithner, C.; Theisen, L. Application of time-resolved vanadium-51 2D NMR for quantitation of kinetic exchange pathways between vanadate monomer, dimer, tetramer, and pentamer. *J. Am. Chem. Soc.* **1990**, *112*, 2901–2908. [CrossRef]

19. Levina, A.; McLeod, A.I.; Pulte, A.; Aitken, J.B.; Lay, P.A. Biotransformations of Antidiabetic Vanadium Prodrugs in Mammalian Cells and Cell Culture Media: A XANES Spectroscopic Study. *Inorg. Chem.* **2015**, *54*, 6707–6718. [CrossRef]

20. Saklayen, M.G. The Global Epidemic of the Metabolic Syndrome. *Curr. Hypertens. Rep.* **2018**, *20*, 12. [CrossRef]

21. Giovannucci, E.; Harlan, D.M.; Archer, M.C.; Bergenstal, R.M.; Gapstur, S.M.; Habel, L.A.; Pollak, M.; Regensteiner, J.G.; Yee, D. Diabetes and Cancer: A consensus report. *Diabetes Care* **2010**, *33*, 1674–1685. [CrossRef] [PubMed]

22. Tschöp, M.H. Metabolic Disease Therapies. *Cell Metab.* **2017**, *26*, 579–583.

23. Lee, C.-W.; Wong, L.L.-Y.; Tse, E.Y.-T.; Liu, H.-F.; Leong, V.Y.-L.; Lee, J.M.-F.; Hardie, D.G.; Ng, I.O.-L.; Ching, Y.-P. AMPK Promotes p53 Acetylation via Phosphorylation and Inactivation of SIRT1 in Liver Cancer Cells. *Cancer Res.* **2012**, *72*, 4394. [CrossRef] [PubMed]

24. Koundouros, N.; Poulogiannis, G. Phosphoinositide 3-Kinase/Akt Signaling and Redox Metabolism in Cancer. *Front. Oncol.* **2018**, *8*, 160–169. [CrossRef]

25. Zheng, W.; Feng, Q.; Liu, J.; Guo, Y.; Gao, L.; Li, R.; Xu, M.; Yan, G.; Yin, Z.; Zhang, S.; et al. Inhibition of 6-phosphogluconate Dehydrogenase Reverses Cisplatin Resistance in Ovarian and Lung Cancer. *Front. Pharmacol.* **2017**, *8*, 421–432. [CrossRef]

26. Rehder, D. Implications of vanadium in technical applications and pharmaceutical issues. *Inorg. Chim. Acta* **2017**, *455*, 378–389. [CrossRef]

27. Crans, D.C.; Henry, L.; Cardiff, G.; Posner, B.I. Developing Vanadium as an Antidiabetic or Anticancer Drug: A Clinical and Historical Perspective. In *Essential Metals in Medicine: Therapeutic Use and Toxicity of Metal Ions in the Clinic*, 1st ed.; Carver, P.L., Ed.; De Gruyter: Berlin, Germany; Boston, MA, USA, 2019; pp. 203–230.

28. Jakusch, T.; Kiss, T. In vitro study of the antidiabetic behavior of vanadium compounds. *Coord. Chem. Rev.* **2017**, *351*, 118–126. [CrossRef]

29. Thompson, K.H.; Orvig, C. Vanadium in diabetes: 100 years from Phase 0 to Phase I. *J. Inorg. Biochem.* **2006**, *100*, 1925–1935. [CrossRef]

30. Thompson, K.H.; Lichter, J.; LeBel, C.; Scaife, M.C.; McNeill, J.H.; Orvig, C. Vanadium treatment of type 2 diabetes: A view to the future. *J. Inorg. Biochem.* **2009**, *103*, 554–558. [CrossRef]

31. Rehder, D. Transport, Accumulation, and Physiological Effects of Vanadium. In *Detoxification of Heavy Metals*; Sherameti, I., Varma, A., Eds.; Springer: Berlin/Heidelberg, Germany, 2011; pp. 205–220.

32. Gruzewska, K.; Michno, A.; Pawelczyk, T.; Bielarczyk, H. Essentiality and toxicity of vanadium supplements in health and pathology. *J. Physiol. Pharmacol.* **2014**, *65*, 603–611.

33. Sánchez Lombardo, I.; Sánchez Lara, E.; Pérez Benítez, A.; Mendoza, Á.; Bernès, S.; González Vergara, E. Synthesis of Metforminium(2+) Decavanadates-Crystal Structures and Solid-State Characterization. *Eur. J. Inorg. Chem.* **2014**, *2014*, 4581–4588. [CrossRef]

34. Treviño, S.; González-Vergara, E. Metformin-decavanadate treatment ameliorates hyperglycemia and redox balance of the liver and muscle in a rat model of alloxan-induced diabetes. *New J. Chem.* **2019**, *43*, 17850–17862. [CrossRef]

35. Treviño, S.; Sánchez-Lara, E.; Sarmiento-Ortega, V.E.; Sánchez-Lombardo, I.; Flores-Hernández, J.Á.; Pérez-Benítez, A.; Brambila-Colombres, E.; González-Vergara, E. Hypoglycemic, lipid-lowering and metabolic regulation activities of metforminium decavanadate (H$_2$Metf)$_3$ [V$_{10}$O$_{28}$]·8H$_2$O using hypercaloric-induced carbohydrate and lipid deregulation in Wistar rats as biological model. *J. Inorg. Biochem.* **2015**, *147*, 85–92. [CrossRef] [PubMed]

36. Durif, A.; Averbuch-Pouchot, M.T.; Guitel, J.C. Structure d'un Decavanadate d'Hexasodium Hydrate. *Acta Crystallogr. Sect. B Struct. Sci.* **1980**, *36*, 680–682. [CrossRef]

37. Howarth, O.W.; Jarrold, M. Protonation of the decavanadate(6-) ion: A vanadium-51 nuclear magnetic resonance study. *J. Chem. Soc. Dalton Trans.* **1978**, *5*, 503–506. [CrossRef]

38. Goddard, J.B.; Gonas, A.M. Kinetics of the dissociation of decavanadate ion in basic solutions. *Inorg. Chem.* **1973**, *12*, 574–579. [CrossRef]

39. Schwarzenbach, G.; Geier, G. 97. Die Raschacidifierung und -alkalisierung von Vanadaten. *Helv. Chim. Acta* **1963**, *46*, 906–926. [CrossRef]

40. Pascale, F.; Bedouet, L.; Baylatry, M.; Namur, J.; Laurent, A. Comparative Chemosensitivity of VX2 and HCC Cell Lines to Drugs Used in TACE. *Anticancer Res.* **2015**, *35*, 6497–6503.

41. Mohamadi, M.; Yousef Ebrahimipour, S.; Torkzadeh-Mahani, M.; Foro, S.; Akbari, A. A mononuclear diketone-based oxido-vanadium(IV) complex: Structure, DNA and BSA binding, molecular docking and anticancer activities against MCF-7, HPG-2, and HT-29 cell lines. *RSC Adv.* **2015**, *5*, 101063–101075. [CrossRef]

42. Dhahagani, K.; Mathan Kumar, S.; Chakkaravarthi, G.; Anitha, K.; Rajesh, J.; Ramu, A.; Rajagopal, G. Synthesis and spectral characterization of Schiff base complexes of Cu(II), Co(II), Zn(II) and VO(IV) containing 4-(4-aminophenyl)morpholine derivatives: Antimicrobial evaluation and anticancer studies. *Spectrochim. Acta A Mol. Biomol. Spectrosc.* **2014**, *117*, 87–94. [CrossRef]

43. Correia, I.; Adão, P.; Roy, S.; Wahba, M.; Matos, C.; Maurya, M.R.; Marques, F.; Pavan, F.R.; Leite, C.Q.F.; Avecilla, F.; et al. Hydroxyquinoline derived vanadium(IV and V) and copper(II) complexes as potential anti-tuberculosis and anti-tumor agents. *J. Inorg. Biochem.* **2014**, *141*, 83–93. [CrossRef] [PubMed]

44. Amuthan, V.D.A.; Chandrashekara, S.S.; Venkata, R.; Richard, L. Cytoprotective Activity of Neichitti (Vernonia cinerea) in Human Embryonic Kidney (HEK293) Normal Cells and Human Cervix Epitheloid Carcinoma (HeLa) Cells against Cisplatin Induced Toxicity: A Comparative Study. *J. Clin. Diagn. Res.* **2019**, *13*, KC01–KC06. [CrossRef]

45. Fujita, H.; Hirose, K.; Sato, M.; Fujioka, I.; Fujita, T.; Aoki, M.; Takai, Y. Metformin attenuates hypoxia-induced resistance to cisplatin in the HepG2 cell line. *Oncol. Lett.* **2019**, *17*, 2431–2440. [CrossRef] [PubMed]

46. Cheng, M.; Li, N.; Wang, N.; Hu, K.; Xiao, Z.; Wu, P.; Wei, Y. Synthesis, structure and antitumor studies of a novel decavanadate complex with a wavelike two-dimensional network. *Polyhedron* **2018**, *155*, 313–319. [CrossRef]

47. Li, Y.-T.; Zhu, C.-Y.; Wu, Z.-Y.; Jiang, M.; Yan, C.-W. Synthesis, crystal structures and anticancer activities of two decavanadate compounds. *Transit. Met. Chem.* **2010**, *35*, 597–603. [CrossRef]

48. Missina, J.M.; Gavinho, B.; Postal, K.; Santana, F.S.; Valdameri, G.; de Souza, E.M.; Hughes, D.L.; Ramirez, M.I.; Soares, J.F.; Nunes, G.G. Effects of Decavanadate Salts with Organic and Inorganic Cations on Escherichia coli, Giardia intestinalis, and Vero Cells. *Inorg. Chem.* **2018**, *57*, 11930–11941. [CrossRef]

49. Wang, Q.; Liu, T.-T.; Fu, Y.; Wang, K.; Yang, X.-G. Vanadium compounds discriminate hepatoma and normal hepatic cells by differential regulation of reactive oxygen species. *J. Biol. Inorg. Chem.* **2010**, *15*, 1087–1097. [CrossRef]

50. Hao, J.; Huang, K.; Chen, C.; Liang, Y.; Wang, Y.; Zhang, X.; Huang, H. Polydatin Improves Glucose and Lipid Metabolisms in Insulin-Resistant HepG2 Cells through the AMPK Pathway. *Biol. Pharm. Bull.* **2018**, *41*, 891–898. [CrossRef]

51. Ferretti, A.C.; Hidalgo, F.; Tonucci, F.M.; Almada, E.; Pariani, A.; Larocca, M.C.; Favre, C. Metformin and glucose starvation decrease the migratory ability of hepatocellular carcinoma cells: Targeting AMPK activation to control migration. *Sci. Rep.* **2019**, *9*, 2815. [CrossRef]

52. Hardie, D.G. Molecular Pathways: Is AMPK a Friend or a Foe in Cancer? *Clin. Cancer Res.* **2015**, *21*, 3836–3840. [CrossRef]

53. Lu, B.; Ennis, D.; Lai, R.; Bogdanovic, E.; Nikolov, R.; Salamon, L.; Fantus, C.; Le-Tien, H.; Fantus, I.G. Enhanced sensitivity of insulin-resistant adipocytes to vanadate is associated with oxidative stress and decreased reduction of vanadate (+5) to vanadyl (+4). *J. Biol. Chem.* **2001**, *276*, 35589–35598. [CrossRef]

54. Tardif, A.; Julien, N.; Chiasson, J.-L.; Coderre, L. Stimulation of glucose uptake by chronic vanadate pretreatment in cardiomyocytes requires PI 3-kinase and p38 MAPK activation. *Am. J. Physiol. Endocrinol. Metab.* **2003**, *284*, E1055–E1064. [CrossRef] [PubMed]

55. Žižić, M.; Miladinović, Z.; Stanić, M.; Hadžibrahimović, M.; Živić, M.; Zakrzewska, J. 51V NMR investigation of cell-associated vanadate species in Phycomyces blakesleeanus mycelium. *Res. Microbiol.* **2016**, *167*, 521–528. [CrossRef] [PubMed]

56. Zhao, Q.; Chen, D.; Liu, P.; Wei, T.; Zhang, F.; Ding, W. Oxidovanadium(IV) sulfate-induced glucose uptake in HepG2 cells through IR/Akt pathway and hydroxyl radicals. *J. Inorg. Biochem.* **2015**, *149*, 39–44. [CrossRef]

57. Seale, A.P.; de Jesus, L.A.; Park, M.C.; Kim, Y.S. Vanadium and insulin increase adiponectin production in 3T3-L1 adipocytes. *Pharmacol. Res.* **2006**, *54*, 30–38. [CrossRef] [PubMed]

58. Yraola, F.; García-Vicente, S.; Marti, L.; Albericio, F.; Zorzano, A.; Royo, M. Understanding the mechanism of action of the novel SSAO substrate $(C_7NH_{10})_6(V_{10}O_{28})\cdot 2H_2O$, a prodrug of peroxovanadate insulin mimetics. *Chem. Biol. Drug. Des.* **2007**, *69*, 423–428. [CrossRef]

59. Marín-Peñalver, J.J.; Martín-Timón, I.; Sevillano-Collantes, C.; Del Cañizo-Gómez, F.J. Update on the treatment of type 2 diabetes mellitus. *World J. Diabetes* **2016**, *7*, 354–395. [CrossRef]

60. Skelin, M.; Rupnik, M.; Cencic, A. Pancreatic beta cell lines and their applications in diabetes mellitus research. *Altex* **2010**, *27*, 105–113. [CrossRef]

61. Zhang, A.Q.; Gao, Z.Y.; Gilon, P.; Nenquin, M.; Drews, G.; Henquin, J.C. Vanadate stimulation of insulin release in normal mouse islets. *J. Biol. Chem.* **1991**, *266*, 21649–21656.

62. Proks, P.; Ashfield, R.; Ashcroft, F.M. Interaction of vanadate with the cloned beta cell K(ATP) channel. *J. Biol. Chem.* **1999**, *274*, 25393–25397. [CrossRef]

63. Gogg, S.; Chen, J.; Efendic, S.; Smith, U.; Ostenson, C. Effects of phosphotyrosine phosphatase inhibition on insulin secretion and intracellular signaling events in rat pancreatic islets. *Biochem. Biophys. Res. Commun.* **2001**, *280*, 1161–1168. [CrossRef] [PubMed]

64. Chen, J.; Ostenson, C.G. Inhibition of protein-tyrosine phosphatases stimulates insulin secretion in pancreatic islets of diabetic Goto-Kakizaki rats. *Pancreas* **2005**, *30*, 314–317. [CrossRef] [PubMed]

65. Liu, H.-K.; Green, B.D.; McClenaghan, N.H.; McCluskey, J.T.; Flatt, P.R. Long-Term Beneficial Effects of Vanadate, Tungstate, and Molybdate on Insulin Secretion and Function of Cultured Beta Cells. *Pancreas* **2004**, *28*, 264–268. [CrossRef] [PubMed]

66. Zhao, Y.; Ye, L.; Liu, H.; Xia, Q.; Zhang, Y.; Yang, X.; Wang, K. Vanadium compounds induced mitochondria permeability transition pore (PTP) opening related to oxidative stress. *J. Inorg. Biochem.* **2010**, *104*, 371–378. [CrossRef]

67. Hosseini, M.J.; Shaki, F.; Ghazi-Khansari, M.; Pourahmad, J. Toxicity of vanadium on isolated rat liver mitochondria: A new mechanistic approach. *Metallomics* **2013**, *5*, 152–166. [CrossRef]

68. Soares, S.S.; Gutiérrez-Merino, C.; Aureliano, M. Mitochondria as a target for decavanadate toxicity in Sparus aurata heart. *Aquat. Toxicol.* **2007**, *83*, 1–9. [CrossRef]

69. Crans, D.C.; Bunch, R.L.; Theisen, L.A. Interaction of trace levels of vanadium(IV) and vanadium(V) in biological systems. *J. Am. Chem. Soc.* **1989**, *111*, 7597–7607. [CrossRef]

70. Griffin, E.; Levina, A.; Lay, P.A. Vanadium(V) tris-3,5-di-tert-butylcatecholato complex: Links between speciation and anti-proliferative activity in human pancreatic cancer cells. *J. Inorg. Biochem.* **2019**, *201*, 110815. [CrossRef]

71. Aureliano, M.; Tiago, T.; Gândara, R.M.C.; Sousa, A.; Moderno, A.; Kaliva, M.; Salifoglou, A.; Duarte, R.O.; Moura, J.J.G. Interactions of vanadium(V)–citrate complexes with the sarcoplasmic reticulum calcium pump. *J. Inorg. Biochem.* **2005**, *99*, 2355–2361. [CrossRef]

72. Sánchez-Lombardo, I.; Baruah, B.; Alvarez, S.; Werst, K.R.; Segaline, N.A.; Levinger, N.E.; Crans, D.C. Size and shape trump charge in interactions of oxovanadates with self-assembled interfaces: Application of continuous shape measure analysis to the decavanadate anion. *New J. Chem.* **2016**, *40*, 962–975. [CrossRef]

73. Student, A.K.; Hsu, R.Y.; Lane, M.D. Induction of fatty acid synthetase synthesis in differentiating 3T3-L1 preadipocytes. *J. Biol. Chem.* **1980**, *255*, 4745–4750. [PubMed]

Publisher's Note: MDPI stays neutral with regard to jurisdictional claims in published maps and institutional affiliations.

MDPI
St. Alban-Anlage 66
4052 Basel
Switzerland
Tel. +41 61 683 77 34
Fax +41 61 302 89 18
www.mdpi.com

Inorganics Editorial Office
E-mail: inorganics@mdpi.com
www.mdpi.com/journal/inorganics